P. Dubin, J. Bock, R. Davis, D.N. Schulz, C. Thies (Eds.)

Macromolecular Complexes in Chemistry and Biology

With 196 Figures and 39 Tables

Springer-Verlag
Berlin Heidelberg New York
London Paris Tokyo
Hong Kong Barcelona Budapest

Editors

Prof. Paul Dubin, Dept. of Chemistry, Indiana-Purdue University, 402 N. Blackford Street, Indianapolis, IN 46202-3274, USA

Dr. J. Bock
Exxon Research & Engineering Company, Route 22, East, Annandale, NJ 08801, USA

Prof. Richie Davis
Dept. of Chemical Engineering, VPI & SU, Blacksburg, VA 24061, USA

Dr. Donald N. Schulz
Exxon Research & Engineering Company, Route 22, East, Annandale, NJ 08801, USA

Prof. Curt Thies
Biological Transport Laboratory, School of Engineering and Applied Scence, Washington University, St. Louis, MO 63130-4899, USA

Library of Congress Cataloging-in-Publication Data
Macromolecular complexes in chemistry and biology/P. Dubin ... [et al.], eds. p. cm.
Includes bibliographical references.
ISBN-13: 978-3-642-78471-2 **e-ISBN-13: 978-3-642-78469-9**
DOI: 10.1007/ 978-3-642-78469-9
1. Polymer solutions. 2. Polyelectrolytes. 3. Complex compounds.
QD381.9.S65M32 1994 547.7'0454--dc20 93-38677 CIP

Softcover reprint of the hardcover 1st edition 1994

SPIN: 10076493 3020 5 4 3 2 1 0 - Printed on acid-free paper

ALLE·ZEIT·WACH
1842

List of Authors

Dr. Roland Audebert
Université P. et M. Curie, ESPCI Lab. Physico-Chimie Macromolélulaire,
10 rue Vauquelin, F-75231 Paris Cedex 05, France

Dr. Jan Bock
Exxon Research and Engineering Company, Route 22, East Annandale,
NJ 08801, USA

Dr. Jeffrey Bodycomb
Dept. of Mechanics and Materials Science, Rutgers University, P.O. Box 909,
Piscataway, NJ 08855-0909, USA

K.D. Branham
Dept. of Polymer Science, The University of Southern Mississippi, Hattiesburg,
MS 39406-0076, USA

Prof. Diane J. Burgess
Dept. of Pharmaceutics, University of Illinois, 833 S. Wood St., M/C 880
Chicago, IL 60614, USA

Dr. Slavomir Bystricky
Institute of Chemistry, Slovak Academy of Sciences, Dubravska cesta 9, 842 38
Bratislava, Slovakia

Dr. L.-S. Choi
Dept. of the Navy, Chemistry Division, Code 6120, Naval Research Laboratory,
Washington, DC 20375-5320, USA

Dr. Herbert Dautzenberg
Max Planck Inst. für Kolloid-und Grenzflächenforschung, Kantstr. 55, D-14513
Teltow-Seehof, Germany

Dr. Edith Dellacherie
Laboratoire de Chimie-Physique Macromoleculaire, URA CNRS 494, ENSIC,
B.P. 451 F-54001 Nancy Cedex, France

Dr. A.J. Dualeh
Ciba-Geigy Corp., Ardsley, New York 10502, USA

Prof. Paul. L. Dubin
Dept. of Chemistry, Indiana-Purdue University, N. Blackford Street, Indianapolis, IN 46205-2810, USA

Prof. Wilmer K. Fife
Indiana University-Purdue University at Indianapolis, Dept. of Chemistry, N. Blackford Street, Indianapolis, IN 46205, USA

Prof. D. Frugier
Université P. et M. Curie, ESPCI Lab. Physico-Macromoléculaire 10, rue Vauquelin F-75231 Paris-Cedex 05, France

Dr. Gérard Gebel
CEA-CENG, DRFMC/SESAM/Laboratoire de Physico-Chimie Moléculaire, B.P. 85X F-38041 Grenoble Cedex, France

Prof. Charles Glatz
Dept. of Chemical Engineering, Iowa State University, 231 Sweeney Hall, Ames, IA 50011, USA

Prof. Masanori Hara
Department of Mechanics and Materials Science, Rutgers University P.O. Box 909 Piscataway, NJ 08855-0909, USA

Prof. Theo E. Hogen-Esch
Loker Hydrocarbon Research, Inst. & Dept. of Chemistry, University of Southern California, 205 LHI, USC Univ. Park Campus, Los Angeles, CA 90089-1661, USA

Dr. Patrick Hubert
Laboratoire de Chimie-Physique Macromoléculaire, URA CNRS 494, ENSIC, B.P. 451 F-54001 Nancy Cedex, France

Dr. F.S. Hwang
Loker Hydrocarbon Research, Inst. & Dept. of Chemistry, University of Southern California, 205 LHI, USC Univ. Park Campus, Los Angeles, CA 90089-1661, USA

Dr. Ilias Iliopoulos
Université P. et M. Curie, CNRS URA 278, ESPCI Lab. Physico-Macromoléculaire 10, rue Vauquelin, F-75231 Paris Cedex 05, France

Prof. Victor A. Kabanov
Faculty of Chemistry, Dept. of Polymer Science, Moscow State University, Leninskie Gory, Moscow V-234, Russia

Dr. Oh-Kil Kim
Dept. of the Navy, Chemistry Division, Code 6120, Naval Research Laboratory, Washington, DC 20375-5320, USA

Dr. J. Koetz
Institut für Polymerchemie, Kantstr. 55, D-14513 Teltow-Seehof, Germany

Prof. Etsuo Kokufuta
Institute of Applied Biochemistry, University of Tsukuba, Tsukuba, Ibaraki 305, Japan

Dr. K.-J. Linow
Institut für Polymerchemie, Kantstr. 55, D-14513 Teltow-Seehof, Germany

Dr. Benoit Magny
Université P. et M. Curie, CNRS URA 278, ESPCI Lab. Physico-Macro-moléculaire, 10, rue Vauquelin, F-75231 Paris Cedex 05, France

Dr. Anna Malovikovú
Institute of Chemistry, Slovak Academy of Sciences, Dubravska cesta 9, 842 38 Bratislava, Slovakia

Dr. J.J. Maurer
Exxon Research and Engineering Co., Route 22, East, Annandale, NJ 08801, USA

Prof. C.L. McCormick
Dept. of Polymer Science, The University of Southern Mississippi Hattiesburg, MS 39406-0076, USA

Dr. B. Philipp
Max Planck Institut für Kolloid-und Grenzflächenforschung, Kantstr. 55, D-14513 Teltow-Seehof, Germany

G. Rother
Max Planck Institut für Kolloid-und Grenzflächenforschung, Kantstr. 55, D-14513 Teltow-Seehof, Germany

Dr. Donald N. Schulz
Polymers and Fluids Lab., Exxon Research and Engineering Company, Route 22, East Annandale, NJ 08801, USA

Dr. J.-Y. Shieh
Department of Chemical Engineering, Iowa State University, 231 Sweeney Hall, Ames, IA 50011, USA

Prof. Carol A. Steiner
Dept. of Chemical Engineering, The City College of CUNY, Convent Avenue and 140 St. New York, NY 10031, USA

Dr. Shinji Takeoka
Dept. of Polymer Chemistry, Waseda University, Tokyo 169, Japan

Prof. Eihun Tsuchida
Dept. of Polymer Chemistry, Waseda University, Tokyo 169, Japan

Dr. P.L. Valint Jr.
Bausch and Lomb Company, 1400 N. Goodman Street, Rochester, New York 14692, USA

Dr. Ramesh Varadaraj
Exxon Research and Engineering Co., Corporate Research Route 22, East Annandale, NJ 08801, USA

Dr. C.G. Varelas
Dept. of Chemical Engineering, Mackay School of Mines, University of Nevada at Reno, Nevada, USA

Dr. J. Xia
Life Technologies Inc, Gaithersburg, MD 20884-9980, USA

Y.-X. Zhang
Loker Hyrocarbon Research, Inst. & Dept. of Chemistry, University of Southern California, 205 LHI, USC Univ. Park Campus, Los Angeles, CA 90089-1661, USA

Preface

Throughout most of this century, progress in Polymer Science took place by way of studies of dilute polymer solutions and of the solid state. More recently, attention has been focused on the semi-dilute regime. The elucidation of transient network and temporal gel states under conditions of moderate concentration by numerous hydrodynamic and optical techniques is now well established and a solid theoretical underpinning for these phenomena is in place. However, in most of the systems studied, interpolymer interactions are relatively weak forces, because in solutions of typical polymers, strong interpolymer interactions lead to bulk phase separation. However, polymers that form highly associating structures in equilibrium solution now constitute an important class of synthetic macromolecules, and it is evident that strong intermacromolecular interactions – without accompanying phase separation – are prevalent in biopolymer systems as well. This behavior is generally characteristic of amphiphilic polymers, which contain both hydrophilic and hydrophobic repeat units. It is also characteristic of systems in which two macromolecular species are complementary, such as hydrogen-bond-donating and hydrogen-bond-accepting polymers, or polycations and polyanions.

Interest in associating polymer systems currently derives from both applied and fundamental questions. Synthetic water-soluble polymers bearing hydrophobic groups exhibit special rheological properties that make them well-suited as thickeners and viscosity-modifiers. Drag-reduction effects may arise in such polymer solutions, and may also be observed in polymer complexes based on hydrogen-bonding. Viscosity-modification has also been the motivation for many detailed studies of ionomer solutions, in which association via ion-pairing leads to aggregate formation in non-aqueous solvents. Complexes in which one component is a biological polymer also represent technological possibilities. These include protein separation via interactions with synthetic polyions, and modifications of the properties of DNA via complexation.

In this variety of associating polymer systems, a few central questions emerge as universal themes. One concerns the structure of multipolymer aggregates. In any system with strong interpolymer forces, one could anticipate interpolymer collapse, gelation, and phase separation as three reasonable consequences of these forces. Why then do many systems exhibit stable equilibrium aggregate states? In other words, the question "why do chains associate" must always be followed by another, often more difficult query: "how do they know when to stop?" If the association is at least partly open-ended, characterization of these complexes poses a set of related challenges. Since most techniques probe structure in a limited dimensional range (e.g. short range for absorption and fluorescence spectroscopy, medium range – long range for scattering methods, and long-range for rheology and viscometry) the use of a variety of experimental techniques becomes especially valuable.

This book originated from a symposium on "Soluble Polymer Complexes" sponsored by the Polymer Division of the American Chemical Society. To the editors of this volume it seemed clear that polymeric materials that can form ordered aggregates or complexes carry the potential for some novel commercial applications in both polymer chemistry and biotechnology. Consideration of the properties of these materials also raises some interesting scientific questions. These questions, which were formerly the basis for only hypothesis and speculation, appear now to be ideal subject matter for modern instrumental methods such as fluorescence probe studies and dyamic and total intensity scattering. Despite the fact that polymer complexes are of interest to a broad range of specialists, including chemical engineers, synthetic chemists, polymer physicits and biological chemists, the literature in this area is rather fragmented. Two previous books attempted to collect the material in this field. *Intermacromolecular Complexes*, by E. Tsuchida and K. Abe (Springer, 1982) surveyed the literature on both soluble and insoluble complexes and offer extensive descriptions of many specific polymer systems. *Microdomains in Polymer Solutions*, edited by one of us (PD) in 1985 (Plenum) contained reports about amphiphilic polymers and polymer-surfactant systems, about ordering in polyelectrolytes, and about association and aggregation in synthetic and natural polymers. Several new developments have taken place since the publication of those texts: the growth of biotechnology and the related interest in technological applications of polymer-protein interactions; the expansion of fluorescence and scattering techniques for the study of intra- and interpolymer association; and the study of ionomers in solution. These new developments reinforce two points about associating macromolecular systems: (1) they are important in both natural and abiotic systems, and (2) contrary to earlier thought, they are neither so inherently irreversible nor so heterogeneous as to preclude fundamental study.

In organizing the contributions to this volume, the editors first considered that the fundamental driving forces for association provide the best guide to sub-topical arrangement. Thus, we first discuss *hydrophobically* associating polymers, and secondly, *Coulombic* polyelectrolyte complexes. In the case of *Complexes involving proteins*, the subtleties of protein structure préclude the specification of a single type of interaction, but the motivation for grouping these papers together is clear, especially given the focus on protein separations. Lastly, we consider studies of the association of *Ionomers in organic media*, which occupies an important position between the long-standing literature on weakly ionic polymers in the bulk state and investigations of charged polymers in water.

The first seven chapters deal with complexes formed by the association of hydrophobic groups on water-soluble polymer backbones. These polymers pose a synthetic problem because of the incompatibility of the two monomer types. Characterization is also challenging, especially for polymers with low hydrophobe content.

Schulz, Bock and Valint begin by reviewing synthetic routes, through both direct copolymerization and post-polymerization reaction. NMR, UV and pyrolysis-GC methods of characterization are discussed. In the second chapter, *Varadaraj, Branham, McCormick and Bock* report on the use of emission and absorption probes to characterize the microstructure of hydrophobically associating polymers. The results suggest that interpolymer domains are more polar and less organized than conventional micelles. In chapter three, *Bock, Varadaraj, Schulz and Maurer* describe the unusual properties of hydrophobically associating water-soluble polymers. Examples of such properties are enhanced viscosification, shear thickening rheology, decreased sensitivity to salts, and increased mechanical stability. In chapter four, *Magny, Illiopoulos and Audebert* describe the dilute-solution viscometry and fluorescence of random copolymers of sodium acrylate and alkyl (octyl to octadecyl) acrylamide. Upon salt addition, a transition from nonaggregated to aggregated coils occurs. Interchain association and polyelectrolyte effects are superimposed in apparent molecular weight changes. *Varelas, Dualeh and Steiner* discuss, in chapter five, hydrogel networks formed from hydrophobically modified cellulose. Cluster formation is found to depend upon solvent composition, and the number of side-chains per cluster in turn influences the rheology of the network. The hydrophobic microdomains so formed act as solubilization sites for apolar solutes. In chapter six, *Fife* reviews the role of macromolecular association on the polymerization of alkylvinylpyridinium ions. For example, Coulombic interactions control copolymerization with anionic monomers, while hydrophobic forces influence copolymerization with water insoluble monomers. Lastly, *Zhang, Hwang and Hogen-Esch* describe, in chapter seven, the synthesis, characterization and viscosity behavior of "fluorophobe" modified polyacrylamide and cellulose derivatives. These polymers are found to be more efficient viscosity-enhancers than conventional alkyl hydrophobic polymers.

Soluble complexes involving polyelectrolytes are the subject of chapters seven through thirteen. These complexes display structural complexity and sensitivity to environment resulting from their ionic groups. The consequent structures range from colloidal droplets to semicrystalline fibers, in dilute solution, to gels and conducting solids at high concentrations. Two main classes of polyelectrolyte complexes (PEC) are discussed: those governed principally by Coulombic forces, and those strongly influenced by hydrogen bonding.

Dautzenberg, Koetz, Linow, Philipp and Rother describe in chapter eight the use of light scattering to probe the structure of PECs formed from oppositely charged polyacrylamides, thereby addressing what has been a long-standing problem. In chapter nine, *Frugier and Audebert* explore the solution behavior of oppositely charged polyacrylamides of low charge density, with special emphases on the effects of salt and polymer MW on phase separation. Their results point out the need for improved theories in this area. The kinetics of polyelectrolyte complexation and the role of PECs in biology are addressed in chapter ten by *Kabanov*. It is suggested that DNA complexed with polycations may be a "synthetic virus" with exciting applications in transgenic experiments. *Bystricky and Malovikova* examine the role of stereochemistry in a study of the complexation of D- and L-polylysine with anionic polysaccharides; their results in chapter may bear on the understanding of molecular recognition in biological systems. *Tsuchida and Takeoka* review, in chapter twelve, PEC formation – in both solution and solid states – between polyethylene oxide and polyacrytic or polymethacrylic acid. The intriguing application of the solid complexes in electrochromic devices is discussed. Interactions via hydrogen bonds are also the subject of *Kim and Choi's* work in chapter thirteen. Structures formed by high MW polyacrylic acid under high shear are elucidated by the use of fluorescent dyes.

The five subsequent chapters are devoted to protein-polymer complexes. These species are viewed in some cases as providing novel routes to protein separation, but in other situations the protein complex itself may have unique value.

In chapter fourteen, *Hubert and Dellacherie* review the features and limitations of several protein separation techniques based on complex formation. In one case, the hydrodynamic volume of the complex may be the key to enhanced separation. Complexation also can change the protein charge density, lead to the technique of affinophoresis. Two-phase polymeric aqueous systems allow the protein to exhibit preferential partitioning. In a fourth approach to protein separation, insoluble protein-polymer complexes may be isolated and then redissolved. In chapter fifteen, *Xia and Dubin* provide a review of "polyelectrolyte-protein" complexes, emphasizing the different experimental techniques that have been employed and discussing some of the molecular models put forward. *Shieh and Glatz* in chapter sixteen also describe similar phase separation methods for the case of lysozyme purification by precipitation with polyacrylic acid. The authors discuss the critical pH for precipitation, the stoichiometry of the complex, the precipitate's particle size, and the overall protein yield. The phenomenon of coacervation is reviewed by *Burgess* in chapter seventeen from both theoretical and experimental perspectives. The system of gum-arabic and serum albumin is the subject of experimental studies in microcapsule formation. In chapter eighteen, *Kokufuta* summarizes the complexation of proteins with strong polycations and polyanions in salt-free solution. Of special interest is the ability of precipitated proteins to display enzymatic activity.

The final section of the book deals with association of weakly ionic polymers (ionomers) in non-aqueous solvents. This area of study may bridge the gap between the fields of polyelectrolytes and ionomers. In the former case, the charged moieties are dissociated and ionized and their presence leads to chain expansion and interchain repulsion. In the latter case, the low dielectric environment in the solid state polymer precludes dissociation, and the ionic residues interact attractively through dipole-dipole forces, leading to intermacromolecular cohesion. Thus the ways these two systems behave are very different.

Ionomer solutions may display both types of behavior, depending on the dielectric constant of the medium. This question is specifically addressed in chapter nineteen by *Gebel*. The properties of lightly sulfornated polystyrene (LSPS) in polar and nonpolar solvents are discussed in terms of polyelectrolyte and ionomer behavior, respectively. In these two cases, rheological and scattering data are rationalized according to different structural models. Similar analyses are carried out for new data on perfluorinated ionomers. In chapter twenty, *Bodycomb and Hara* use dynamic and static light scattering to study the dependence of the aggregation of LSPS in toluene on ion content and ionomer concentration. They conclude that very large species can form at increased ion content.

We close this preface with expressions of appreciation for the support of the initial symposium from the American Chemical Society, for professional guidance from the staff of Springer, and for the extended cooperation of the contributors.

November 1993

Paul Dubin
J. Bock
Richie Davis
Donald N. Schulz
Curt Thies

Table of Contents

Part I. Hydrophobically Associating Polymers

1	**Synthesis and Characterization of Hydrophobically Associating Water-Soluble Polymers** *D.N. Schulz, J. Bock, and P.L. Valint Jr.*	3
1.1	Introduction	3
1.2	Synthesis	3
1.3	Copolymerization	4
1.4	Postpolymerization Modification	6
1.5	Characterization-Hydrophobe Incorporation	10
1.6	Summary	12
1.7	References	12
2	**Analysis of Hydrophobically Associating Copolymers Utilizing Spectroscopic Probes and Labels** *R. Varadaraj, K.D. Branham, C.L. McCormick, and J. Bock*	15
2.1	Introduction	15
2.2	Emission Probes and Labels	16
2.3	Absorption Probes	27
2.4	Concluding Remarks	30
2.5	References	30
3	**Solution Properties of Hydrophobically Associating Water-Soluble Polymers** *J. Bock, R. Varadaraj, D.N. Schulz, and J.J. Maurer*	33
3.1	Introduction	33
3.2	Polymer Systems	34
3.3	Solubility Characteristics	35
3.4	Solution Stability	37
3.5	Solution Rheological Properties	41
3.6	Summary	49
3.7	References	50

4 **Aggregation of Hydrophobically Modified Polyelectrolytes in Dilute Solution: Tonic Strength Effects**
B. Magny, I. Iliopoulos, and R. Audebert 51
4.1 Introduction 51
4.2 Experimental 52
4.3 The Polymers 53
4.4 Dilute Solution Viscosity 54
4.5 Fluorescence Spectroscopy 56
4.6 Intrinsic Viscosity 58
4.7 Concluding Remarks 61
4.8 References 61

5 **Microdomain Composition in Two-Phase Hydrogels**
C.G. Varelas, A.J. Dualeh, and C.A. Steiner 63
5.1 Introduction 63
5.2 Background 64
5.2.1 Microphase Separation in Polymers 64
5.2.2 Importance of Graft Polymers 65
5.3 Results 65
5.3.1 Hydrogels from Surfactant Solutions 65
5.3.2 Hydrogels from Ethanol/Water Solutions 67
5.3.3 Solute Uptake by Gels 69
5.4 Conclusions 69
5.5 References 69

6 **Molecular Association and Polymerization of 1-Alkyl-4-vinylpyridinium Ions**
W.K. Fife 71
6.1 Complexation in 1-Alkyl-4-vinylpyridinium Ions and Related Polymers 71
6.2 Homopolymerization of 1-Alkyl-4-vinylpyridinium Ions 77
6.3 Copolymerization of 1-Alkyl-4-vinylpyridinium Ions 85
6.4 Conclusion 91
6.5 References 91

7 **Fluorocarbon-Modified Water Soluble Polymers**
Y.-X. Zhang, F.S. Hwang, and T.E. Hogen-Esch 95
7.1 Introduction 95
7.2 Experimental 97
7.3 Results and Discussion 101
7.4 Conclusions 115
7.5 References 116

Part II. Polyelectrolyte Complexes

8 Static Light Scattering of Polyelectrolyte Complex Solutions
H. Dautzenberg, J. Koetz, K.-J. Linow, B. Philipp, and G. Rother 119
8.1 Introduction 119
8.2 Survey of Static Light Scattering Studies on PEC Solutions 120
8.3 Interpretation of Light Scattering Experiments 122
8.4 Experimental 124
8.4.1 Materials 124
8.4.2 Methods of Investigation 125
8.5 Results and Discussion 126
8.6 Conclusion 133
8.7 References 133

9 Interaction Between Oppositely Charged Low Ionic Density Polyelectrolytes: Complex Formation or Simple Mixture?
D. Frugier and R. Audebert 135
9.1 Introduction 135
9.2 Material and Techniques 138
9.2.1 Polymer Synthesis 138
9.2.2 Polymer Characterization 138
9.2.3 Other Techniques 139
9.3 Phase Diagram 139
9.3.1 Phase Diagram Representation 139
9.3.2 Influence of the Charge Density 139
9.3.3 Influence of the Ionic Strength 140
9.3.4 Influence of the Molecular Weight of the Samples 141
9.3.5 Phase Diagram and Complex Formation 141
9.4 Polymer-Polymer Affinity and Phase Diagram 144
9.5 Conclusion 148
9.6 References 148

10 Basic Properties of Soluble Interpolyelectrolyte Complexes Applied to Bioengineering and Cell Transformations
V.A. Kabanov 151
10.1 Introduction 151
10.2 Kinetic and Equilibrium Properties of Interpolyelectrolyte Complexes 153
10.3 Interpolyelectrolyte Complexes as Protein Carriers 161
10.4 Complexes of DNA with Synthetic Polycations for Cell Transformation 167
10.5 Conclusion 173
10.6 References 173

11 Conformation Presumption for Polysaccharide-Polylysine Complexation
S. Bystricky and A. Maloviková 175
11.1 Introduction 175
11.2 Complex Formation 175
11.3 Pectate-Polylysine Interaction 176
11.4 Polyguluronate Rich Alginate-Polylysine Interaction 178
11.5 Polymannuronate Rich Alginate-Polylysine Interaction 180
11.6 Conclusion 181
11.7 References 182

12 Interpolymer Complexes and their Ion-Conduction
E. Tsuchida and S. Takeoka 183
12.1 Introduction 183
12.2 Classification of Interpolymer Complexes 184
12.3 Formation of Interpolymer Complexes from PAA with POE . 186
12.4 Thermodynamics of Interpolymer Complexes from PAA (or PMMA) with POE 192
12.5 Selective and Substitution Interpolymer Complexation 195
12.6 Solid Properties of a Hydrogen-Bonding Complex 200
12.7 Ion Conduction and Solid Polymer Electrolytes 202
12.8 Ion Conduction of Hydrogen-Bonding Complexes 206
12.9 References 211

13 Fluorescence Probe Studies of Poly(acrylic acid) Interchain Complexation Induced by High Shear Flow and Influence of Cationic Surfactants on the Complexation
O.-K. Kim and L.-S. Choi 215
13.1 Introduction 215
13.2 Experimental 216
13.2.1 Materials 216
13.2.2 Flow Processing 217
13.2.3 Fluorescence Measurements 217
13.3 Results and Discussion 217
13.3.1 Drag Reduction (DR) and PAA Conformation 217
13.3.2 Local Chain Rigidity 220
13.3.3 Hydrophobic Association 221
13.3.4 Hydrophobe-Assisted Rigidity 222
13.4 References 226

Part III. Biopolymer Systems

14 Water-Soluble Biospecific Polymers for New Affinity Purification Techniques
P. Hubert and E. Dellacherie 229

14.1 Introduction 229
14.2 Discrimination on the Basis of High Molecular Weight 231
14.2.1 Biospecific Ultrafiltration 231
14.2.2 Biospecific Gel Filtration 232
14.3 Discrimination on the Basis of High Density of Charges: Affinophoresis 236
14.4 Discrimination on the Basis of Surface Tension Properties: Affinity Partition 237
14.5 Discrimination on the Basis of Reversible Solubility: Affinity Precipitation 241
14.6 Advantages and Drawbacks of Techniques Involving Water-Soluble Biospecific Polymers 243
14.7 References 245

15 Protein-Polyelectrolyte Complexes
J. Xia and P.L. Dubin 247
15.1 Introduction 247
15.2 Investigation Methods 247
15.3 Factors Influencing Protein-Polyelectrolyte Complexation and Structures of the Protein-Polyelectrolyte Complexes 260
15.4 Protein Separation by Polyelectrolytes 265
15.5 Enzymes in Polyelectrolyte Complexes 268
15.6 Conclusion 269
15.7 References 270

16 Precipitation of Proteins with Polyelectrolytes: Role of Polymer Molecular Weight
J.-Y. Shieh and Ch.E. Glatz 273
16.1 Introduction 273
16.2 Materials and Methods 274
16.3 Results and Discussion 276
16.4 Conclusions 283
16.5 References 284

17 Complex Coacervation: Micro-Capsule Formation
D.J. Burgess 285
17.1 Introduction and Terminology 285
17.2 Simple Coacervation 286
17.3 Complex Coacervation 287
17.4 Theory of Complex Coacervation 289
17.5 Coacervation as a Method of Microencapsulation 291
17.6 Materials and Methods 294
17.7 Results 296
17.8 Conclusions 299
17.9 References 299

18 **Complexation of Proteins with Polyelectrolytes in a Salt-Free System and Biochemical Characteristics of the Resulting Complexes**
E. Kokufuta ... 301
18.1 Introduction ... 302
18.2 Experimental Section ... 303
18.3 Results and Discussion ... 308
18.4 Conclusions and Topics for Future Research ... 323
18.5 References ... 324

Part IV. Ionomers in Solution

19 **Ionomer Solutions: Polyelectrolyte or Ionomer Behavior**
G. Gebel ... 329
19.1 Introduction ... 329
19.2 Sulfonated Polystyrene Ionomer Solutions in Nonpolar Solvents 331
19.3 Sulfonated Polystyrene Ionomer Solutions in Polar Solvents .. 337
19.4 Perfluorinated Ionomer Solutions ... 340
19.5 Conclusion ... 344
19.6 References ... 345

20 **Scattering Studies of Ionomer Aggregates in Nonpolar Solvents**
J. Bodycomb and M. Hara ... 347
20.1 Introduction ... 347
20.2 Experimental ... 350
20.3 Light Scattering Analysis ... 350
20.4 Results and Discussion ... 352
20.5 References ... 356

Subject Index ... 357

Part I

Hydrophobically Associating Polymers

1 Synthesis and Characterization of Hydrophobically Associating Water-Soluble Polymers

D.N. Schulz, J. Bock and P.L. Valint Jr.

Hydrophobically associating water-soluble polymers are of interest because of their unusual properties in solution; e.g. enhanced viscosification efficiency, shear-thickening rheology, shear stability, and salt stability. However, such macromolecules are difficult to prepare and analyze. This chapter reviews synthetic routes designed to overcome the problems of mixing and reacting oil-soluble and water-soluble reagents/monomers. Both direct copolymerization and post-polymerization methods are described. It also presents methods for determining the amounts of hydrophobe incorporated by NMR, UV, and degradation/GC.

1.1 Introduction

Hydrophobically associating water-soluble polymers are of interest because of their unusual properties in solution. Even small levels of hydrophobic groups on water-soluble polymer backbones can have a profound effect on aqueous solution rheology. For example, solutions of such polymers show enhanced viscosification efficiency, shear-thickening rheology, as well as improved shear and salt stability [1–19].

Nevertheless, hydrophobically associating water-soluble polymers present challenges to both the synthetic and characterization chemists. Hydrophobically associating water soluble polymers are often difficult to synthesize because of problems associated with mixing oil-soluble and water-soluble reagents/monomers. Similarly, since only small levels of hydrophobe are needed for rheology control, characterization of hydrophobe contents often stretch the detection limits of conventional polymer characterization methods.

This chapter reviews recent advances in the synthesis and characterization of hydrophobically modified water-soluble polymers. The synthetic part describes both direct synthesis via copolymerization and post-polymerization methods. The characterization section presents methods for measuring/estimating hydrophobe incorporation.

1.2 Synthesis

There are basically two ways to incorporate hydrophobic moieties into water-soluble polymer chains; i.e. direct copolymerization of hydrophobic and water

Macromolecular Complexes
in Chemistry and Biology
Eds.: Dubin/Bock/Davis/Schulz/Thies

soluble monomers and post-polymerization functionalization. Each method has its advantages and disadvantages.

1.3 Copolymerization

Direct copolymerization of both oil-soluble and water-soluble monomers is difficult because of the phase incompatibility of hydrocarbon and water soluble monomers. Thus, simply dispersing the hydrophobic monomers in the aqueous phase with the hydrophilic monomer leads to a heterogeneous product with low hydrophobe incorporation.

Use of mutual solvents (e.g. alcohols) or solvent mixtures to dissolve both the water and oil-soluble monomers also has some serious limitations [4]. For example, even though a common solvent can be found for the monomers, it is unlikely that the polymer will also be soluble in such a medium. Thus, the polymer will precipitate before it has built up a sufficient molecular weight for use as a viscosifier. In addition, most of the possible water miscible solvents (e.g. alcohols, ether, acetone) are chain transfer agents for free radical polymerization, and their presence leads to low molecular weight products.

Better direct polymerization methods involve the use of micellar copolymerization or use of surfactant macromonomers. In the micellar copolymerization method [8, 13], an oil-soluble acrylamide; (e.g. C_8-alkyl acrylamide) is solubilized in the aqueous phase by appropriate surfactant(s) and copolymerized with water-soluble acrylamide (AM) monomer (Eq. 1.1). The resulting product is a copolymer of acrylamide and C_8-alkyl acrylamide (RAM).

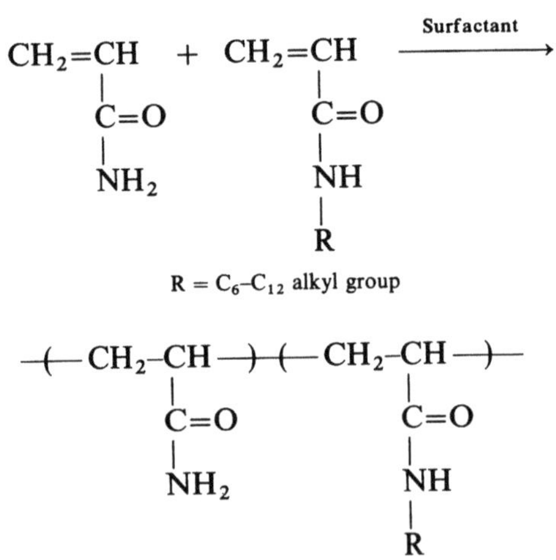

RAM (1.1)

Typical surfactants for neutral or anionically charged hydrophobic polymers are sodium dodecyl sulfate or nonionic ethoxylates. For cationic polymers, either cationic or nonionic surfactants are used. Fluorinated surfactants are needed to micellize fluorophobe (fluorinated hydrophobe) monomers [16]. The surfactant type and concentration are chosen to provide solubilization of the hydrophobic monomer so that there are only a few hydrophobic monomer groups per micelle. This reduces the tendency for blocky incorporation and consequently yields a more random copolymer.

In the surfactant macromonomer method, [9–11, 17] a water-soluble (or dispersible) surfactant, nonylphenoxy poly(ethylene oxide), is built into an acrylate monomer. The surfactant macromer has both a hydrophilic head group,$-(CH_2CH_2O)-_n$, and a hydrophobic tail ($-C_6H_4-R$). Thus, no external surfactant is needed to solubilize the hydrophobic unit during copolymerization. This macromonomer copolymerizes easily to form a hydrophobically associating polyacrylamide (PAM-Surf) (Eq. 1.2).

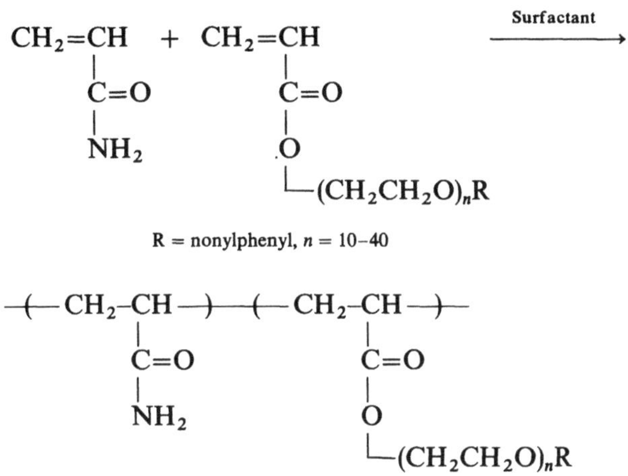

(1.2)

PAM-Surf

A comparison of typical recipes for synthesizing RAM and PAM-Surf polymers is shown in Table 1.1 [20].

Table 1.1. Synthesis of hydrophobically associating PAM's typical recipes[a]

RAM (*Micellar Pzn*)	PAM-Surf (*Surfactant Macromonomer*)
○ 470.7 g H_2O	○ 500 g H_2O
15.85 g Sodium, dodecyl sulfate	30.0 g Acrylamide
14.76 g Acrylamide	1–5 g Nonyl Phenoxy
0.288 g Octyl acrylamide	(etheroxy) acrylate
○ Above heated to 50 °C	○ Above heated to 50 °C
○ 0.01 g $K_2S_2O_8$ Added	○ 0.01 g $K_2S_2O_8$ Added

[a] Reprinted with permission from Schulz DN In: Shalaby SW, McCormick CL, Butler GB (eds): Water-soluble polymers: synthesis, solution properties, and applications, ACS Symposium Series 467, Am Chem Soc, Washington, DC, p 60

1.4 Post Polymerization Modification

A number of hydrophobically modified water-soluble polymers have been prepared via post-polymerization reaction. For example, Straus [21] has prepared polysoaps by the quarternization of poly(2-vinylpyridine) with *n*-dodecyl bromide (Eq. 1.3).

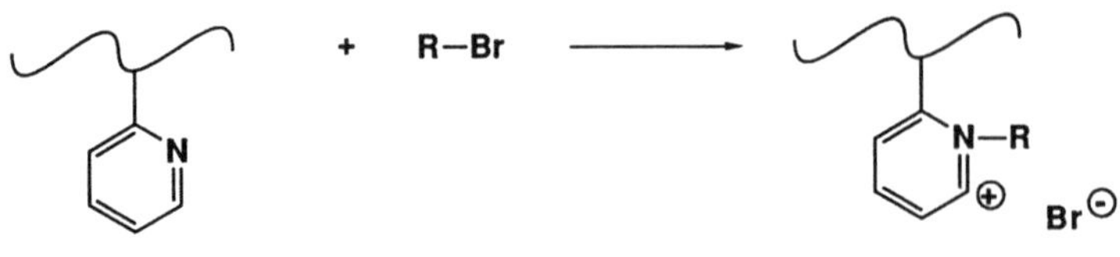

(1.3)

Polysoaps are highly substituted hydrophobically modified polymers that exhibit abnormally compact coils because of the overwhelming effects of intramolecular association, and micellization.

Straus [22–24] has modulated the effects of association by varying the number of soap molecules, while holding the total linear charge density the same. This condition has been achieved by quaternizing poly(4-vinylpyridine) to various extents with highly hydrophobic *n*-dodecyl bromide and completing the process with the less hydrophobic ethyl bromide to form hydrophobic polyelectrolytes (**1**).

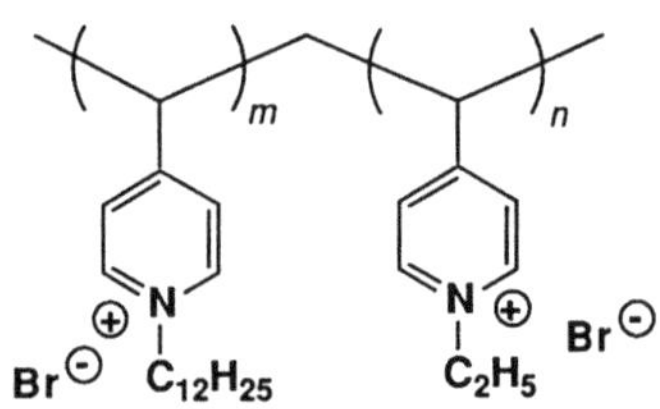

1

Hydrophobically modified hydroxyethyl cellulose (HEC) has also been synthesized by post-polymerization reaction. HEC is reacted with low levels of certain reactive hydrophobes; e.g. long chain alkyl epoxides, alkyl halides, acyl halides, isocyanates or anhydrides [1, 2, 25]. (Scheme 1.1 [3]). To handle the incompatibility of the hydrophobic reagent and hydrophilic polymer, the reaction is run in solution with a common solvent or in slurry media. Slurry methods are preferred because of their lower reactor viscosities. A typical slurry process involves swelling HEC with isopropanol, adding NaOH in H_2O, followed by addition of a long chain epoxide; such as 1,2-epoxydecane, 1,2-epoxydodecane.

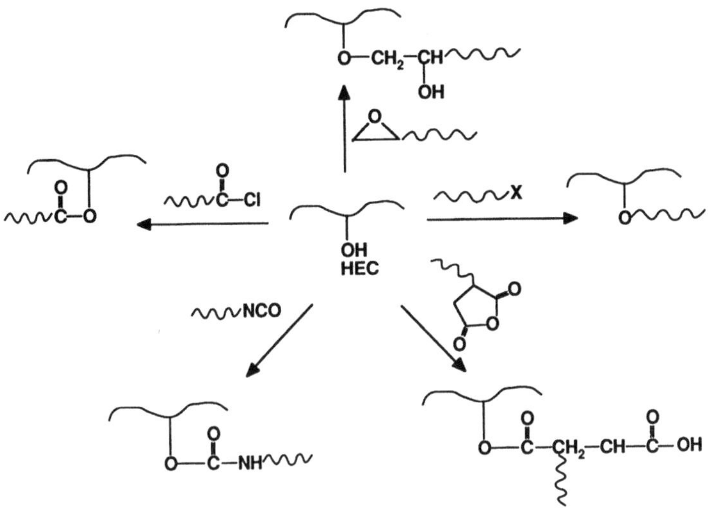

Scheme 1.1

Recently, Hogen-Esch [26] has modified HEC with fluorocarbon tosylates and epoxides (Eq. 1.4, 1.5).

$$\mathrm{Cel[OCH_2CH_2ONa] + TSOCH_2(CF_2)_\mathit{n}CF_3 \rightarrow}$$
$$\mathrm{Cel[OCH_2CH_2OCH_2(CF_2)_\mathit{n}CF_3]} \qquad (1.4)$$
$$\mathrm{Cel[OCH_2CH_2ONa] + \underset{\quad O}{CH_2{-}CH}{-}CH_2OCH_2(CF_2)_\mathit{n}CF_3 \rightarrow \rightarrow}$$
$$\mathrm{Cel[OCH_2CH_2OCH_2CH(OH)CH_2OCH_2(CF_2)_\mathit{n}CF_3]} \qquad (1.5)$$

Fluorocarbon-modified HEC is a more potent viscosifier than hydrophobe-modified HEC because fluorocarbons can micellize at lower concentrations than alkyl hydrophobes. Moreover, the fluorocarbon epoxide synthetic route gives fluorocarbon modified HEC polymers with a better balance of solubility and viscosification efficiency than does the fluorocarbon tosylate method.

In addition, hydrophobically modified ethoxylated urethanes (HEUR) are prepared by hydrophobe capping of urethane prepolymers. Comb, block and star HEUR's have been synthesized. (Schemes 1.2–1.4) [27]. Such polymers are very effective associative thickeners for urethane paints and coatings.

Iliopoulous [28, 29] has developed a novel method for hydrophobic post-polymerization modification of poly(acrylic acid) by long chain amines in an aprotic solvent, 1-methyl-2-pyrrolidone (MPD), in the presence of *N*,*N*-dicyclohexylcarbodiimide (DCC) (Eq. 1.6).

Treatment with base yields hydrophobically-modified poly(sodium acrylate) (Eq. 1.6).

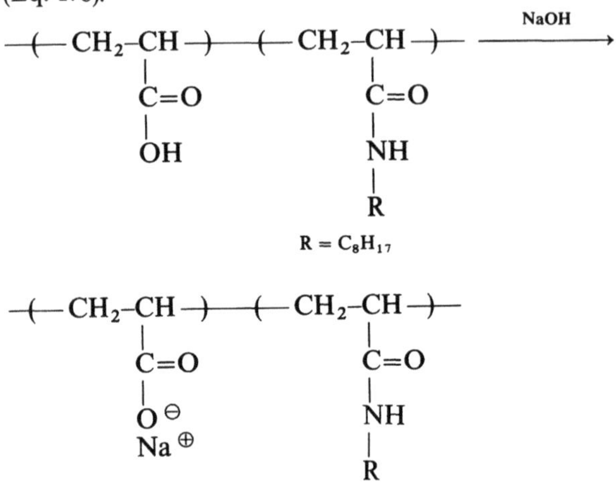

(1.6)

This synthetic method yields a product with a random distribution of alkyl groups and an unchanged backbone $\overline{DP}$.

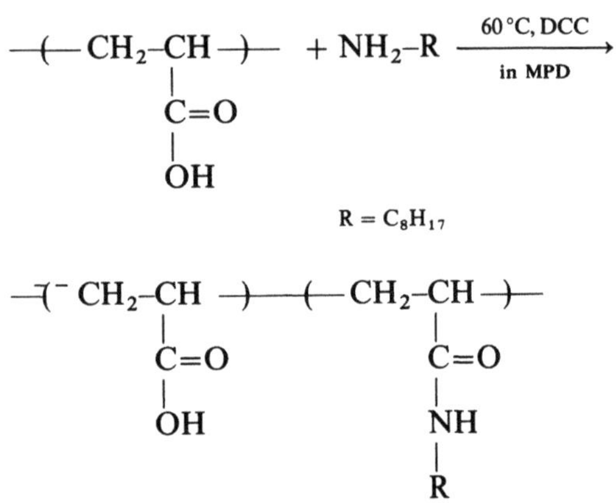

(1.7)

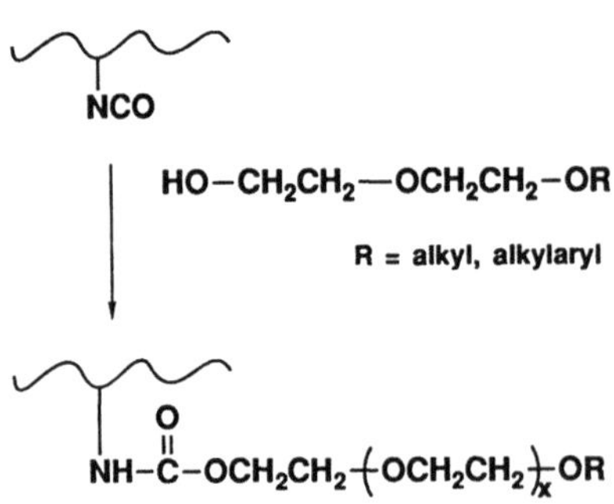

Comb HEUR

Scheme 1.2

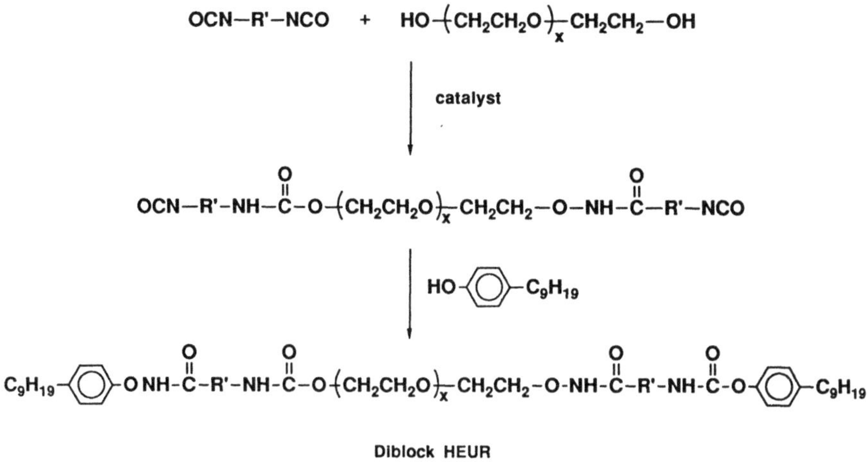

Scheme 1.3

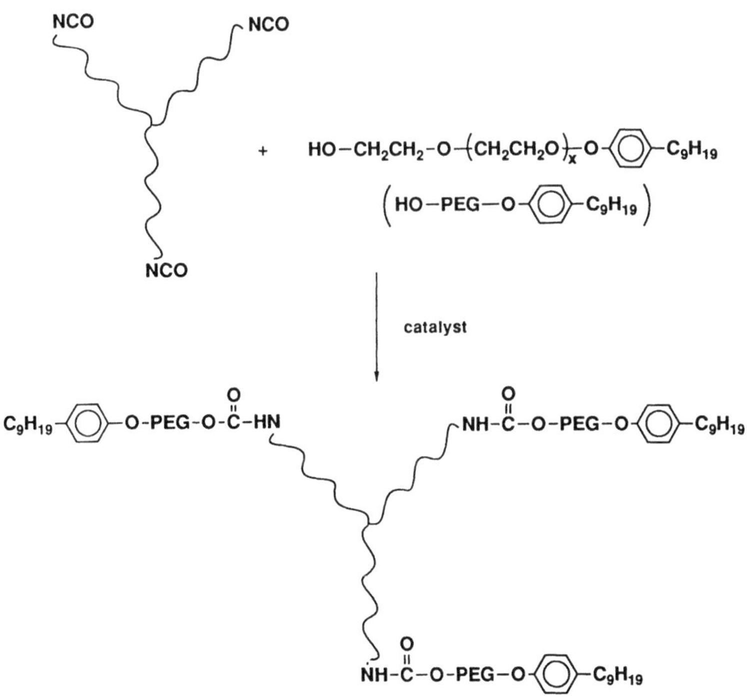

Scheme 1.4

As seen from the above examples, most of these hydrophobe modification processes involve introducing a hydrophobe onto a water soluble polymer (WSP) backbone. However, Bock and Valint [30] have reversed the process; they have added solubilizing ionic moieties to a hydrophobic block copolymer, poly(*t*-butylstyrene-*b*-styrene), backbone by selective sulfonation of the polystyrene block (Scheme 1.5). The resulting material is a novel hydrophobically associating block copolymer.

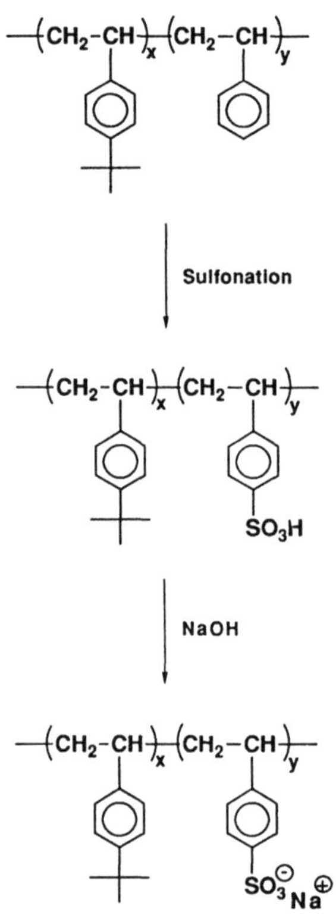

Scheme 1.5

1.5 Characterization–Hydrophobe Incorporation

Evidence for hydrophobe incorporation is easy to observe in the solution properties (e.g. viscosity) of water soluble polymers containing such groups. It is also possible to develop inferential relationships between levels of hydrophobic monomer or reagent charged to the reaction vessel and the final solution viscosity of the polymer. However, detailed analytical data on hydrophobe incorporation, especially for simple alkyl hydrophobes has proven difficult to obtain. Among the characterization techniques that have been used are NMR,

UV, degradation/GC and recently the use of ^{14}C labeled monomers. (The latter method will be the subject of a future report.)

NMR determinations are not sensitive enough to detect low levels of hydrophobe (i.e. 1–2%). Yet, such low levels are all that is needed for viscosification if the backbone molecular weight is high (i.e. $> 10^6$–10^7 g/mol).

On the other hand, ^1H and ^{13}C NMR [31] has been successfully used to determine the hydrophobe level for hydrophobically modified poly(sodium acrylates) of lower molecular weights (i.e. $< 10^5$ g/mol) and high (up to 30%) hydrophobe contents, provided the pH = 8 and the alkyl group is kept small ($< C_8$). A deaggregating but nonprecipitating solvent pair (i.e. D_2O/CD_3OD, to 70/30) is used to improve resolution.

The sequence distribution [31] of hydrophobic comonomer in hydrophobic poly(sodium acrylates) has also been obtained for low molecular weight/high hydrophobe materials. These polymers have been found to obey Bernoullian statistics.

On the other hand, UV is a probe that is sensitive enough to measure low levels of hydrophobes, provided that there is a UV active chromophore in the hydrophobic segment. Incorporation levels of *N*-aryl substituted acrylamide (**2**) and nonyl phenoxy (etheroxy) acrylates (**3**) into polyacrylamide backbones have been determined using such UV methods [14, 17]. The absorbtivity of **2** has been measured by comparison of the absorptivity of the hydrophobic unit in the polymer to that of the model compound (**4**). Assuming the absorptivity of the hydrophobe in the polymer is similar to that of the model, polymer absorptivity to the model gives the weight fraction of the hydrophobe in the polymer.

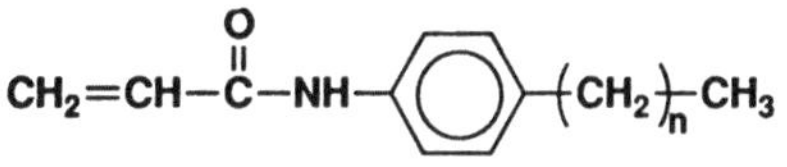

n = 1,3

2

n = 10-40

3

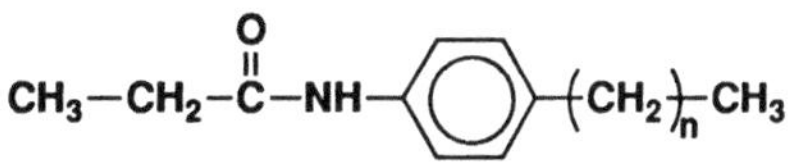

n = 1,3

4

In turn, the absorptivity of (**3**) in the copolymer has been determined using the monomer itself as a model [16]. The amount of **3** in the copolymer is calculated after correction for low level acrylamide absorption. Absorbance vs. concentration plots for both **3** and **4** followed Beer's Law.

Finally, degradation/GC have been used to determine the hydrophobe content in certain hydrophobe modified natural polymers. For example, the hydrophobe content of hydrophobe modified HEC have been determined by exhaustive cleavage of the polymer by aqueous hydroiodic acid followed by GC analysis of the iodoalkane formed [3].

1.6 Summary

Despite their utility, hydrophobically associating water soluble polymers are difficult to synthesis and characterize. This chapter reviews the synthetic routes developed to overcome problems associated with mixing oil-soluble and water-soluble reagents/monomers. It also describes methods for determining the amount of hydrophobe incorporated via NMR, UV, and degradation/GC.

Acknowledgments. The authors wish to thank A. Ozbun for typing this manuscript and P. Wright for drawing the chemical structures. Appreciation is also extended to the Exxon Research and Engineering Company for permission to publish this work.

1.7 References

1. Landoll LM (1980) U.S. Patent 4 228 277
2. Landoll LM (1982) J Polym Sci Chem Ed 20:443
3. San AC, Landoll LM (1989) In: Glass JE (ed Polymers in aqueous media: performance through association. Advances in Chemistry Series 223, Am Chem Soc, Washington, DC, p 343
4. Emmons WD, Stevens TE (1985) U.S. Patent 4 395 524
5. Evani S (1984) U.S. Patent 4 432 801
6. Turner SR, Siano DB, Bock J (1985) U.S. Patent 4 502 182
7. Turner SR, Siano DB, Bock J (1985) U.S. Patent 4 521 580
8. Turner SR, Siano DB, Bock J (1985) U.S. Patent 4 528 348
9. Schulz DN, Maurer JJ, Bock J (1984) U.S. Patent 4 463 151
10. Schulz DN, Maurer JJ, Bock J (1984) U.S. Patent 4 463 152
11. Maurer JJ, Schulz DN, Bock J (1986) U.S. Patent 4 571 926
12. Costein VG, King T (1985) U.S. Patent 4 541 935
13. Bock J, Valint PL, Pace SJ, Siano DB, Schulz DN, Turner SR (1988) In: Stahl GA, Schulz DN (eds) Water soluble polymers for petroleum recovery. Plenum, New York, p 147
14. Valint Jr. PL, Bock J, Schulz DN (1989) In: Glass JE (ed) Polymers in aqueous media: performance through association. Advances in Chemistry Series 223, Am Chem Soc, Washington, DC, p 399
15. Bock J, Siano DB, Valint Jr. PL, Pace SJ (1989) In: Glass JE (ed) Polymers in aqueous media: performance through association. Advances in Chemistry Series 223, Am Chem Soc, Washington, DC, p 412
16. Zhang YX, Da AH, Hogen-Esch TE, Butler GB (1991) In: Shalaby SW, McCormick CL, Butler GB (eds) Water soluble polymers: synthesis, solution properties and application. ACS Symposium Series 467, Am Chem Soc, Washington, DC, p 159

17. Schulz DN, Kaladas JJ, Maurer JJ, Bock J, Pace SJ, Schulz WW (1987) Polymer 28:2110
18. McCormick CL, Johnson CB (1988) In: Stahl GA, Schulz DN (eds) Water soluble polymers for petroleum recovery. Plenum, New York, p 161
19. McCormick CL, Johnson CB (1989) In: Glass JE (ed) Polymers in aqueous media: performance through association, Advances in Chemistry Series 223, Am Chem Soc, Washington, DC, p 437
20. Schulz DN In: Shalaby SW, McCormick CL, Butler GB (eds) Water soluble polymers; synthesis, solution properties and application. ACS Symposium Series 467, Am Chem Soc, Washington, DC, p 60
21. Strauss UP (1951) J Polym Sci 6:649
22. Strauss UP, Gershfeld NL (1954) J Phys Chem 58:747
23. Strauss UP, Gershfeld NL (1958) J Phys Chem 60:577
24. Strauss UP (1989) In: Glass JE (ed) Polymers in aqueous media: performance through association. ACS Advances in Chemistry Series 223, Am Chem Soc, Washington, DC, ch. 16
25. Hoy KL, Hoy RC (1984) U.S. Patent 4426485
26. Hwang F, Hogen-Esch TE (1991) Am Chem Soc Div of Polym Chem Prepr 32:591
27. Karunsena A, Brown GB, Glass JE (1989) In: Glass JE (ed) Polymers in aqueous media: performance through association. ACS Advances in Chemistry Series 223, Am Chem Soc, Washington, DC, ch. 26
28. Wang TK, Iliopoulos I, Audebert R (1988) Polym Bull 20:577
29. Wang TK, Iliopoulos I, Audebert R (1989) In: Shalaby SE, McCormick CL, Butler GB (eds) Water soluble polymers: synthesis, solution properties, applications. ACS Advances in Chemistry Series 467, Am Chem Soc, Washington, DC, p 218
30. Valint Jr. PL, Bock J (1988) Macromolecules 21:175
31. Mogny B, Lafuma F, Iliopoulos I (1992) Polymer, in press

2 Analysis of Hydrophobically Associating Copolymers Utilizing Spectroscopic Probes and Labels

R. Varadaraj, K.D. Branham, C.L. McCormick, and J. Bock

Over the past decade much of the research on hydrophobically associating polymers have focused on synthesis and rheological property characterization. In comparison, relatively fewer studies have addressed microstructural details. Aqueous solutions of hydrophobically associating polymers have the potential to form unique microheterogeneous systems. Exploring microstructural features on these soluble polymer complexes and differentiating them from well-characterized common microheterogeneous systems like micelles, microemulsions, and vesicles is important and can add value to the efforts in synthesis and rheological property characterization. The spectroscopic probe approach has been used extensively to study a variety of microheterogeneous systems with varying degrees of success. In this chapter the use of emission and absorption probes to study hydrophobically associating polymers in aqueous media is described.

2.1 Introduction

Water-soluble copolymers possessing intra- or intermolecular modes of hydrophobic associations have been the focus of considerable research over the past decade in both academic and industrial laboratories. Impetus for the research has resulted from the utility of these water-based, microheterogeneous systems in applications including surface coatings, personal care and oil field formulation, flocculation, rheology modification, and colloid stabilization [1].

In this report we focus on major classes of copolymers with hydrophobic, *intermolecular* associations in aqueous media. These copolymers are termed "associative thickeners" and display high viscosities at polymer concentrations above the critical overlap concentration (C^*) (Fig. 2.1). Importantly, properties can be controlled by synthetic manipulation of hydrophobic/hydrophilic sequences in the copolymer microstructure. The rheological properties are, of course, also a function of concentration, temperature, added electrolytes, shear stress, etc. A challenge to polymer chemists is determining key features of these soluble polymer complexes which differentiate them from well characterized microheterogeneous systems such as micelles and vesicles. Since associating polymers form hyprophobic domains in aqueous solutions, many fundamental questions may be asked: how hydrophobic are the associated regions? what are the size and nature of these aggregates? is there a relationship between hydrophobic microstructure and macroscopic properties? how may we perturb or trigger these hydrophobic associations through addition of salt or surfactants, or through changes in pH or temperature? In this chapter the role of

Macromolecular Complexes
in Chemistry and Biology
Eds Dublin/Bock/Davis/Schulz/Thies

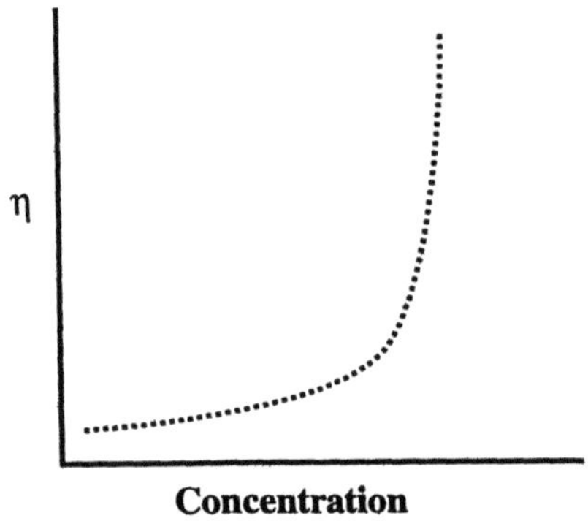

Fig. 2.1. Viscosity as a function of concentration for an associative thickener

emission probes and labels, as well as absorption probes, in studying such questions is described.

2.2 Emission Probes and Labels

Fluorescence emission studies on hydrophobically associating polymers using a variety of probes and labels appear in the literature. The majority of these studies have been performed on three distinct classes of polymers: 1) hydrophobically-modified water-soluble cellulosics, 2) polyethylene oxide (PEO) polymers end-capped with hydrophobic groups and, 3) hydrophobically-modified polyacrylamides. Each type of polymer is generated by quite different synthetic methods, resulting in different polymer microstructure. Knowledge of polymer microstructure is often an important factor in interpreting fluorescence data with respect to the macroscopic organization of the hydrophobic aggregates.

2.2.1 Hydrophobically-Modified Cellulosics

Fluorescence studies of pyrene-labeled hydroxypropyl cellulose (HPC-Py) were performed by Winnik and coworkers [2]. Hydroxypropyl cellulose (M_n 36 000) was reacted with sodium hydride to give the alkoxide anion, which was subsequently reacted with 4-(1-pyrenyl)butyl tosylate to yield HPC-Py having the label attached through hydrolytically stable ether linkages (Fig. 2.2). The pyrene labels were assumed to be distributed randomly along the semirigid polymer backbone for HPC-Py/216 (1 pyrene per 216 glucose units) and HPC-Py/26 (1 pyrene per 26 glucose units).

Evidence for intermolecular aggregation was presented in terms of the degree of excimer fluorescence to that of monomer fluorescence. The ratio of the fluorescence intensity of "excited dimer" or excimer, formed by overlap of two pyrene labels, relative to monomer fluorescence (I_E/I_M) may be used as an indicator of polymer–polymer interactions, providing intramolecular excimer formation can be differentiated from intermolecular excimer formation. (I_E/I_M)

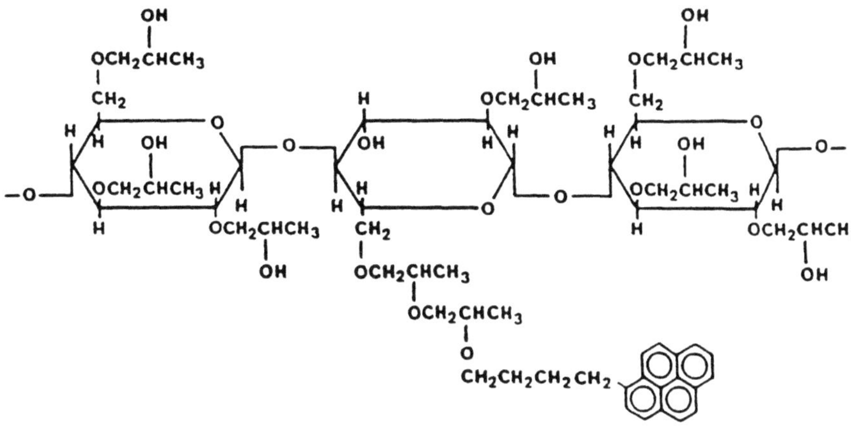

Fig. 2.2. Idealized structure of pyrene-labeled (hydroxypropyl)cellulose

was measured as a function of Py-HPC concentration, as well as a function of methanol content in methanol/water mixtures. Enhancement of I_E/I_M with decreasing methanol content demonstrated that hydrophobic associations between the pyrenyl groups occurred in aqueous solutions. Measurements of I_E/I_M against polymer concentration (HPC-Py/26) were also reported for pure water. These measurements, indicating the presence of both intra- and intermolecular associations of the pyrenyl groups, are shown in Fig. 2.3. The value of I_E/I_M at 0.2 ppm demonstrated that some intramolecular excimer formation occurs. As the polymer concentration was increased to 200 ppm (Fig. 2.3a), an increase in I_E/I_M was noted, consistent with interpolymer associations. Next, a dilution experiment was performed (Fig. 2.3b) in which HPC-Py/26 was diluted with unlabeled HPC. The total polymer concentration was maintained at 130 ppm. As the concentration of labeled polymer (and thus the concentration of pyrene label) decreased, I_E/I_M decreased. This was interpreted as an indication that the labels were not driving the associations, but that associations were predominantly due to hydrophobic associations through other parts of the structure.

Excitation spectra of HPC-Py/26 in water revealed a red-shift in the excimer species relative to monomer. This, along with the lack of a detectable rise time in the fluorescence decay profile of the excimer species, indicated that the excimers exist predominantly as ground-state aggregates. This result is a very significant indication of the nature of the aggregates as well as the association mechanism.

Temperature studies on similar HPC-Py polymers were also performed [3]. Unmodified HPC was found to have an LCST of 42.5 °C, measured by turbidimetry. Correlation of fluorescence and turbidimetric data for two samples, HPC-Py/438 and HPC-Py/56, were reported. The temperature dependent behavior of the polymer containing smaller amounts of label was very similar to HPC. However, HPC-Py56 behaved differently, showing an LCST at 44 °C. I_E/I_M measurements indicated the presence of aggregates at room temperature,

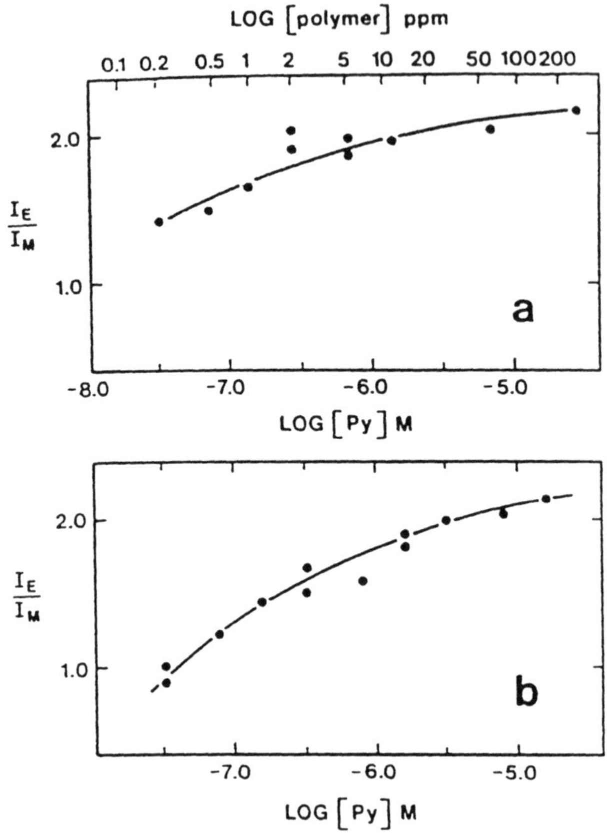

Fig. 2.3 a,b. Plots of I_E/I_M as a function of log [HPC-Py/26]: **a** in water solution; **b** in water containing unlabeled HPC

as noted above. As the temperature was increased to approximately 40 °C, I_E/I_M increased suggesting the formation of dimers and higher aggregates. This was mirrored by a concomitant increase in the solution turbidity with a maximum at 40 °C, also indicating the formation of polymeric aggregates below the LCST. As the temperature was raised above 40 °C, I_E/I_M decreased, indicating the break-up of pyrenyl label associations due to increased thermal energy. This trend continued up to the LCST at 44 °C, where I_E/I_M reached a plateau value. At approximately the onset of the plateau temperature of I_E/I_M, the turbidity of the solution increased rapidly, indicating the onset of precipitation and thus the LCST of the polymer. A summary of this behavior is presented in Fig. 2.4. These data show a correlation between the molecular level (via fluorescence measurements) and the 1000 Å level (via turbidimetric measurements). Other evidence for interpolymer associations, at the 10–100 Å level was reported [4] through energy transfer measurements of HPC-Py polymers and HPC polymers labeled in a similar fashion with fluorene chromophores.

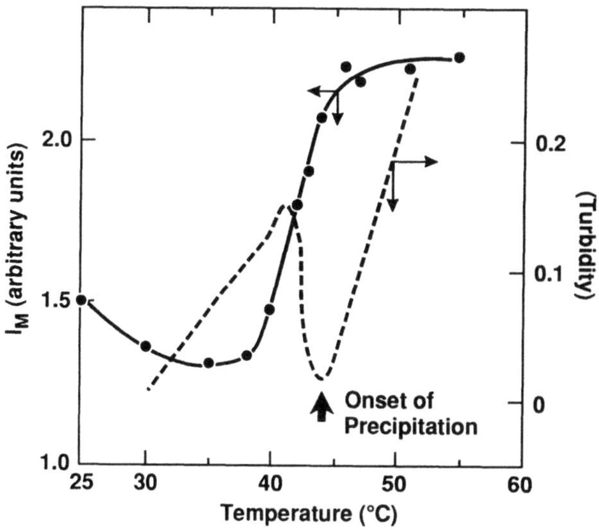

Fig. 2.4. Pyrene monomer emission intensity (I_M) as s function of temperature for a solution of HPC-Py/56 in water (0.12 g/l) and average turbidity for a solution of HPC-Py/56 in water (0.40 g/l) as a function of temperature

2.2.2 Hydrophobically-Modified PEO Copolymers

The second class of polymers, end-capped PEO polymers, have been studied using both probe and label experiments. In the first study, Frank and coworkers [5] reported fluorescence studies on PEO polymers of varied molecular weight end-labeled with pyrene. The polymers (Fig. 2.5) in this study had M_w values of 4800, 9200 and 11 000.

I_E/I_M values were measured as a function of polymer concentration for the three samples (Fig. 2.6). The initial plateau regions of the curves are indicative of intramolecular excimer formation due to end-to-end cyclization of the flexible PEO chains. The portions of the curves with positive slopes show increasing I_E/I_M with increasing polymer concentration due to intermolecular excimer formation. The break in the curves occurs at approximately the same molar concentration of polymer, regardless of molecular weight. This indicates definitively, that intermolecular excimer formation is driven by hydrophobic associations of the pyrenyl groups and not polymer chain overlap. This observation was supported by measurements of I_E/I_M as a function of concentration in water/methanol mixtures of varying methanol contents. I_E/I_M remained constant over the concentration range observed, indicating the excimer formation was intramolecular in these solvent systems.

In the second study Wang and Winnik [6] reported pyrene probe fluorescence experiments on PEO polymers end-capped with *n*-hexadecyl groups through urethane linkages. These polymers are examples of hydrophobically-modified ethoxylated urethane (HEUR) associative thickeners (Fig. 2.7). Two

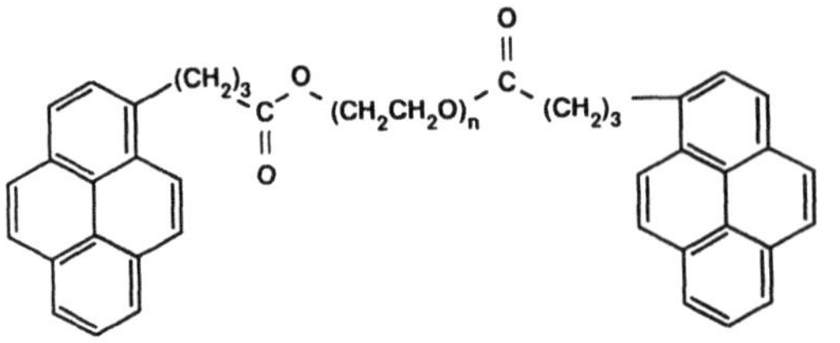

Fig. 2.5. Structure of end-labeled PEO polymers

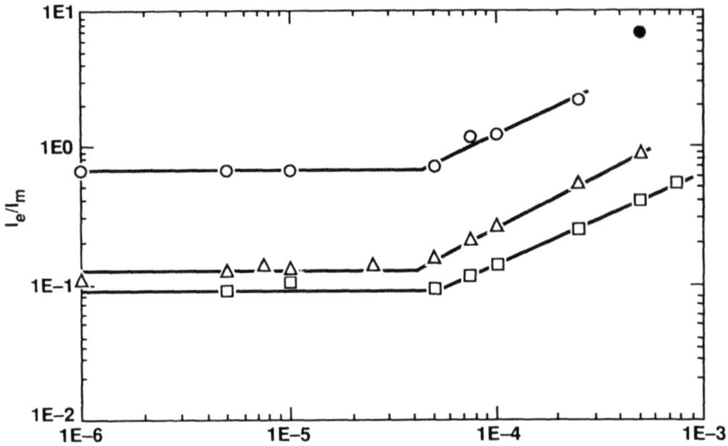

Fig. 2.6. Log-log plot of I_E/I_M ratios as a function of polymer concentration in deionized water for Py-PEO-Py samples of different molecular weights at 298 K: (○) Py-PEO-Py(4800); (△) Py-PEO-Py(9200); (+) Py-PEO-Py(11200); (●) phase separation occurred

$C_{16}H_{33}O$ - DI - $[(OCH_2CH_2)_xO$ - DI$]_n$ - $OC_{16}H_{33}$

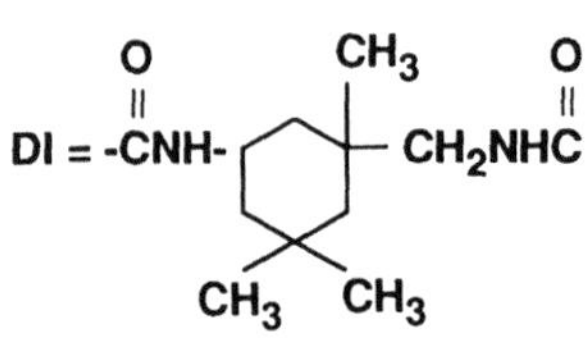

Fig. 2.7. Structure of PEO polymer end-capped with *n*-hexadecyl groups

samples were examined, having molecular weights (M_w) of approximately 40 000 (PEO-C16/40) and 47 000 (PEO-C16/47), respectively. The M_n of each polymer was calculated from respective molecular weight distributions and found to be approximately 25 000 for both polymers.

The above study utilized pyrene as an environmental probe. The steady-state fluorescence spectrum of pyrene possesses five distinct vibronic bands. The third vibronic band at approximately 383 nm is allowed and is insensitive to solvent polarity. The first vibronic, however, is forbidden and yields enhanced values of

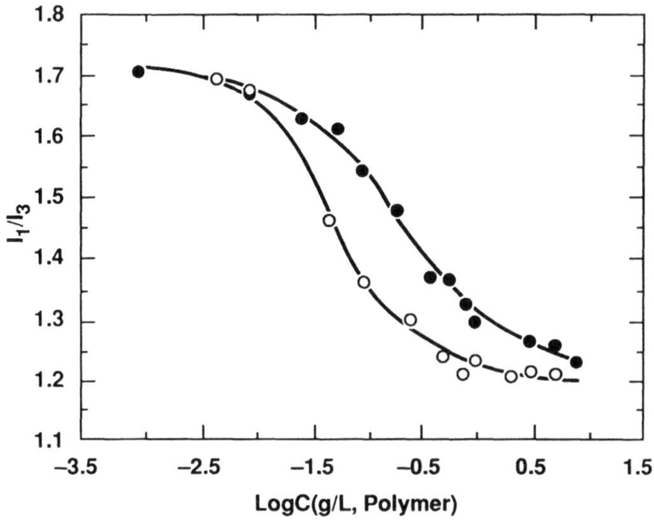

Fig. 2.8. Plot of fluorescence intensity ratio I_1/I_3 as a function of polymer concentration: (○) PEO-C16/40; (●) PEO-C16/47

fluorescence intensity in polar solvents and/or environments. Therefore, the ratio of the intensity of the first band to that of the third (I_1/I_3) can be used as an indication of the local environment that the pyrene probe experiences [7]. In hydrocarbon solvents the value of I_1/I_3 is approximately 0.6, approximately 1.1 to 1.2 in common micelles, and in water the values have been reported to be approximately 1.6 [8] or 1.9 [9]. This apparent discrepancy for the I_1/I_3 value in water has not been accounted for in the literature [10]. However, in the present discussion it is the change in the value of I_1/I_3, under varied conditions, which is the important factor, rather than the absolute value.

Plots of I_1/I_3 as a function of polymer concentration from 0.001 g/l to 10 g/l (Fig. 2.8) show several features. At low polymer concentration the pyrene probe was solubilized in an environment close to that of water; at high polymer concentration the environment resembled that of common micelles. This conclusion was supported by fluorescence lifetime measurements. At polymer concentrations above 0.1 g/l, the pyrene probes had average lifetimes of 300 ns, typical of pyrene in SDS micelles. Critical aggregation concentrations for each polymer were also estimated from the plots in Fig. 2.8 by drawing tangents to each curve passing through their respective inflection points. The intersections of these tangents with the horizontal axis were taken as the critical aggregation concentration for each polymer. These values were reported to be 0.1 g/l for PEO-C16/40 and 0.2 g/l for PEO-C16/47.

Excimer formation was also found to occur with the onset of aggregation. Figure 2.9 shows a plot of I_E/I_M as a function of log polymer concentration for each polymer. The maxima of the curves occur at approximately the critical aggregation concentrations. Assuming that at the maxima of the curves there

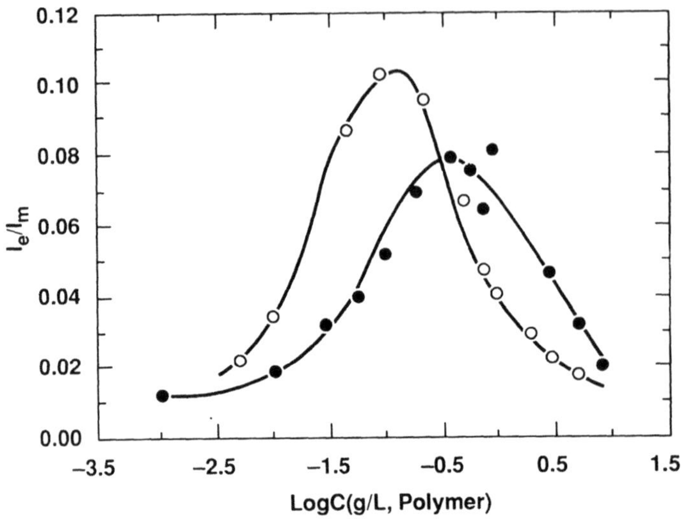

Fig. 2.9. Plot of I_E/I_M as a function of polymer concentration: (○) PEO-C16/40; (●) PEO-C16/47

are exactly two pyrene molecules per micellar aggregate, mean aggregation numbers of hexadecyl groups were estimated by Poisson statistics [11]. These values were reported to be 13 for PEO-C16/40 and 23 for PEO-C16/47, values which are of the same magnitude as those of hexadecyl-EO nonionic surfactants.

The authors of the above study point out that the onset of aggregation occurs at a critical concentration, but that the "association transition" is broad, possibly a result of the polydispersity of the samples. They also note that both end groups of one polymer molecule may participate in a single micellar cluster, but at sufficiently high polymer concentration "bridging interactions" of polymer chains between two micellar clusters result in gelation and a large increase in the solution viscosity.

2.2.3 *Hydrophobically-Modified Polyacrylamides*

The third class of polymers consists of hydrophobically-modified polyacrylamides. These polymers are usually synthesized using the micellar polymerization technique first reported by Turner and coworkers [12] in which small amounts (usually less than 2 mole percent) of an alkyl or aryl hydrophobic comonomer are incorporated into the acrylamide backbone. Polymerizations are carried out in water, using a surfactant (SDS) to solubilize the hydrophobic comonomer. Acrylamide polymerizes very efficiently in water and molecular weights of these hydrophobically-modified acrylamides may reach several million [13, 14]. Also this polymerization method appears to introduce a "blocky" copolymer microstructure where the hydrophobic groups residing in the micelles are incorporated into the growing polymer chain in runs separated by longer segments of acrylamide [15, 16, 17]. Numerous rheological studies describe

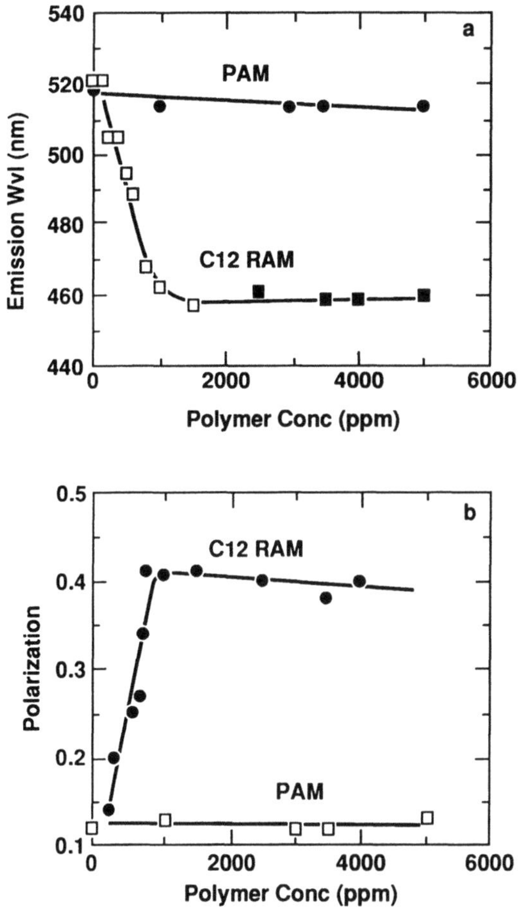

Fig. 2.10 a. The emission wavelength of 10^{-5} M ANS in solutions of varying concentration of polyacrylamide (PAM) and a polyacrylamide copolymerized with 1 mol% dodecylacrylamide (C12RAM). **b** The polarization of fluorescence of 10^{-5} M ANS from the same solutions of PAM and the hydrophobic copolymer with dodecylacrylamide

the associative behavior of this class of hydrophobically associating polymer [12, 13, 18]. Also, both probe and label spectroscopic studies have appeared in the literature recently [14–17, 19].

Siano and coworkers [14] report the use of 8-anilino-1-naphthalenesulfonic acid (ANS) as a probe of environment polarity in polyacrylamides containing 1 mole percent *n*-dodecylacrylamide (estimated from feed ratio) in 2 wt% NaCl. ANS, an amphoteric probe is selectively solubilized at the water-hydrophobe interface. The fluorescence emission maximum is shifted from 520 nm, in aqueous environments to approximately 462 nm in hydrophobic environments [18]. Studies of the polymer described above and a homopolymer of acrylamide were accomplished by varying polymer concentration up to 5000 ppm at constant ANS concentration. In Fig. 2.10 a, a plot of the emission wavelength maximum of ANS vs polymer concentration indicated an increasingly hydrophobic environment with increasing copolymer concentration. Above approximately 1000 ppm, emission wavelength remained constant, indicating that ANS must be bound in a hydrophobic region of the polymer. No spectral changes occurred for ANS in homo-polyacrylamide solutions, indicating that the dye did not bind to the unmodified polymer.

Further insight into the nature of the hydrophobic aggregates of this polymer was gained through fluorescence polarization experiments. Exciting ANS with polarized light and measuring the polarization of the emitted light allows determination rigidity or microviscosity of the hydrophobic aggregates in the copolymer. If the dye is held in a rigid, viscous environment, the emitted light retains much of the polarization of the exciting source. If the dye is unbound and free to move, Brownian motions allow the molecule to reorient during its excited-state lifetime and much of the polarization is lost. The polarization has been found to have a value of 0.12 in water or 2% NaCl and the theoretical maximum is about 0.45 in rigid media.

Figure 2.10 b shows the emission polarization of ANS in solutions of the homopolyacry-lamide and the hydrophobically-modified copolymer. The data for the copolymer indicated that above approximately 1000 ppm, the ANS molecules are in a rigid, viscous microenvironment. Below 1000 ppm, the polarization increases with copolymer concentration. The break in this curve is at approximately the same concentration as the one in the emission wavelength in Fig. 2.10 b. No change in polarization is evident with the homo-polyacrylamide over the entire concentration range. Both of the experiments are consistent with the hypothesis of aggregation of the polymer chains through the hydrophobic dodecyl groups.

Flynn and Goodwin [19] employed the micellar polymerization technique to make copolymers of acrylamide and dodecylmethacrylate (0.2 to 1.1 mole percent in feed). Pyrene probe fluorescence studies were reported. I_1/I_3 measurements were made for polymer concentrations over the range of approximately 0.1 g/dl to 1.0 g/dl in 0.1 M NaCl. Figure 2.11 shows I_1/I_3 as a function of polymer concentration for homo-polyacrylamide and copolymers containing approximately 0.4 and 0.2 mole percent dodecyl methacrylate. The pyrene probe experienced an aqueous environment over the entire concentration range for the polyacrylamide sample. For the copolymers, however, the probe was solubilized in an increasingly hydrophobic environment with increasing polymer concentration. At high polymer concentration I_1/I_3 was approximately 1.08, similar to the value for pyrene in common micelles. The authors also noted that although the probes were solubilized in hydrophobic microdomains, this process appears to be different than the micellization process encountered for common surfactants, where the probe maintains its aqueous value up to the CMC of the surfactant. At the surfactant CMC, a sharp break in the curves is observed and the I_1/I_3 drops to a lower value. With the copolymers, the change was continuous, but not as abrupt. The authors suggested partitioning of the pyrene probe between the hydrophobic microdomains and the bulk solution to explain this observation. Alternatively, the gradual change in the I_1/I_3 may indicate that the polarity of the hydrophobic aggregates changed with increasing polymer concentration.

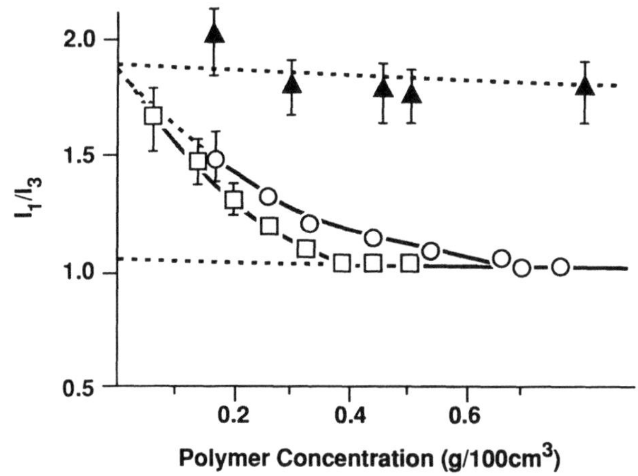

Fig. 2.11. Change in I_1/I_3 ratio with concentration of polymers: 0.4 ( ); 0.2 (○); 0.0 (Δ)

I_1/I_3 and viscosity measurements were also compared for the above copolymers. Plots of I_1/I_3 and reduced viscosity are shown as a function of concentration (Fig. 2.12). These data indicate that the pyrene probe is in a micelle-like environment at the critical overlap concentration for each polymer, demonstrating that viscosification of the polymer solution occurs through inter-polymer hydrophobic associations.

Labeled acrylamide copolymers have been reported by McCormick and coworkers [16]. Micellar and solution polymerization techniques were used to synthesize copolymers of acrylamide and a pyrene labeled monomer, [1-(β-aminoethyl) sulfonamido-1-pyrene]acrylamide (APS). In these systems the APS served as both a photophysical label and hydrophobic comonomer.

The polymer polymerized by the micellar technique (surfactant polymer) contained 0.25 mole% APS, while the solution polymerized (solution polymer, DMF/H_2O 57:43 volume fraction) polymer contained 0.35 mole%. Steady-state emission spectra of the two polymers, shown in Fig. 2.13, indicated that at constant polymer concentration, I_E/I_M was higher for the surfactant polymer even though it contained less of the pyrene-sulfonamide label. This was indicative of a higher local label concentration in the surfactant polymer and thus a blocky copolymer microstructure.

Rheological studies also indicated differences in behavior as a function of concentration for these two polymers. The plots of relative and reduced viscosities for the copolymer prepared in surfactant were similar to the profile in Fig. 2.1, indicating that it behaved as a hydrophobically associating polymer in deionized water and 2% NaCl. The solution polymer, in contrast, was shown to behave as a polymeric microsuspension in these solvents.

Correlation of photophysical data with rheological data was possible with the surfactant polymerized system. Plots of I_E/I_M and reduced viscosity as a function of log polymer concentration indicated interpolymer associations.

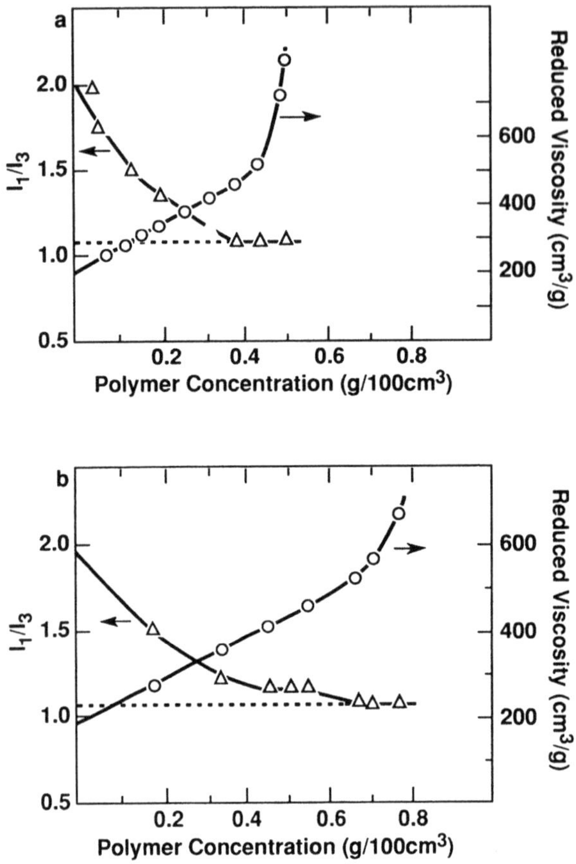

Fig. 2.12. Comparison of reduced viscosity and I_1/I_3: **a** polymer 0.4 and **b** polymer 0.2

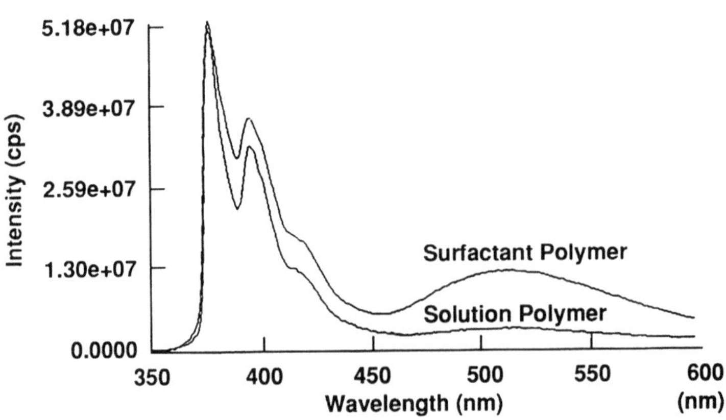

Fig. 2.13. Steady-state fluorescence emission spectra of pyrene-sulfonamide-labeled polyacrylamides in water

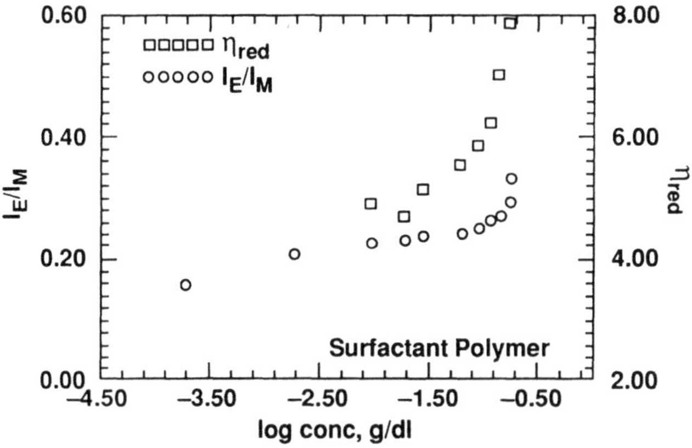

Fig. 2.14. RED and I_E/I_M as a function of log polymer concentration of the surfactant polymerized labeled polyacrylamide in water

These data appear in Fig. 2.14. C^* of the polymer is indicated by upward curvature in both sets of data. This correlation verifies the existence of microscopic interactions which affect macroscopic solution properties. Again it is possible that two or more labels or "blocks" of labels may participate in one micellar aggregate, bridging interactions of some polymer chains between these aggregates must be present to result in solution viscosification. Other information regarding the nature of the aggregates was obtained from excitation and time-resolved fluorescence studies. A 4 nm red-shift of the excimer excitation spectrum relative to monomer and the lack of a resolvable rise time in the fluorescence decay profile of the excimer were cited as strong evidence for ground-state (static) aggregation of the pyrene labels above C^*. Significantly, fluorescence lifetime and quenching studies on these labeled systems indicate a very hydrated, open structure. Clearly, the intermolecular domains of these model associative thickeners are neither as "hydrophobic" in nature nor as organized as micelles.

2.3 Absorption Probes

The spectral shift in the long wavelength absorption band of certain types of dye molecules has been used as an indicator of the micropolarity of microheterogeneous systems. Methyl Orange, 1-methyl-8-oxyquinolinium betaine, orange red, phaeophytin, 4-hexadecyl-1-[4-(4-oxyphenyl)buta-1,3-dien-yl]pyridinium merocyanine, 1-methyl-4-carboxymethoxy-pyridinium iodite, and 2,6-diphenyl-4-(2,4,6-triphenyl-1-pyridino) phenoxide have been successfully utilized to study hydrophobic microdomains and interfaces [20–26].

Among the absorption probes investigated, the 2,6-diphenyl-4-(2,4,6-triphenyl-1-pyridino)phenoxide is claimed to be the most sensitive to solvent polarity. Reichardt et al. [26] were the first to synthesize and conduct detailed spectroscopic investigations on this probe and have published extensive literature on its solvatochromic properties. For example, the solvatochromic band at 810 nm in diphenyl ether shifts to 515 nm in methanol. This 295 nm shift in the long wavelength absorption maximum band, which occurs when the solvent is changed from nonpolar to polar, is the highest reported solvatochromic shift for a water insoluble dye. The sensitivity of the dye to small changes in solvent polarity has been used by Dimroth and Reichardt [26] to establish an empirical scale of solvent polarity, called the E_T scale. The E_T scale provides an empirical characterization of the solvent polarity. The 2,6-diphenyl-4-(2,4,6-triphenyl-1-pyridino)phenoxide dye itself has been referred to as the $E_T(30)$ dye.

Recognizing its potential for probing hydrophobic microdomains, $E_T(30)$ has been used to investigate the hydrophobic microdomains of a variety of microheterogeneous systems like micelles, microemulsions and phospholipid bilayers [27, 28]. The rule of thumb with the $E_T(30)$ probe is that a bathochromic shift in the long wavelength absorption band $\{\lambda(\max)\}$ occurs with increasing hydrophobic character of the microenvironment of the dye.

Varadaraj et al. [29] have demonstrated that the $E_T(30)$ probe can be used as an extrinsic probe to study the hydrophobic microdomains formed due to polymer association in the hydrophobically associating copolymer system acrylamide–alkylacrylamide. The probe was solubilized in the aqueous polymer solution and absorption spectra recorded. Results are given in Table 2.1.

The $E_T(30)$ probe is insoluble in polyacrylamide (PAM) and the hydrolyzed polyacrylamide (HPAM) solutions since hydrophobic microdomains do not exist in these systems. A 580 nm absorption peak was observed for the hydro-

Table 2.1. Spectroscopic characteristics of $E_T(30)$ and pyrene probes in RAM and PAM polymer solutions

Medium/Polymer	λmax (nm)	I_1/I_3	τ (ns)
Water	± 5 nm		± 5 ns
PAM	no peak	1.78	124
RAM	580	1.57	97
			275
HPAM	no peak	1.84	123
HRAM	580	1.85	151
			290
2% NaCl			
PAM	no peak	1.88	129
RAM	568	1.66	97
			283
HPAM	no peak	1.91	129
HRAM	578	1.85	115
			216

[polymer]: 500 ppm; Temperature: 25 °C; C_8 acrylamide at 1 mol% incorporation in RAM and HRAM

phobically associating n-C_8 alkyl polyacrylamide (RAM) and its partially hydrolyzed analog, HRAM, in water. The λ(max) value for $E_T(30)$ in micelles of typical anionic and nonionic surfactants is about 500 nm. These observations indicated that the hydrophobic microenvironment in these systems is more hydrophobic than that normally encountered in micelles of surfactants.

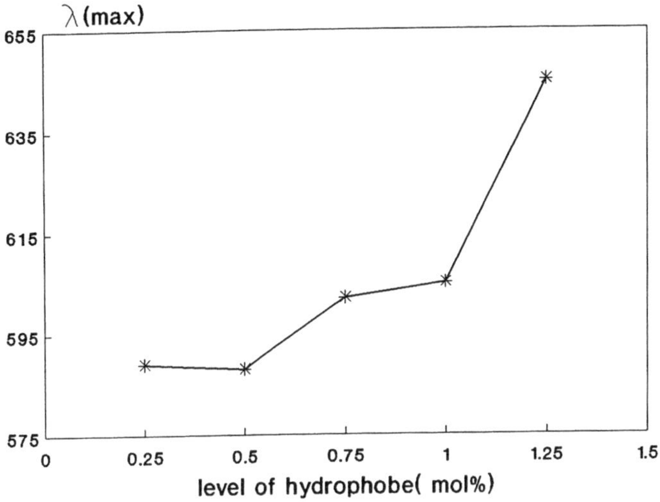

Fig. 2.15. Wavelength of maximum absorption for $E_T(30)$ probe for 2000 ppm HRAM polymer in 2% NaCl solution (hydrophobe: n-C_8 acrylamide)

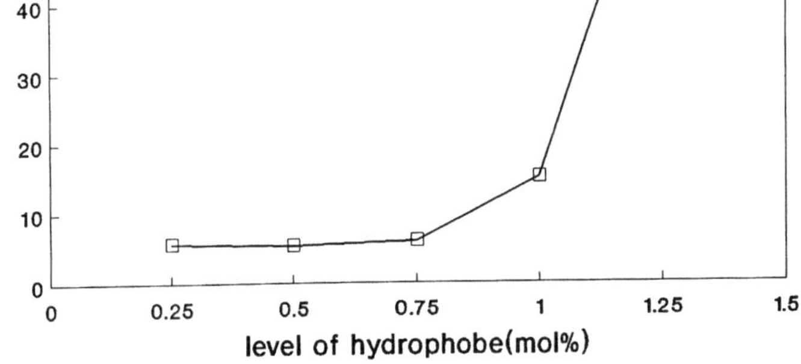

Fig. 2.16. Low shear solution viscosity for 2000 ppm HRAM polymer in 2% NaCl solution (hydrophobe: n-C_8 acrylamide)

The effect of electrolyte addition on the hydrophobic microdomain has also been investigated. Addition of NaCl to the polymer solution of RAM causes a 12 nm blue shift in λ(max) while the change for HRAM is negligible and within the error limit. The explanation advanced is that salt addition is causing a reduction in hydrodynamic volume of the polymer. This squeezes the probe from the core and pushes it toward the interface which is more hydrophilic. Interestingly, the I_1/I_3 data obtained from using pyrene as an emission probe generates the same information.

The $E_T(30)$ has also been used to explore polymer association as a function of the hydrophobe level. As shown in Fig. 2.15, the increase in λ(max) tracks the increase in hydrophobe content in the polymer. This indicates increased hydrophobe character for the microdomains and stronger polymer association with increasing hydrophobe levels. It is noteworthy that the increase in λ(max) dovetails the increase in solution viscosity (Fig. 2.16). Thus, the dye probe provides microscopic domain information which complements the macroscopic solution viscosity characterization of the polymer system.

2.4 Concluding Remarks

In this chapter we have attempted to show how emission and absorption probes and labels can be used to obtain information on the hydrophobic microdomains formed by the aggregation of hydrophobically associating polymers in aqueous media. Knowledge of polymer microstructure and how this may effect the response of the probe or label is often important in the interpretation of spectroscopic data. While the mechanism(s) of associative thickening are still to be fully elucidated, hydrophobic domains can and have been detected by spectroscopic probes and labels. Critical to designing associative thickeners with improved rheological properties will be the development of new techniques for probing microstructure and macroscopic organization.

2.5 References

1. a. McCormick CL, Bock J, Schulz DN (1989) Encyclopedia of Polymer Science and Engineering. John Wiley, 17:730; b. Polymers as Rheology Modifiers (1991) Schulz DN, Glass JE (eds) ACS Symposium Series #462, American Chemical Society, Washington DC; c. Water Soluble Polymers (1991) Shalaby B, McCormick (eds) ACS Symposium Series #467, American Chemical Society, Washington DC; d. Polymers in Aqueous Media (1989) Glass JE (ed) Advances in Chemistry Series #223; American Chemical Society, Washington DC
2. Winnik FM, Winnik MA, Tazuke S, Ober CK (1987) Macromolecules 20:38
3. Winnik FM (1987) Macromolecules 20:2745
4. Winnik FM (1989) Macromolecules 22:734
5. Char K, Frank CW, Gast AP, Tang WT (1987) Macromolecules 20:1833
6. Wang Y, Winnik MA (1990) Langmuir 6:1437
7. Kalyanasundarum K, (1987) Photochemistry in microheterogeneous systems. Academic Press
8. Kalyanasundarum K, Thomas JK (1977) J Am Chem Soc 99:2059
9. Dong DC, Winnik MA (1984) Canadian Journal of Chemistry 62:2560

10. Dong DC, Winnik MA (1982) Photochemistry and Photobiology 35:17
11. Thomas JK (1984) The Chemistry of excitation at interfaces. American Chemical Society, Washington DC
12. Turner SR, Siano DB, Bock J, U.S. Patent 4520182.
13. Bock J, Siano DB, Valint PL, Pace SJ (1989) In: Glass JE (ed) Polymers in aqueous media, Advances in Chemistry Series #223; American Chemical Society, Washington, DC
14. Branham KD, Middleton JC, McCormick CL (1991) Polymer Preprints 32(1):106
15. McCormick CL, Ezzell SE, Hoyle CE, Creed D, Macromolecules, accepted for publication
16. Dowling KC, Thomas JK (1990) Macromolecules 23:1059
17. McCormick CL, Nonaka T, Johnson CB (1988) Polymer 29:731
18. Reference 14 and references therein
19. Flynn CE, Goodwin JW (1991) In: Schulz DN, Glass JE (eds) Polymers as rheology modifiers. ACS Symposium Series #462, American Chemical Society, Washington, DC
20. Takagishi T, Nakota Y, Kuroki N (1974) J Poly Sci 12:807
21. Ueda M, Schelly ZA (1989) Langmuir 5:1005
22. De-min Z, Schelly ZA (1992) Langmuir 8:48
23. Hochapfel A, Biodart M, Laurent M (1988) J Colloid and Interface Sci 121:289
24. Warr GG, Evans DF (1988) Langmuir 4:217
25. Kido J, Hiyoshi M, Endo C, Nagai K (1991) J. Colloid and Interface Sci 142:326
26. a. Reichardt C, Gorner EH (1983) Leibigs Ann Chem 721; b. Dimroth K, Reichardt C (1969) Liebigs Ann Chem 727:93; c. Reichardt C, Dimroth K (1968) Fortschr Chem Forsch 11:1
27. Johnson BP, Gabrielsen B, Matulenko M, Dorsey JG (1986) Anal Lett 198:939
28. a. Zachariasse KA, Van Phuc N, Kozankiewicz B (1981) J Phys Chem 85:2676; b. Varadaraj R, Bock J, Valint P, Brons N (1990) Langmuir 6:1376
29. a. Varadaraj R, Bock J, Brons N, Pace S (1991) Polymer Preprints 32:575; b. Varadaraj R, Bock J, Brons N, Pace S (1993) J. Phys. Chem (In Press)

3 Solution Properties of Hydrophobically Associating Water-Soluble Polymers

J. Bock, R. Varadaraj, D.N. Schulz, and J.J. Maurer

The unique and technologically useful solution properties of hydrophobically associating polymers are a prime motivation for their study. Advances in synthesis have enabled the preparation of a variety of polymer architectures and chemical compositions which contain hydrophobic functionality. Sufficient structural variations are available to allow the development of basic structure, property, and performance relationships. Hydrophobically associating polymers impart unusual rheological properties to aqueous fluids. Enhanced viscosification, shear thickening or dilatant rheology, and viscosity which is less sensitive to ionic strength of the fluid have been found. In addition, improved mechanical and thermal stability have been observed. Under high shear or extensional deformation experienced during injection into porous media or spray through nozzles, the associating polymer solutions exhibit less permanent loss in viscosity. This chapter will focus on the solution properties of hydrophobically associating polymers and the relationships between these properties of polymer structure. The synthesis and structural characterization of these polymers is the subject of a separate chapter.

3.1 Introduction

It has been recognized that the formation of amphiphilic aggregates in water is due to the strong affinity of the polar groups to water and the tendency of alkyl chains to avoid contact with water. Hydrogen bonding with the polar groups tends to alter the structure of water and to "squeeze out" the hydrocarbon chains which are then held together by nonpolar Van der Waals forces. This effect, known as the *hydrophobic effect* [1–3] has been a useful concept to describe association phenomena of small amphiphiles, such as surfactants. Similar concepts should be applicable to water-soluble polymers containing nonpolar hydrocarbon chains, although steric constraints and excluded volume effects will add complexity.

The solution properties of polymeric amphiphilic molecules are strongly influenced by molecular aggregation and have been a subject of widespread research [4–19]. These polymers contain a major amount of polar, water-soluble groups and a minor amount of nonpolar hydrocarbon chains. The amphiphilic nature of these structures causes polymer aggregation in aqueous media which results in unique and useful solution properties. This chapter describes the solution properties of hydrophobically associating water-soluble polymers. In addition, the relationship between these properties and performance attributes in a variety of technological applications will be described.

Macromolecular Complexes
in Chemistry and Biology
Eds Dubin/Bock/Davis/Schulz/Thies

Details of the synthesis and structural characterization of the polymers are described in another chapter.

3.2 Polymer Systems

Hydrophobically associating polymers are water soluble polymers containing a small amount of hydrophobic functionality. The conformation of the polymer in solution is influenced by a variety of structural parameters. Polymer architecture, such as random or blocky arrangement of the hydrophobic groups will have a significant influence on polymer conformation and interactions in solution. The presence of hydrophobic functionality can result in inter- or intramolecular association or some combination of both. The relative amounts of these types of association will affect the conformation of the molecules in solution and in turn the properties of the solution. While many chemically different systems, such as hydrophobically modified cellulose [19], urethanes [15], and others [11] have been described in the literature, we will focus on only a few critical structures. To illustrate the relationships between structure and the observed solution phenomena, three types of hydrophobically associating polymers will be discussed as shown in Fig. 3.1. These are:

(1) Copolymers with random hydrophobic monomers,
(2) Copolymers with random surfactant monomers,
(3) Block copolymers.

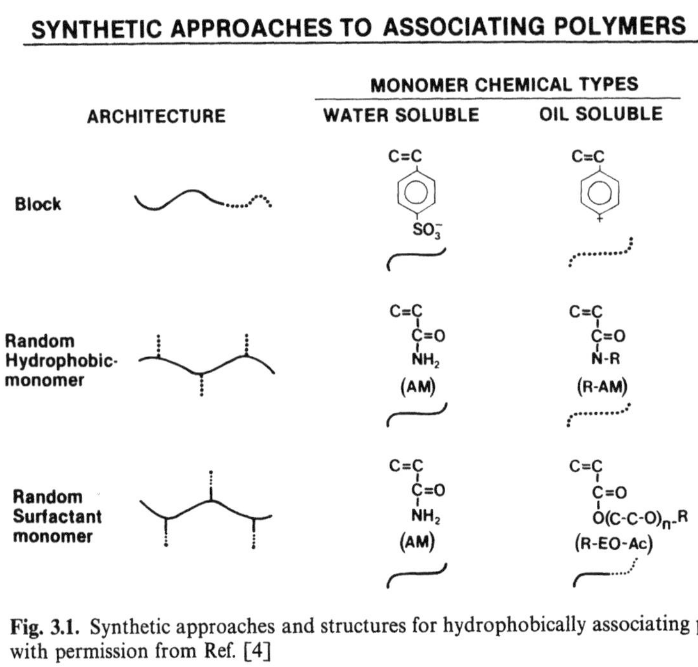

Fig. 3.1. Synthetic approaches and structures for hydrophobically associating polymers. Reprinted with permission from Ref. [4]

The synthesis of these polymers is the subject of another chapter and thus details will not be reviewed here. However, it is important to understand some of the differences in structure to be able to understand the properties of these polymers in solution. The block systems consisted of anionically polymerized styrene and *tert*-butyl styrene blocks [7]. The styrene units were sulfonated and following neutralization with base, the polymers were rendered water soluble. These polymers rely on ionic functionality for water solubility. The hydrophobic unit is a chain of *tert*-butyl styrene groups and thus is quite bulky.

The two random architectures are derived from free radical polymerization of acrylamide with either an alkylacrylamide [5] or a nonylphenoxy-poly(etheroxy) acrylate monomer [6, 13]. Type (1) systems require the micellar polymerization to enable incorporation of the hydrophobic monomer. The backbones of these polymers are built from nonionic, water-soluble monomers and thus are quite different than the type (3) block copolymers in terms of solubility characteristics.

3.3 Solubility Characteristics

Solubility characteristics of high molecular weight polymers can be problematic. Long dissolution times and 'fish eyes' in the final solution are often observed. Associating polymers have additional complexity. The solution history of the polymer can affect inter- and intramolecular associations which influence dissolution rates. Preparing solutions above the polymer overlap concentration, C^*, requires patience and care. The design of the polymer structure can greatly affect solubility and dissolution characteristics.

In general, the solubility characteristics of a hydrophobically associating polymer are influenced largely by the architecture and chemical composition of the polymer. Examples include the chemical architecture and amount of ionically charged and uncharged water-soluble, polar groups as well as the hydrophobe content and type. The higher the hydrophobe content and the longer the hydrocarbon chain, the greater the extent of aggregation and the less soluble is the polymer in water. As observed with conventional surfactants, a critical chain length of about 6 to 8 carbons is required for the onset of aggregation; however, as the carbon length approaches 16 to 18, there is a drastic decrease in solubility. The critical micelle concentration, CMC, is the concentration of amphiphilic molecules where aggregates or micelles begin to form. The CMC depends on the architecture of the molecule, relative amount of hydrophilic to lypophilic functionality, chain length and structure of the lypophile or hydrophobe, as well as physicochemical parameters such as temperature, pH, ionic strength of the water, etc. It is anticipated that these variables will also influence the nature of assembly and the solubility characteristics of hydrophobically associating polymers.

The solubility characteristics of hydrophobically associating copolymers of *p*(alkyl acrylamide-*co*-acrylamide), RAM, have been studied by a number of groups [4, 5, 9, 10, 14, 18, 20]. RAM polymers with molecular weights of several million and hydrophobe levels of up to about 1.25 mol% n-C_8 have been found to be soluble. As the hydrophobe chain length increases, the amount of hydrophobe must be decreased to maintain solubility. Lowering molecular weight improves solubility and thus can be used to counter-balance an increase in hydrophobe content or chain length.

Another way to improve the solubility characteristics of hydrophobically associating polymers is through incorporation of water-soluble, ionic monomers into the polymer. Carboxylate functionality has been introduced into RAM polymers by either copolymerization with acrylic acid salt or by a post-polymerization partial hydrolysis of the acrylamide groups. Incorporating about 20 mol% sodium acrylate functionality, significantly improves solubility of these HRAM polymers. Sulfonate groups can be introduced by copolymerizing with a sulfonate monomer such as vinyl sulphonate or 2-acrylamido-2-methylpropane sulfonate, AMPS. We call these polymers SRAM.

The effect of monovalent and divalent salts on the solubility of these hydrophobically associating polymers (HAPs) is similar to that of ionic surfactants. An increase in salt content decreases solubility. With increasing salinity, the hydrocarbon chains are forced into closer proximity to the point where the subtle balance between hydrophobic associative forces and hydrophilic hydration forces breaks down, and phase separation results. Divalent cations have a larger effect on decreasing polymer solubility than do alkaline earth or monovalent cations. This is particularly so when the polymer contains anionic functionality such as acrylate or sulfonate. Another interesting phenomena occurs in mixed salts with certain polymer compositions when the ratio of divalent to monovalent cation is varied. A window of solubility is observed similar to that found with anionic surfactant solutions.

Another interesting aspect of associating polymers is the effect of polymer concentration on solubility. Increasing polymer concentration causes the system to pass from a dilute to semi-dilute to gel phase. These transitions can be quite sharp. The dilute region is the region below C^* (the overlap concentration) the semi-dilute region is above C^*, and the gel phase is one where a network structure predominates. The behavior of the system in the gel state is currently under study.

The effect of surfactants on the solubility of hydrophobically associating polymers is a relatively less explored and fairly complex area [20–22]. Since surfactants themselves aggregate above their critical micelle concentrations and provide hydrophobic regions in the micellar structures, it is to be expected that the solubilities of hydrophobically associating polymers should increase in surfactant solutions. It has been observed that even below CMC there is a surfactant-polymer interaction wherein the hydrophobic chains of the hydrophobically associating polymer and the hydrophobic tails of the surfactants interact to increase solubility. A systematic study addressing the interaction between hydrophobically associating polymer structure and surfactant type and concentration is in progress.

3.4 Solution Stability

It is well established that the stability of the aqueous polymer solutions can be influenced by one or more of a complex set of factors including polymer composition, salt environment, temperature, pH, oxidative effects, bacteria, etc. [11, 15, 16, 24]. The influence of these factors has been considered extensively in studies of acrylamide-based polymers and biopolymers to control mobility in Chemically Enhanced Oil Recovery, CEOR, processes. With respect to polyacrylamide (PAM) and partially hydrolyzed polyacrylamide (HPAM), hydrolysis is a major factor influencing solution stability.

3.4.1 Chemical Stability

Extensive hydrolysis has been observed during high temperature aging (one to four weeks at 93 °C) of HPAM polymers (Fig. 3.2). Hydrolysis of two types of associating polymers (RAM and HRAM) also occurs during high temperature (96 °C), anaerobic aging of these systems in mixed salt (3% NaCl + 0.3% $CaCl_2$) environment. Thus, hydrolytic instability is a major issue regarding the development of improved acrylamide-based polymers for use in high temperature environments.

A comprehensive evaluation of the influence of comonomer type and level and the initial degree of hydrolysis (τ_i) on hydrolytic stability and phase behavior of a series of acrylamide-based copolymers was conducted. The comonomers examined were AMPS, *N*-vinyl pyrrolidone (NVP) and dimethyl

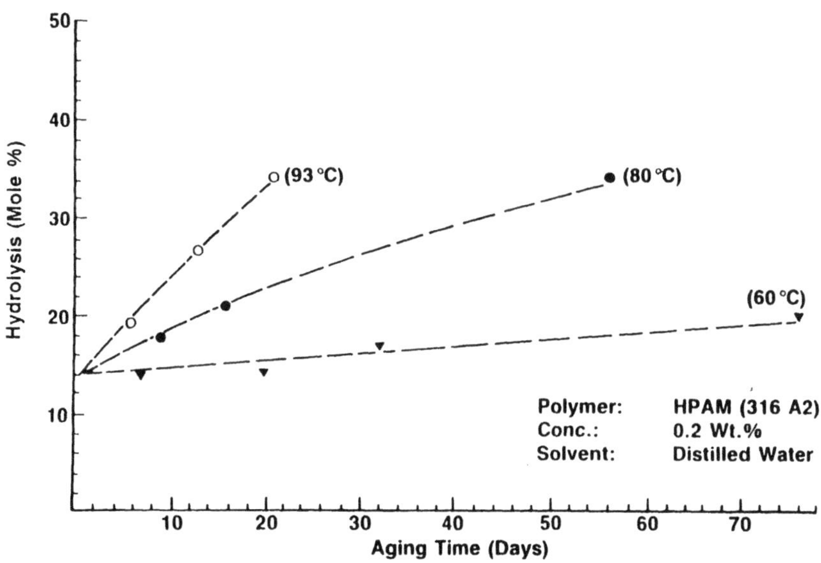

Fig. 3.2. Influence of temperature on the hydrolysis of HPAM

acrylamide (DMA). Several key aging conditions – temperature, brine composition, and oxygen content – were also explored with regard to their influence on hydrolytic stability and/or phase behavior. The major general conclusions from these studies are as follows:

(1) At low comonomer levels (~ 30 mol %), the relative hydrolytic stability of these copolymers at 93 °C, in 3.5% NaCl, was AMPS > NVP ≫ DMA ≫ PAM as shown in Fig. 3.3.
(2) At high comonomer level (> 50 mol %), both the AMPS and the NVP comonomer lead to significantly improved thermal hydrolytic stability in the test protocols (93°, 3.5% NaCl, aerobic).
(3) The DMA copolymers were the least stable of the three comonomer systems examined at all levels (30–60 mol %). However, DMA copolymers had improved hydrolytic stability in comparison with PAM polymers. The high degree of hydrolysis observed suggests that the DMA comonomer may be undergoing hydrolysis at these aging conditions.
(4) Complex hydrolysis characteristics were noted for several of these polymers, in particular for the 30 mol % copolymers, which suggests that multiple hydrolysis mechanisms may be operative.
(5) All of the 30 mol % polymers exhibited a viscosity loss of > 50% of the initial value in less than one year of anaerobic (< 2 ppb O_2) aging at 96 or 120 °C.
(6) Oxygen content had little or no effect on solution stability. Brine composition had no effect up to about 30–50 days. At longer aging periods complex

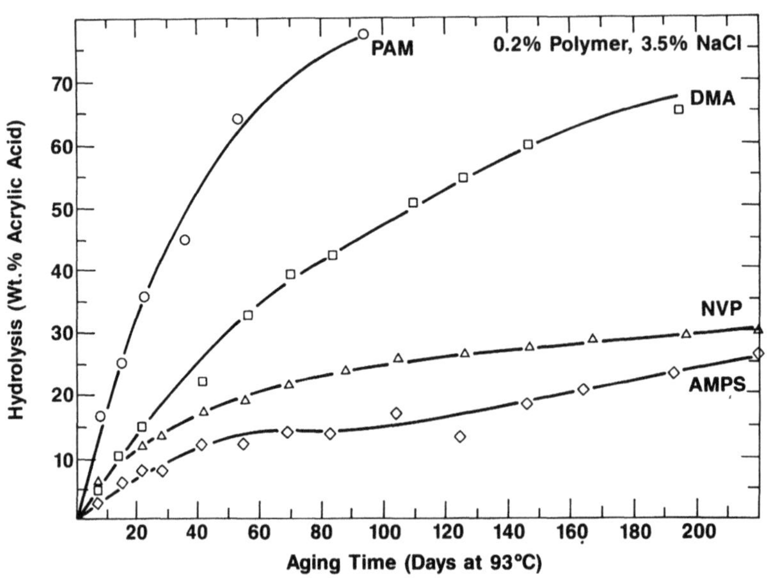

Fig. 3.3. Influence of comonomer type on hydrolytic stability

hydrolysis characteristics were detected which vary slightly with brine composition and polymer composition. These effects again suggest that multiple hydrolysis mechanisms may be operative.

Through proper design of backbone monomer composition and the use of a more hydrolytically stable amide linkage to the hydrophobic functionality, thermally stable hydrophobically associating polymers can be obtained. This is another area worthy of further study.

3.4.2 Physical Stability

The response of a solution of hydrophobically associating polymer to changes in ionic strength or salt content is very important for many applications. Since the hydrophobic association or aggregation is enhanced with increasing salt, as in classical micellar solutions, unusual solution rheology as a function of salt content can be anticipated. A complex balance of competing effects needs to be considered to explain the observed response of viscosity to increasing salt content. As shown in Fig. 3.4, the viscosity of a HRAM polymer first decreases as small mounts of salt are added to the aqueous solution. However the extent of viscosity loss is significantly less than that observed for a solution of a convention polyelectrolyte of similar composition (without the hydrophobic groups) and molecular weight. Charge screening of the ions in the polymer by the counter ions of the salt cause a decrease in electrostatic repulsion of the polyions and a collapse of the chain. The hydrophobic associations apparently

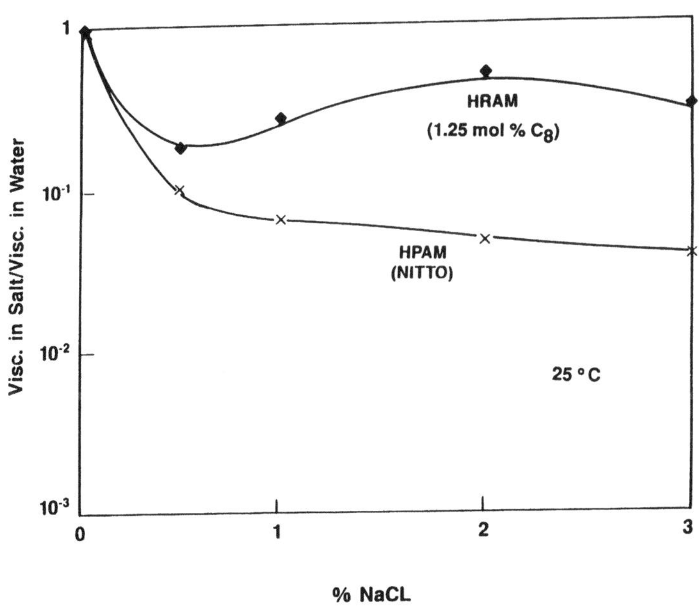

Fig. 3.4. Viscosity of HRAM as a function of added salt

restrain the collapse or reduce the hydrodynamic size of the coils caused by ionic screening.

As the salt concentration is further increased, hydrophobic interactions are enhanced and solution viscosity can be observed to increase. This is quite unusual and very important from an application viewpoint. This observation suggests that by properly balancing the amount of charged groups and the hydrophobic functionality, a polymer solution with isoviscous response over a broad salt concentration range could be obtained.

Another unusual aspect of the physical stability of associating polymer solutions is their response to shear. Above the overlap concentration, associating polymer solutions can exhibit a shear thickening or dilatant regime. While the relaxation time of the polymer coils in this regime can be relatively long, the solution will exhibit a stable or steady state viscosity at a fixed shear rate. This suggests that the morphology of the polymer in solution is dependent on shear rate but will reach a stable value given sufficient time.

3.4.3 Mechanical Stability

In many applications, polymer solutions are subjected to high shear or extensional forces. For example, during injection of displacement fluids for chemically enhanced oil recovery, CEOR, the fluid is subjected to high extensional forces which can cause mechanical breakage of the polymer backbone. Looking at the change in solution viscosity after flow through a Berea sandstone core provides a measure of mechanical stability. As shown in Fig. 3.5, the PAM and HRAM polymer solutions of comparable starting viscosity maintain that viscosity at high flow rates. Replacing the high MW polymer with lower MW polymer having associating groups results in solutions with improved mechanical stability. The associations under high stress can provide a mechanism for stress

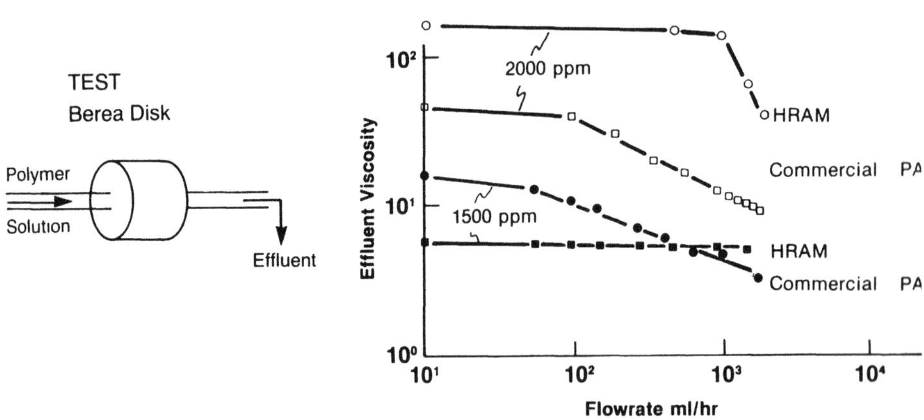

Fig. 3.5. Mechanical stability of HRAM and PAM as measured by flow through a Berea sand-stone core at flow rates of 10–10^3 ml/h

release without permanent damage to the polymer backbone. Once the stress is removed, the associations can reform without loss in solution viscosity. These associating polymer solutions could thus be considered as 'smart' fluids. The solutions can sense the external shear force and intelligently respond by altering the associations to avoid permanent destruction.

3.5 Solution Rheological Properties

3.5.1 Copolymers with Random Hydrophobic Monomers

The rheology of dilute polymer solutions has been used extensively to gain insight into the structure and conformation of polymers in solution [10–12, 15, 16]. In many technologies such as paints, lubricants, displacement fluids for enhanced oil recovery, and water-based metal working fluids, polymers are used to control solution rheology [12]. Dilute solution rheology provides information on the dynamics of individual polymer chains. The intrinsic viscosity, $[\eta]$, provides a measure of the molecular weight of a polymer through a relationship such as the Mark–Houwink–Sakurada equation, $[\eta] = kM^a$. Prior studies of the dilute solution properties of polyacrylamide systems [11] provide a basis to compare hydrophobically modified systems.

The introduction of hydrophobic groups into a water soluble polymer complicates the dilute solution viscometrics. The intrinsic viscosity of both the RAM and HRAM solutions decreases and the Huggins' constant increases as the amount of hydrophobe on the polymer increases. The effect is more pronounced beyond a certain level of hydrophobe which depends on the amount of charged groups in the polymer. HRAM polymers containing approximately 18 mol % carboxylate groups require a higher hydrophobe content to exhibit changes in dilute solution properties relative to uncharged RAM polymers. One explanation for this phenomenon is that the solubilizing effect of the $-COO^-$ is offset by the decreasing solubilizing effect of the hydrophobic substitutes. The increase in the Huggins' constant indicates a decrease in solvent quality with increasing hydrophobe content. Thus a Huggins' constant greater than the usual 0.3–0.8 found in conventional polymers [4] appears to be a good indicator of hydrophobic associations. These observations are consistent with the dilute solution behavior of ionomers (hydrocarbon soluble polymers containing a small amount of ionic groups) in a hydrocarbon solvent [25–26].

The presence of charge in the form of anionic acrylate functionality on the polymer has an opposite effect to that of hydrophobic functionality on the dilute solution properties. Increasing the amount of sodium acrylate groups on the polymer increases the intrinsic viscosity and reduces the Huggins' constant for a given hydrophobe level. Thus acrylate functionality seems to improve the solvent quality and open up the polymer coil in dilute solution. The fact that charged groups and hydrophobic groups have opposite effects can be used to modulate the solubility characteristics and control the associative properties of these polymers.

Another way to modulate the properties of water soluble polymers is to change the solvent quality by adding salt. The intrinsic viscosity decreases and the Huggins' constant increases with increasing ionic strength or salt level for a series of HRAM polymers with about 18 mol % sodium acrylate and varying *N*-n-octylacrylamide content. As the hydrophobe content increases, the influence of salinity in the water becomes stronger and again there appears to be a critical hydrophobe level required to cause a change in intrinsic viscosity and Huggins' constant. As before, the Huggins' constant appears to be a sensitive indicator of the presence of hydrophobe on the polymer. If the Huggins' constant is an indication of associative effects, then these interactions are present even in dilute solution where non-associating polymer molecules are believed to be isolated and acting independently. The addition of salt to the solution enhances the associations which may be predominantly intra-molecular in this concentration regime. Further study of the structure and conformation of these hydrophobe containing molecules in dilute solution by a scattering probe would be useful in interpreting the macroscopic rheological information.

Semi-Dilute Viscometrics. Solution viscometrics at concentrations above the overlap concentration indicate dramatic effects due to the associative nature of the hydrophobic groups in the polymer. As shown by the reduced viscosity–concentration profiles of Fig. 3.6 (large graph), the introduction of only 1.0 mol % *N*-n-octylacrylamide to polyacrylamide can increase the viscosification efficiency dramatically. Increasing the hydrophobe level to 1.25 mol% further increases solution viscosity. At a concentration of 2000 ppm, the presence of the hydrophobe has resulted in over an order of magnitude increase in viscosity. This is in contrast to the behavior of these polymers in dilute solution as shown in the inset diagram in Fig. 3.6. Above the chain overlap concentra-

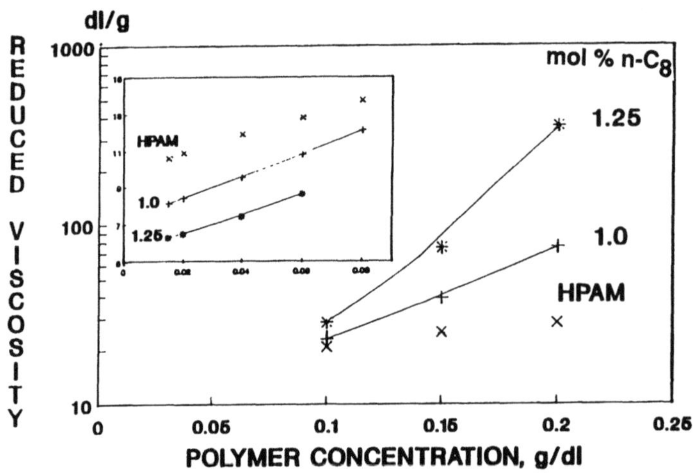

Fig. 3.6. Reduced viscosity vs concentration profile from n-C_8 HRAM 1.0, 1.25 mol % hydrophobe compared with HPAM. Reprinted with permission from Ref. 9

tion, inter-molecular associations increase the resistance to flow or solution viscosity. This may indicate an increase in the hydrodynamic radius of the coils due to inter-molecular associations as well as increased pseudo entanglements.

The conventional approach to control hydrodynamic radii of polymer coils in solution is through molecular weight. To study this effect, different molecular weights of HRAM polymers containing 1.0 mol % *N*-n-octylacrylamide and 18 mol % sodium acrylate were prepared. Viscosity at 1.3 s^{-1} shear rate was determined on solutions of these polymers in 2% NaCl as a function of polymer concentration. As shown in Fig. 3.7, polymers with higher intrinsic viscosity (labels on the curves) corresponding to higher molecular weight, exhibit higher solution viscosity at a given polymer concentration. Thus, while hydrophobic associations contribute to the viscosification efficiency, molecular weight of the polymer remains an important structural parameter for solution viscosity control. Quantification of these relationships would be useful in designing polymers for viscosity control. One approach would involve testing the validity of conventional scaling laws to describe the polymer concentration and molecular weight dependence of the zero shear viscosity of hydrophobically associating polymers. However, data covering a broad range of these variables would be needed.

A further complication in understanding the rheology of hydrophobically associating polymers is the unique response to shear rate and solvent quality (e.g. salt content). As shown in Fig. 3.8, the viscosity can be independent (Newtonian), decrease (pseudoplastic) or even increase (dilatant) with shear rate depending on the polymer concentration and shear rate. As the hydrophobe level and polymer concentration increase beyond a critical value, a shear rate region in

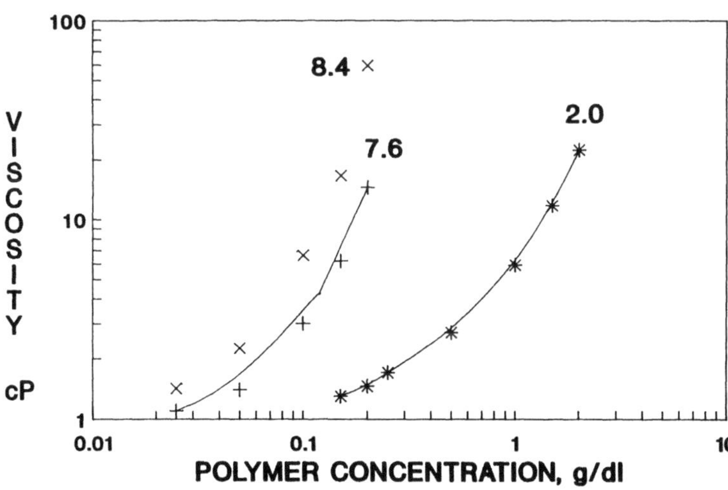

Fig. 3.7. Viscosity concentration profile for C_8 HRAM's at 1 mol % C_8 and 18 mol % sodium acrylate with different intrinsic viscosities (molecular weights). Reprinted with permission from Ref. 9

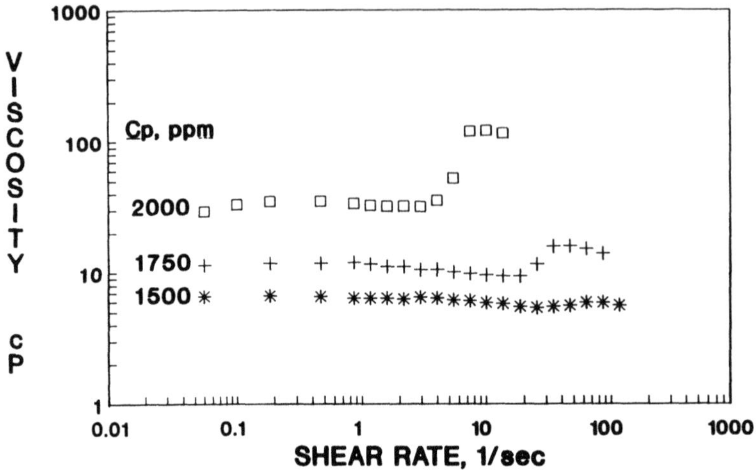

Fig. 3.8. Effect of shear rate on viscosity for HRAM polymer with 1.25 mol % n-Eq. Reprinted with permission from Ref. 9

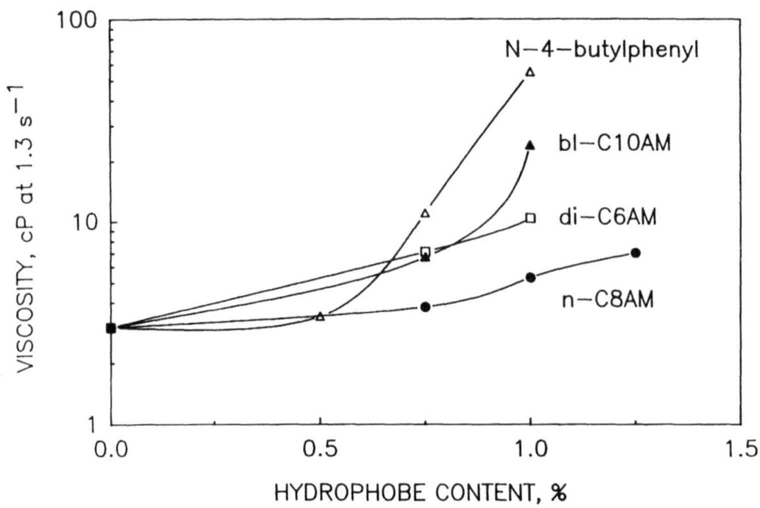

Fig. 3.9. Effect of hydrophobe type on solution viscosity. n-C_8AM is *N*-n-octylacrylamide, diC_6AM is *N*, *N'*-n hyxylacrylamide, bi-C_{10}AM is a bilinear alkyl acrylamide with C_{10} branches, *N*-4-butyl phenyl is a *N*-4-butylphenylacrylamide. Reprinted with permission from Ref. 9

which viscosity increases with shear is observed. Below and above this shear rate region, the viscosity behaves quite normally with shear and the time required to reach a steady state or equilibrium condition is relatively short. However, within the boundaries of the shear regime for shear thickening, polymer relaxation times increase significantly. An explanation of the molecular behavior of the system for shear thickening involves a change in the relative amount of intra-

and inter-molecular associations with shear. As the polymer chains are extended under a shear field, the amount of inter-molecular associations increase which results in enhanced viscosity. This explanation has been used to explain the shear thickening behavior of ionically associating polymers in hydrocarbon solvents [25, 26].

The above studies relate to hydrophobically associating polymers containing a particular type of functionality – namely, *N*-n-octylacrylamide. Other hydrophobe structures have been studied. The length of the hydrophobe group, the nature of hydrophobe branching and the presence of phenyl group can have a significant effect on polymer properties. Figure 3.9 shows the effect of hydrophobe structure on solution viscosity as a function of hydrophobe content in the polymer. Increasing hydrophobe chain length and total hydrocarbon in the hydrophobic group results in an increase in solution viscosity at high hydrophobe content (e.g. diC_6AM, $biC_{10}AM$). In addition, the presence of a phenyl group in the hydrophobe significantly increases solution viscosity.

3.5.2 Copolymers with Random Surfactant Monomers

Surfactant macromonomers such as nonylphenoxypoly(etheroxy) acrylates, R-EO-Ac (Fig. 3.1), are water soluble or water dispersible. The extent of water solubility depends upon the hydrophile-lipophile balance (HLB) of the surfactant and the temperature. Surfactant monomers, wherein R = nonylphenyl and $n = 10$, 20, and 40 have been synthesized.

Such surfomers can be easily copolymerized with acrylamide (AM) to form water soluble copolymers, i.e. poly(AM-*co*-R-EO-Ac), which contain low levels (< 5 mol %) of surfactant group. Unlike the copolymerization of acrylamide with other hydrophobic monomers (e.g. long chain alkylacrylamides), large amounts of external surfactant are not needed to solubilize the hydrophobic monomer. Rather, conventional free radical solution methods can be used to form high molecular weight copolymers [13].

Aqueous Solution Properties. Acrylamide copolymers containing even small quantities of surfomer units (i.e. < 0.5 mol %) also exhibit interesting solution properties. For example, below C^* the surfomer copolymers show lower intrinsic viscosities $[\eta]$ and elevated Huggins constants, k_H (Table 3.1). As with the RAM polymers, below C^* the chains interact intramolecularly to collapse the coil, lowering $[\eta]$ and raising the slope of the reduced viscosity concentration plot, i.e. k_H [13].

Above C^* these polymers show substantially higher viscosities than unsubstituted polyacrylamide. In this regime the viscosity depends upon the level and type of surfomer, as well as conventional polymerization variables, such as the ratio of monomer concentration to the square root of the initiator concentration, $[M]/[I]^{1/2}$. Of course, the $[M]/[I]^{1/2}$ ratio is related to the degree of polymerization (DP) (or molecular weight M_w) the polymer for an ideal free radical polymerization by the well-known relationship:

$$M_w \text{ or DP} = \frac{k_p[\mathrm{M}]}{\{F\,k_d\,k_t\}^{1/2}}\,[\mathrm{I}]^{1/2}$$

where k_p = propagation rate constant, F = frequency factor, [M] = monomer concentration, k_d = initiator decomposition constant, k_t = termination rate constant and [I] = initiator concentration.

Above C^*, the solution viscosity of the surfomer copolymers is more sensitive to changes in the level of surfactant macromonomer than to changes in $[\mathrm{M}]/[\mathrm{I}]^{1/2}$. For example, doubling the surfomer level from 0.3 to 0.6 mol % greatly increases the polymer solution viscosity. However, doubling the $[\mathrm{M}]/[\mathrm{I}]^{1/2}$ from 30 to 60 has almost no effect on the viscosity-concentration profile of the surfomer copolymer (Fig. 3.10) [13].

Table 3.1. Viscosity Data[a] for PAM and P(AM-*co*-R-EO-Ac)[c] copolymers

	PAM	P(AM-*co*-R-EO-Ac)	
Viscosity at 3000 ppm (cP)	4.50	33	11.50
Viscosity at 1000 ppm (cP)	1.70	2.00	2.50
$[\eta]$ (dl g^{-1})[b]	7.6	5.1	3.8
k_b^H	0.29	1.27	11.6

[a] In 2% NaCl solution, filtered through a 150 mesh screen and measured at 1.3 s^{-1}.
[b] $[\eta]$ and kH calculated according to the Flory–Huggins relationship, $\eta_{sp/c} = [\eta] + k_H[\eta]^2C + k_H[\eta]^3C^2$.
[c] 0.3 mol % (charge) surfomer, nonylphenylpoly(etheroxy) acrylate of variable poly(etheroxy) chain length.

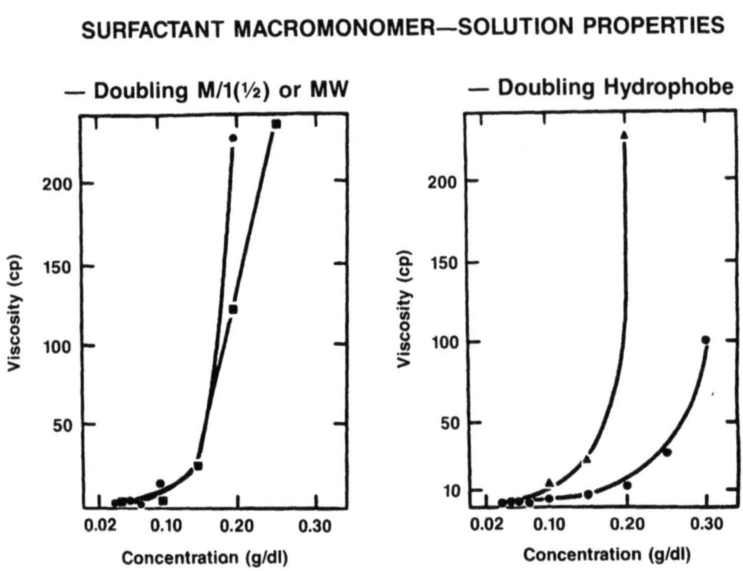

Fig. 3.10. Effects of doubling $[\mathrm{M}]/[\mathrm{I}]^{1/2}$ vs doubling the hydrophobe content in surfomer copolymers

As with other high molecular weight, water-soluble polymers, the mode of polymer purification and isolation can also have a profound effect on polymer solution properties. Polymers purified by the solid polymer route (i.e. precipitation/vacuum drying or dialysis/freeze drying) have lower viscosities than polymers purified by the solution polymer route (i.e. dialysis followed by dilution). This effect is especially pronounced at high $[M]/[I]^{1/2}$ ratios or for high molecular weight polymers, suggesting the loss or breakdown of some polymer upon precipitation and redissolution or the removal of low molecular weight impurities via dialysis [13].

3.5.3 Block Polymers

Valint and Bock [7] compared the dilute solution properties of diblock, random and homopolymer system. A comparison of the reduced viscosity as a function of concentration for the block polymer, D5-TBS-SS3 and random polymer, R5-TBS-SS, having 3.3 mol % *tert*-butylsytrene content and M_w of 9.0×10^5 and 8.3×10^5, respectively, is shown in Fig. 3.11. It can be seen that the random polymer is very similar in rheological behavior to the homopolymer. The *tert*-butylstyrene hydrophobic groups are not aggregating in the random polymer due to the hydrophilicity and steric bulk of the neighboring styrene sulfonate groups. The block polymer exhibited a reduced viscosity about one order of magnitude higher than the homopolymer of similar molecular weight. Thus, the presence of a relatively small amount of hydrophobic groups in a block architecture significantly enhanced the viscosification efficiency of the polymer.

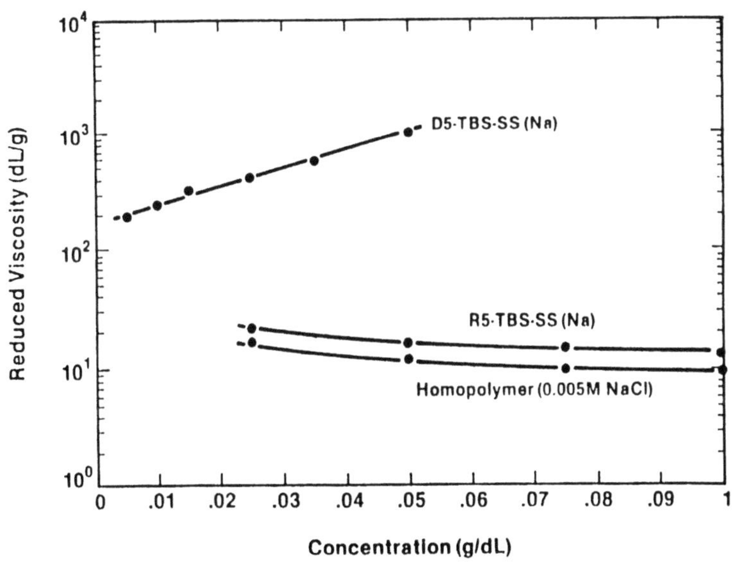

Fig. 3.11. Reduced viscosity plots comparing diblock poly(*tert*-butyl styrene-*co*-sodium styrene sulfonate), D5-TBS-SS(Na), with random poly(*tert*-butyl styrene-*co*-sodium styrene sulfonate), R5-TBS-SS(Na), and homopolymer poly(sodium styrene sulfonate). Reprinted with permission from Ref. 7

The opposite slope of the reduced viscosity concentration profile for the block polymer solution as compared to that of the homopolymer is indicative of fundamental mechanistic differences in viscosification. The decrease of reduced viscosity with increasing polymer concentration for the homopolymers results from a decrease in effective polymer size by coiling. This is due to a reduction of charged group repulsion through shielding of the anionic charge by the proximity of other polymer molecules. With the block polymer, the reduced viscosity increases with increasing concentration because of polymer hydrophobe association. It is envisioned that the *tert*-butylstyrene blocks associate or aggregate due to hydrophobic interactions in the aqueous medium. As the polymer concentration increases, the size and/or number of these aggregates increases. This results in enhanced interaction and, hence, solution viscosity.

In dilute solution, the reduced viscosity of a ionically charged polymer increases upon dilution because of the polyelectrolyte effect [2, 11]. As the ionic centers along the backbone become deshielded by dilution, the polymer hydrodynamic size increases due to electrostatic repulsion. Surprisingly, the polyelectrolyte effect was not observed with the block polymers, most likely due to the association of polymer molecules. Polymer concentrations as low as 50 ppm were used with no apparent upturn in reduced viscosity with decreasing concentration. Apparently, block polymer aggregation is present to very low concentrations which maintains a locally effective shielding of the sulfonate anions within the aggregate. Alternatively, individual block polymer molecules may assume a collapsed coil conformation in dilute solution such that the hydrophobic *tert*-butylstyrene block is surrounded by the styrene sulfonate block to isolate it from the aqueous medium. This type of conformation would counter chain expansion due to charge repulsion of the sulfonate groups with no polyelectrolyte effect as was observed. As of yet we have not made direct measurements of the hydrodynamic size of these polymers or their aggregates in solution using light or neutron scattering techniques, for example, to be able to determine the validity of the above postulates.

To define the effect of molecular weight on intrinsic viscosity, diblock polymers, D5-TBS-SS, with 3.3 mol % *tert*-butylstyrene were prepared having weight average molecular weights from 240 000 to 1.7 million. It can be seen from the plot in Fig. 3.12 that a power law relationship, described by the Mark–Houwink–Sakurada equation ($[\eta] = kM^a$) exists between the molecular weight and intrinsic viscosity for this series of block polymers in distilled water. The exponent, a, was determined to be 1.0 which was in good agreement with $a = 0.93$ reported [16] for polystyrene sulfonate in 0.005 M sodium chloride solution. The lower plot in Fig. 3.12 represents the literature data. Therefore, the influence of molecular weight on the hydrodynamic size of the block polymers in solution was similar to what was observed for the homopolymer analogs and suggests that both polymers behave as rigid rods in solution. However, the block polymers have intrinsic viscosities about an order of magnitude higher than the homopolymers due to the aggregation of several polymer molecules via the association of the hydrophobic blocks.

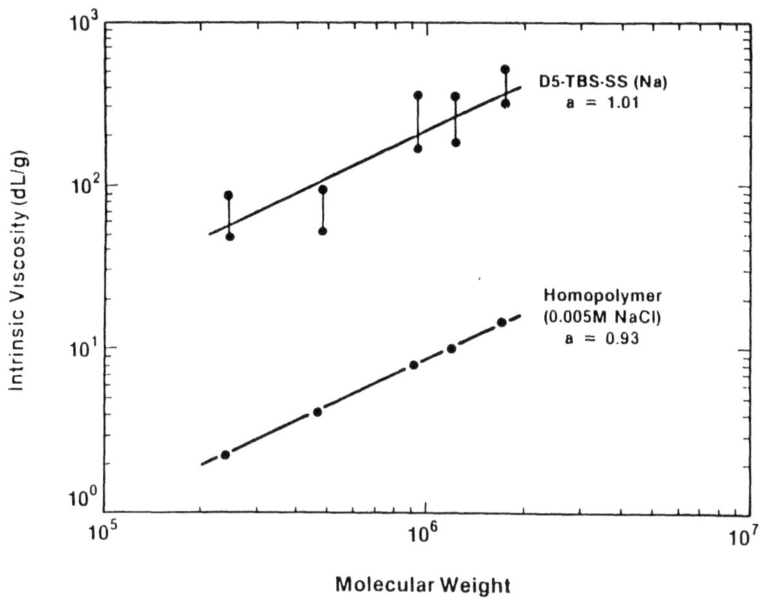

Fig. 3.12. Mark–Houwink–Sakurada plots for diblock, D5-TBS-SS(Na), and homopolymer, sodium polystyrene sulfonate. Reprinted with permission from Ref. 7

Rheological studies of solutions of these polymers in distilled water clearly identified viscosity enhancement due to the presence of the hydrophobic *tert*-butylstyrene endblocks. Comparison with homopolymer and random copolymers of similar composition and molecular weight demonstrated the increased viscosification with the block configuration. The block polymer solutions were also observed to solubilize toluene with a marked increase in the solution viscosity. These results are interpreted as the result of polymer intermolecular association through the hydrophobic blocks similar to micellization of surfactants.

3.6 Summary

Hydrophobically associating water-soluble polymers are of interest because of their unique and technologically useful solution behavior. Enhanced viscosification, shear thickening rheology, maintenance of viscosity in salt, as well as improved mechanical and thermal stability, have been observed for some of these systems. Polymer architecture, such as block or random hydrophobe distribution, is important. The balancing of hydrophobic and charged groups provides additional synthetic design to control solution properties. Such versatile solution properties make hydrophobically associating polymers excellent candidates for use as aqueous rheology modifiers.

3.7 References

1. Tanford C (1973) The hydrophobic effect: Formation of micelles and biological membranes. Wiley, New York
2. Ben-Naim A (1980) Hydrophobic interactions. Plenum, New York.
3. Franks F (1975) Water: A comprehensive treatise, vol. 4. Plenum, New York
4. Bock J, Siano DB, Schulz DN, Turner SR, Valint PL, Pace SJ (1988) In: Stahl GA, Schulz DN (eds) Water soluble polymers for petroleum recovery. Plenum, New York, p 147
5. Turner SR, Siano DB, Bock J (1985) US Patents 4 528 348, 4 520 182, and 4 521 580
6. Schulz DN, Maurer JJ, Bock J (1984) US Patents 4 463 151 and 4 463 152
7. Valint PL, Bock J (1986) Polym Mater Sci Eng 55:361; (1985) US Patents 4 492 785 and 4 548 735; (1988) Macromolecules 21:175
8. Valint PL, Bock J, Schulz DN, (1989) In: Glass JE (ed) Polymers in aqueous media: Performance through association, Advances in chemistry series 223. Am Chem Soc, Washington, DC, p 399; J Polym Mater Sci Eng (1987) 57:484
9. Bock J, Siano DB, Valint PL Jr, Pace SJ (1987) Polymer Mater Sci Eng 57:487; Bock, Siano, Valint, Pace in Ref 15, p. 411
10. Siano DB, Bock J, Myer P, Valint PL Jr (1987) Polym Mater Sci Eng 57:609; In Ref 15, p 425
11. McCormick CL, Bock J, Schulz DN (1989) Encyclopedia of polymer science and engineering. John Wiley, vol 17, p 730
12. Polymers as Rheology Modifiers (1991). Schulz DN, Glass JE (eds) ACS Symposium Series #462. American Chemical Society, Washington, DC
13. Schulz DN, Kaladas JJ, Maurer JJ, Bock J, Pace SJ, Schulz WW (1987) Polymer 28:2110
14. McCormick CL, Nonaka T, Johnson CB (1988) Polymer 29:731
15. Polymers in Aqueous Media (1989) Glass JE (ed) Advances in chemistry series #223. American Chemical Society, Washington, DC
16. Water Soluble Polymers (1991) Shalaby B, McCormick CL (eds) ACS Symposium Series #467. American Chemical Society, Washington, DC
17. Evani S, Rose GD (1987) Polym Mater Sci Eng 57:477
18. Peer WJ (1878) Polym Mater Sci Eng 57:492, and in Ref 15, p 381
19. Landoll LM (1982) Polym Sci, Polym Chem ed 20:443
20. Biggs S, Selb J, Candau F (1992) Langmuir 8:838
21. Dualeh AJ, Steiner CA (1990) Macromolecules 23:251; (1991) 24:112
22. Steiner CA (1991) J Appl Polym Sci 42:1493
23. Illiopoulos I, Wang KT, Audebert R (1991) Langmuir 7:617
24. Kulicke WM, Bose N, Bouldin M (1988) In: Stahl GA, Schulz DN (eds) Water soluble polymers for petroleum recovery. Plenum, New York, p 1
25. Lundberg RD, Duvdevani I (1991) In: Schulz DN, Glass JE (eds) Polymers as rheology modifiers. Am Chem Soc, Washington, DC, p 156 and references therein
26. Witten TA, Cohen MH (1985) Macromolecules 18:1915

4 Aggregation of Hydrophobically Modified Polyelectrolytes in Dilute Solution: Ionic Strength Effects

B. Magny, I. Iliopoulos and R. Audebert

Viscometry and fluorescence spectroscopy are used to study the dilute solution behavior of hydrophobically modified polyelectrolytes. Random copolymers of sodium acrylate and alkylacrylamide are used for this study. The alkylacrylamide content is low, 1 – 3 mol%, and the length of the alkyl group ranges from octyl to octadecyl. Upon addition of NaCl these copolymers exhibit a transition from a nonaggregated coil to an aggregated one. The transition occurs above a critical salt concentration which is dependent on the length and the content of alkyl groups. This behavior is ascribed to the formation of intrachain hydrophobic aggregates. The copolymer with the shortest alkyl chain (octyl) does not exhibit such aggregation at least in the range of salt concentrations studied, i.e. ≤ 15% NaCl.

4.1 Introduction

Amphiphilic polymers are water soluble or water dispersable polymeric materials constituted of well defined hydrophilic and hydrophobic parts. They can have either a high content in hydrophobes, for instance the alternating copolymers of ionic and hydrophobic monomers (polysoaps) [1 – 3], or a low content of very hydrophobic groups, for instance copolymers of acrylamide (99 mol%) and octylacrylamide (1 mol%) [4].

The latter materials, also known as Hydrophobically Modified (or hydrophobically associating) Water Soluble Polymers (HMWSP) exhibit very salient rheological features markedly different from those of the unmodified parent polymers. For example, in semidilute aqueous solution the interchain aggregation of the very hydrophobic groups leads to an increase in the apparent molecular weight of the system inducing enhanced viscosity and reversible shear sensitivity. Of course, the aggregation efficiency and subsequently the rheological behavior depends on the structural characteristics of the HMWSP, notably the content and the nature of the hydrophobic groups and their distribution along the copolymer chain. Other external factors such as ionic strength and pH may play an important role especially if the copolymer is a polyelectrolyte. It is obvious that these materials may be used as thickening additives in many aqueous-based industrial formulations, e.g. latex paints, drilling muds, hydraulic fracturing fluids, flocculents, foods, cosmetics.

The numerous potential applications of these polymers as well as some fundamental questions related to their aggregation mechanisms and to their use

Macromolecular Complexes
in Chemistry and Biology
Eds. Dubin/Bock/Davis/Schulz/Thies

as models for the understanding of the behavior of more complicated biological polymers can explain the increasing interest in this field (see the other contributions in this book and references [5–6].

Our interest was mainly focused on the study of hydrophobically modified polyelectrolytes. We studied especially the synthesis and the rheological behavior of hydrophobically modified poly(sodium acrylate) (HMPAA). Typically the HMPAA we used contained 1–3 mol% of alkylacrylamide units randomly distributed along the polymer chain [7]. The length of the alkyl groups ranged from octyl to octadecyl.

In semidilute solution and upon addition of salt (NaCl) an "anti-polyelectrolyte effect" was observed. That means that the viscosity increased with the ionic strength and was several orders of magnitude higher than that measured in pure water [8–9].

We ascribed this behavior to the aggregation of the alkyl groups which becomes more effective as the ionic strength increases because of the screening of the electrostatic repulsions between the acrylate groups. Of course, the rise of the viscosity and more generally the rheological behavior of these polymers is the result of a subtle balance between the increase in the apparent molecular weight due to the interchain aggregation and the decrease in the coil dimensions due to the screening of the electrostatic repulsions and the intrachain aggregation.

To understand the general rheological behavior of these materials and especially the effect of competition between inter- and intrachain aggregation one needs information about their behavior in very dilute solution; i.e. in conditions where the polymer chains behave as isolated coils. In this work we use viscometry and fluorescence spectroscopy techniques to study the effect of the ionic strength on the dilute solution properties of HMPAA. The results are discussed in terms of a transition from a nonaggregated coil to an aggregated one.

4.2 Experimental

Poly(acrylic acid) was purchased from Polyscience and its average molecular weight given by the supplier was 150000. The modification reaction was described elsewhere [8]. The modified samples had the same polymerization degree as the precursor polymer and they were obtained in the sodium salt form:

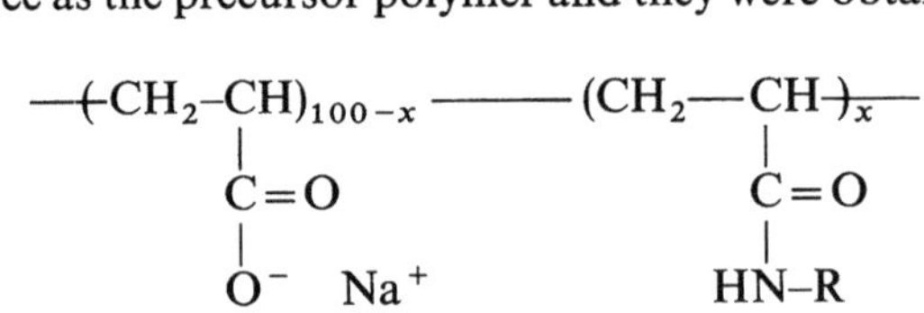

where R is an alkyl chain with n carbon atoms ($n = 8$, 12, 14 or 18) and x the modification degree in mol%. The following sample designation was adopted: 3-C12 contains 3 mol% of dodecyl groups. Because the samples were used in their fully neutralized form (sodium salt), the pH of their solutions was slightly basic (pH$\approx$8.7).

Concentrated polymer stock solutions were prepared under magnetic stirring at least 24 h before use. Final solutions of desired concentration were obtained by addition of a NaCl solution to the appropriate volume of polymer stock solution (NaCl concentration is expressed in g per 100 ml). The polymer concentration used for specific viscosity and for fluorescence measurements was 10^{-2} mol l^{-1} in monomer unit for all the samples. High purity pyrene (supplied by Polyscience) was used as the fluorescence probe at a concentration of $5 \cdot 10^{-7}$ mol l^{-1}.

Viscosity measurements were performed with an automatic (Fica Viscomatic MS) Oswald type capillary viscometer. Fluorescence emission spectra of pyrene were recorded between 355 and 480 nm on a Fica 65 MK II spectrofluorimeter. The excitation wavelength was 334 nm and the band paths were set at 7.5 nm for excitation and at 0.5 nm for emission. The I_1/I_3 ratio was calculated as the ratio of the intensity at 373 nm to that of 384 nm. All the measurements were conducted at 25.0 ± 0.1 °C.

4.3 The Polymers

Hydrophobically associating water soluble polymers can be prepared either by copolymerization of a hydrophilic and a hydrophobic monomer or by modification of a precursor water-soluble polymer. A typical example of materials prepared by the first method is the hydrophobic derivatives of polyacrylamide (PAM) [10–15]. The copolymerization occurs in the presence of a surfactant in order to increase the solubility of the hydrophobic component in the water and the resulting polymers exhibit a rather block structure: the hydrophobic groups have a rather block distribution on the PAM chain [10, 13–15]. The size of the hydrophobic blocks can be controlled by the quantity of the surfactant added (see also the contribution of Prof. McCormick group in this book). Unfortunately this copolymerization technique does not allow the simultaneous control of all the molecular parameters (i.e. polymer molecular weight, degree of modification and alkyl group distribution).

From this point of view the modification method presents some advantages. The degree of polymerization of the modified sample is the same to that of the precursor, provided that neither polymer degradation nor cross-linking occur. The degree of modification can be controlled by the amount of hydrophobic substituent added. Finally, if the modification reaction occurs in the homogeneous solution (the solvent used must solubilize both the water soluble polymer and the hydrophobic substituent) then a random distribution of the alkyl groups on the polymer chain is expected. It must be noted that good control of the molecular parameters can be achieved by the copolymerization technique if it occurs in a common solvent for both the hydrophilic and the hydrophobic monomer [16].

We have used the modification method to obtain hydrophobic derivatives of poly (sodium acrylate) (PAA). In brief, a precursor poly(acrylic acid) (COOH form) was modified by various alkylamines in an aprotic solvent (1-methyl-2-pyrrolidone) and in the presence of *N,N'*-dicyclohexylcarbodiimide [8]

$$-\mathrm{COOH} + \mathrm{RNH_2} \longrightarrow -\mathrm{CONHR}$$

The obtained modified polymers have a random distribution of the alkyl groups [7] and their molecular weight is close to that of the precursor. The alkyl chain length was adjusted by the choice of the alkylamine and the modification degree by the molar ratio of amine to acrylic acid unit (the reaction yield was 100%).

The semidilute solution behavior of these materials was described elsewhere [9]. We want to mention here that in pure water and above a critical polymer concentration, C_p^c, the viscosity of the modified polymers rise sharply and can be several orders of magnitude higher than that of the precursor. The longer the alkyl chain and the higher the modification degree the lower the C_p^c value [9]. Above the C_p^c, the addition of small amounts of NaCl leads to a further enhancement of the viscosity.

4.4 Dilute Solution Viscosity

Typical plots of the specific viscosity, η_{sp}, against the logarithm of salt (NaCl) concentration C_s, are given in Fig. 4.1. As expected, for both the precursor poly(sodium acrylate) and the modified sample 3-C18 (containing 3 mol% of octadecylacrylamide groups), the viscosity decreases monotonically as the NaCl concentration increases. However, the modified polymer exhibits a more pronounced viscosity decrease when the salt concentration is higher than about 0.3%. In the case of the precursor polymer the decrease in viscosity is due only to the screening of the electrostatic interactions. The modified polymer seems to follow the same mechanism at low salt concentrations but apparently at $C_s > 0.3\%$ another mechanism contributes to some additional chain retraction.

The importance of the latter mechanism can be more obvious if the viscosity of the modified polymer is reduced by the viscosity of the precursor. This is shown in Fig. 4.2 when the ratio of the specific viscosity of the modified polymer to that of the precursor, $\eta_{sp,mod}/\eta_{sp,prec}$, is plotted as a function of the NaCl concentration. At low salt concentration, when only the screening of the electrostatic interactions is operative, the ratio $\eta_{sp,mod}/\eta_{sp,prec}$ is constant and very close to 1. Above a critical salt concentration, C_s^c, the modified sample exhibits lower viscosity values than the precursor and then the viscosity ratio decreases linearly with the logarithm of NaCl concentration. This behavior is observed with polymers bearing dodecyl, tetradecyl and octadecyl groups (3 mol%) but not with the one bearing octyl groups (Fig. 4.2). A sample containing 1 mol% of octadecyl groups behaves like the sample 3-C12. The critical salt concentration (C_s^c) decreases when the alkyl chain length or the modification degree increases.

It is very tempting to ascribe this behavior to the hydrophobic aggregation of the alkyl groups. In fact, at the low polymer concentration we used (10^{-2} mol l^{-1} in monomeric units, i.e. of about 0.1%) one can assume that the

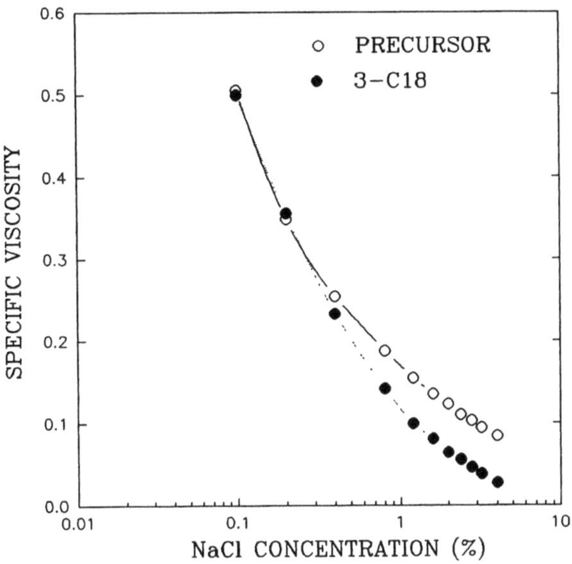

Fig. 4.1. Specific viscosity versus salt concentration for the precursor (○) and a modified polymer 3-C18 (●). Polymer concentration: 10^{-2} unit mol l^{-1}

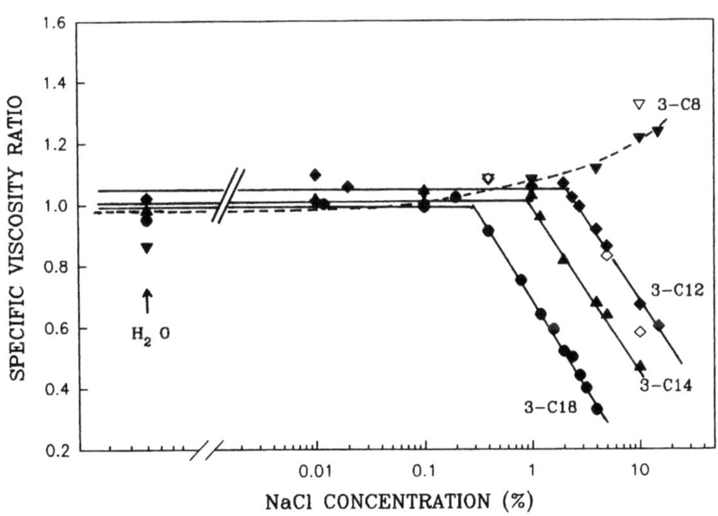

Fig. 4.2. Specific viscosity ratio ($\eta_{sp,mod}/\eta_{sp,prec}$) versus salt concentration for several modified polymers: (●) 3-C18; (▲) 3-C14; (◆) 3-C12; (▼) 3-C8. Polymer concentration as in Fig. 4.1. (◇) and (▽) represent the ratio of intrinsic viscosity $[\eta]_{mod}/[\eta]_{prec}$ for 3-C12 and 3-C8, respectively

polymer chains behave as isolated coils. The coil dimensions are mainly determined by an interplay between two oppositely acting forces: the electrostatic repulsions between charges of the polymer chain and the attractive hydrophobic forces between alkyl groups. At low ionic strength the electrostatic repulsions prevail over the hydrophobic attractions and the modified sample behaves like the precursor. At $C_s > C_s^c$ the electrostatic repulsions are sufficiently reduced and the polymer coil sufficiently contracted to allow the aggregation of the alkyl chains. The hydrophobic attractions are stronger for polymers bearing longer alkyl groups or having a higher content of alkyl groups. This behavior can be compared qualitatively with that of low molecular weight ionic surfactants. At surfactant concentrations lower than their critical micellar concentration (cmc) in pure water, the micellization can be induced by addition of salt because of the decrease of the electrostatic repulsions between the polar heads. The longer the alkyl chain the lower the salt concentration needed to induce micellization [17].

It is also noteworthy that above the aggregation threshold (i.e. above C_s^c) the slope of all the plots reported in Fig. 4.2 is practically the same (0.60 ± 0.05). This probably means that the aggregation mechanism (i.e. number and size of the aggregates and the rate of growth of these aggregates upon salt addition) is the same and is independent of the length of the alkyl group or of the modification degree.

Surprisingly the sample modified with the shorter alkyl chain (3-C8) presents a slight increase in the viscosity ratio as the salt concentration increases. Obviously the octyl groups are too short to induce hydrophobic aggregation as was also put forward by the study of this polymer in semidilute solution [9]. The slight increase in viscosity is due to other effects which will be discussed in the last section of this paper.

4.5 Fluorescence Spectroscopy

Fluorescence spectroscopy is a powerful technique to study the formation of hydrophobic aggregates in water solution. Pyrene is one of the most used fluorescent molecules to probe the formation of these kind of aggregates [18]. Pyrene has been extensively used to determine the micellar size [19] and the cmc [18] in surfactant solutions, but also to study the conformational transitions in amphiphilic polymers [20–21] and the cluster formation between surfactants and polymers [20–22].

The main feature making pyrene a useful probe for the study of hydrophobic microdomains is the sensitivity of the vibrational band structure of its fluorescence emission spectrum to the polarity of its environment [18]. The relative intensity of the peaks of the emission spectrum undergoes significant perturbation when the solvent polarity increases. In particular, the intensity of the first peak (I_1) increases in polar solvents while that of the third peak (I_3) is essentially unaffected. So the ratio I_1/I_3 is sensitive to the solvent polarity and more particularly to the polarity of the microenvironment experienced by the probe. For instance, it has a value of 0.65 in cyclohexane, while in water its value is close to 1.85 [18]. In a surfactant solution pyrene probes the formation of micelles. At concentrations lower than the cmc, the I_1/I_3 ratio is close to that found in pure water but at the cmc a transition is observed and at surfactant concentrations higher than the cmc the I_1/I_3 ratio takes values between those found for aqueous and nonpolar solvents. For example, in $5 \cdot 10^{-2}$ mol l^{-1} sodium dodecylsulfate solution the I_1/I_3value is 1.14 [18].

In Fig. 4.3 two emission spectra of pyrene in a solution of a modified polymer (10^{-2} unit mol l^{-1} of 3-C12) are given. One of them (full line) is obtained in low ionic strength solution (0.1% NaCl) and the ratio I_1/I_3 is found to be significantly the same as that in pure water, i.e. 1.85, indicating no formation of hydrophobic aggregates. When the ionic strength is very high (10% NaCl) the intensity of the third peak increases (dashed line in Fig. 4.3) and the I_1/I_3 value is now 1.45 reflecting the formation of hydrophobic aggregates. These results are in line with the viscosity results shown in Fig. 4.2. In fact, for this modified polymer (3-C12), the critical salt concentration as found by the viscosity measurements and above which the hydrophobic aggregates are formed, is $C_s^c = 2\%$.

In Fig. 4.4 the ratio I_1/I_3 is plotted as a function of NaCl concentration for the same modified polymers as those of Fig. 4.2. It is obvious that for three of the polymers (3-C12, 3-C14 and 3-C18) the pyrene experiences a transition from a high polarity environment ($I_1/I_3 \approx 1.80$) to a lower polarity one ($I_1/I_3 \simeq 1.40$). Such a transition is not observed for the sample 3-C8, while a sample bearing 1 mol% of octadecyl groups (not shown) presents practically the same transition as the 3-C12. These results are in very good correlation with the viscosity results reported in Fig. 4.2. The C_s^c values determined by viscosity are indicated by arrows in Fig. 4.4. The onset of the decrease in the I_1/I_3 ratio fits well with the C_s^c values

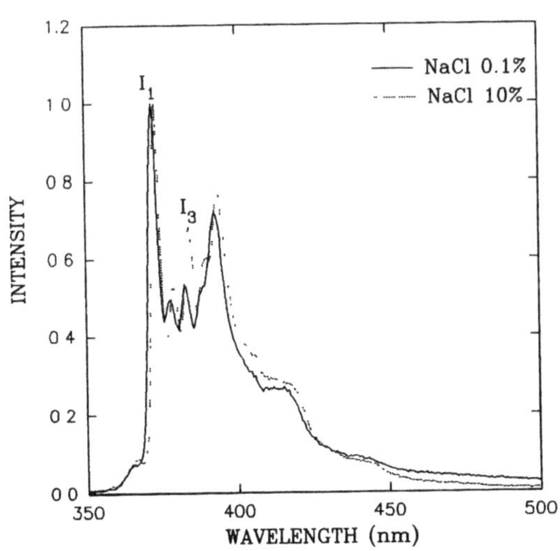

Fig. 4.3. Fluorescence emission spectra of pyrene in aqueous solution of 3-C12, in the presence of NaCl. Salt concentration: 0.1%, solid line; 10% dashed line. Pyrene concentration: 5×10^{-7} mol l^{-1}. Excitation wavelength: 334 nm

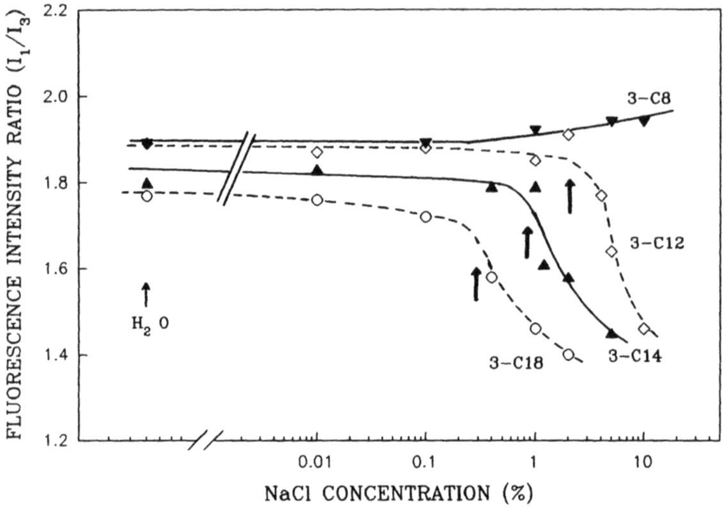

Fig. 4.4. Fluorescence intensity ratio (I_1/I_3) versus salt concentration for the same modified polymers as in Fig. 4.2. Arrows indicate the critical salt concentration as found by the viscosity measurements (Fig. 4.2)

confirming that a transition from a nonaggregated to an aggregated coil occurs and that this transition is due to the formation of hydrophobic microdomains.

At salt concentrations lower than the C_s^c the value of I_1/I_3 depends on the nature of the modified polymer. For the polymer 3-C12 this value is very close to that measured in pure water, but decreases slightly for the 3-C14 and significantly for the 3-C18. It seems that the pyrene experiences a more hydrophobic environment even before the onset of the hydrophobic aggregation. One can ascribe this result to the fact that the hydrophobic pyrene probe lies preferably close to the nonpolar alkyl chain than in the bulk solution.

4.6 Intrinsic Viscosity

Although the viscosity and fluorescence results discussed above were obtained in dilute solution, the polymer concentration was 10^{-2} unit mol l^{-1} (i.e about 0.1%), they can also reflect some contribution of coil/coil interactions. Consequently, a more rigorous approach should be the study of these systems at the infinite dilution limit, for example, by measuring the intrinsic viscosity, $[\eta]$, instead of the specific one. However, such a study presents two main drawbacks. One is that intrinsic viscosity measurements cannot be given at very low salt concentrations, in our case lower than 0.2% NaCl. The other is that these measurements are very time consuming, at least five measurements must be done for each sample at each salt concentration.

To check if there is any significant difference between the results obtained by measuring the specific or the intrinsic viscosity, we performed some intrinsic

Table 4.1. Intrinsic viscosity and Huggins coefficient at low and high ionic strength conditions: 0.4 and 10% NaCl respectively

NaCl (%)	Polymer	$[\eta]$/ml g^{-1}	$\frac{[\eta]_{mod}}{[\eta]_{prec}}$	k	$\frac{k_{mod}}{k_{prec}}$
0.4	Precursor	256	—	0.38	—
	3-C12	278	1.08	0.37	1.0
	3-C8	276	1.08	0.34	0.9
10	Precursor	59	—	0.65	—
	3-C12	34	0.58	6.3	9.7
	3-C8	78	1.32	0.48	0.7

viscosity measurements for the precursor and the modified samples 3-C12 and 3-C8. For salt concentration $\geq 0.4\%$ linear plots of the reduced viscosity versus the polymer concentration were obtained. The values of $[\eta]$ and of the Huggins coefficient, k, in the presence of 0.4% and 10% NaCl are given in Table 4.1. The ratio of the intrinsic viscosity of the modified polymer to that of the precursor $[\eta]_{mod}/[\eta]_{prec}$ is plotted in Fig. 4.2 together with the specific viscosity results. It is obvious that there is not a significant difference between these two kinds of results and consequently we can infer that the specific viscosity measurements depict the polymer coil behavior at the infinite dilution limit.

The intrinsic viscosity measurements also give information about the coil-coil interactions as reflected by the Huggins coefficient. At low ionic strength (0.4% NaCl) $[\eta]$ and k for the two modified samples are close to that of the precursor (Table 4.1) indicating no hydrophobic aggregation. Furthermore k takes values low enough ($\simeq 0.35$) as usually observed for polymers in a good solvent. At high ionic strength (10% NaCl) the sample 3-C12 exhibits hydrophobic aggregation, the ratio $[\eta]_{mod}/[\eta]_{prec}$ takes a low value (0.58) while the Huggins coefficient for this sample increases one order of magnitude: i.e. 6.3 instead of 0.65 for the precursor (Table 4.1 and Fig. 4.5). Such high values of the Huggins coefficient were usually interpreted as increased coil-coil interactions [23]. Similar observations were also reported for other associating polymers [24–25]. The k value for the precursor slightly increases with the salt concentration: from 0.38 at 0.4% NaCl to 0.65 at 10% NaCl. This is in agreement with the general observation that k increases when coil expansion decreases [23] but its value is still between 0.3 and 0.8 as reported for nonassociating polymers.

A very different behavior was observed with the sample 3-C8. In 10% NaCl solution the intrinsic viscosity of this sample exceeds that of the precursor by $\approx 30\%$ (Fig. 4.5 and Table 4.1) in a similar way as the specific viscosity does (Fig. 4.2), and k takes a value typical of nonassociating systems (0.48). Obviously the hydrophobic interactions between octyl groups are not strong enough to induce hydrophobic aggregation.

Until now the solution properties of the copolymer coil were discussed only in terms of electrostatic repulsions between acrylate units and hydrophobic

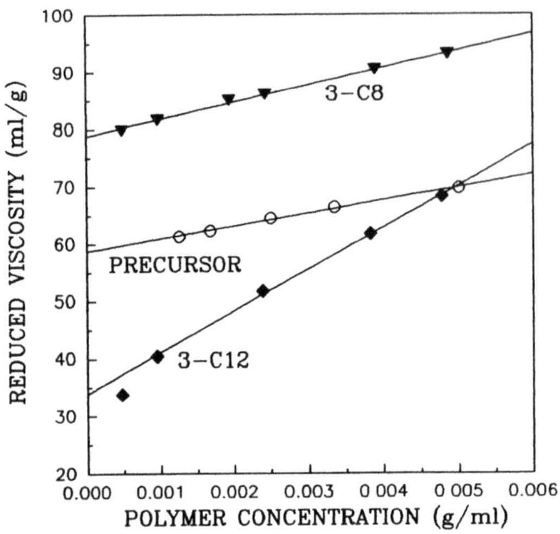

Fig. 4.5. Reduced viscosity versus polymer concentration for the precursor (○) and two modified polymers 3-C12 (◆) and 3-C8 (▼) in 10% NaCl solution

attractions between the alkyl chains. The first kind of interactions are predominant in low ionic strength conditions, the second one becomes important at high ionic strength. Although neglected during our previous discussion interactions between unlike monomers, i.e. between sodium acrylate and alkylacrylamide units, are also present and their contribution to coil expansion and subsequently to the intrinsic viscosity of the modified polymer could be detectable under suitable conditions.

It has been shown that the expansion of a copolymer coil depends largely on the number of heterocontacts, i.e. contacts between unlike monomers [26, 27]. The larger the number of heterocontacts the higher the expansion of the copolymer coil. As a consequence, the intrinsic viscosity of the copolymer is higher than that of the equivalent homopolymers. The importance of this effect increases as the solvent quality decreases [27].

That is happening with the sample 3-C8. The hydrodynamic volume of its coil is influenced by the electrostatic repulsions between the acrylate groups and also the interactions between these groups and the octylacrylamide groups. As shown before, the attractive hydrophobic interaction between octyl groups are negligible for this sample (3-C8), at least under the ionic strength conditions used. In low ionic strength solution, the effect of the electrostatic repulsions on the coil expansion largely prevails over the effect due to heterocontacts and consequently the modified polymer has the same intrinsic viscosity as the precursor. By increasing the ionic strength, the electrostatic repulsions are screened, the solvent quality towards the copolymer decreases, the attractive hydrophobic forces are still negligible and then the heterocontacts effect becomes important. This effect is noticeable even for copolymers with a low content in octylacrylamide units because of the bulkiness of these groups. The intrinsic viscosity of the copolymer becomes larger than that of the precursor.

The heterocontacts effect is operative even for the polymers modified with larger alkyl chains but it is screened by the strong hydrophobic aggregation effect.

4.7 Concluding Remarks

Hydrophobically modified poly(sodium acrylate) exhibit salient rheological behavior in aqueous solution. The fluorescence and viscosity results reported in this study show that in dilute solution a transition from a nonaggregated to an aggregated coil occurs upon addition of salt. The salt concentration at which the aggregation starts, depends on both the alkyl chain length and content. Above this concentration the chain retraction due to the hydrophobic aggregation seems to be only dependent on the salt concentration but independent of the length and the content of the alkyl groups.

Because of the low polymer concentration used, this behavior is believed to be due to the intracoil hydrophobic aggregation. The aggregation coil also exhibits high values of the Huggins coefficient reflecting some inter-coil interactions. Light scattering studies are in progress in our laboratory to determine what the contribution of inter-coil interactions is in the infinite dilution limit.

Acknowledgements. This work was supported by the Société Française HOECHST and the Centre National de Recherche Scientifique. We want to thank Dr R. Zana and Dr L. Bokobza for helpful discussions.

4.8 References

1. (a) Dubin P, Strauss UP (1967) J Phys Chem 71:2757; (b) Dubin P, Strauss UP (1970) J Phys Chem 74:2842
2. Hsu JL, Strauss UP (1987) J Phys Chem 91:6238
3. Binana-Limbelé W, Zana R (1990) Macromolecules 23:2731
4. Bock J, Valint Jr PL, Pace SJ, Siano DB, Schulz DN, Turner SR (1988) In: Stahl CA, Schulz DN (eds) Water soluble polymers for petroleum recovery. Plenum Press, New York, p 147.
5. Glass JE (ed) (1989) Polymers in aqueous media: performance through association. Advances in Chemistry Series 213, Am Chem Soc, Washington, DC
6. Shalaby SW, McCormick CL, Butler GB (eds) (1991) Water soluble polymers. ACS symp Ser 467, Am Chem Soc, Washington, DC
7. Magny B, Lafuma F, Iliopoulos I (1992) Polymer, 33:3151
8. Wang TK, Iliopoulos I, Audebert R (1988) Polym Bull 10:577
9. (a) Wang TK, Iliopoulos I, Audebert R (1989) Am Chem Soc Div of Polym Chem Prepr 30(2):377; (b) Wang TK, Iliopoulos I, Audebert R (1991) In: Shalaby SW, McCormick CL, Butler GB (eds) Water soluble polymers. ACS Symp. Ser. 467, Am Chem Soc, Washington, DC, p. 218
10. Hill A, Candau F, Selb J (1991) Progr Colloid Polym Sci 84:61
11. McCormick CL, Nonaka T, Johnson CB (1988) Polymer 29:731
12. Schulz DN, Kaladas J. Maurer JJ, Bock J, Pace SJ, Schulz WW (1987) Polymer 28:2110
13. Peer WJ (1989) In: Glass JE (ed) Polymers in aqueous media: performance through association. Advances in Chemistry Series 213, Am Chem Soc, Washington, DC, p 381
14. Dowling KC, Thomas JK (1990) Macromolecules 23:1059
15. (a) Ezzel SA, McCormick CL (1989) Am Chem Soc Div Polym Chem Prepr 30(2):340; (b) Ezzel SA, McCormick CL (1991) In: Shalaby SW, McCormick CL, Butler GB (eds) Water Soluble Polymers, ACS Symp Ser 467, Am Chem Soc, Washington, DC, p 130

16. Ringsdorf H, Venzer J, Winnik FM (1990) Am Chem Soc Div Polym Chem Prepr 31(1):568
17. Lindman B, Wennerström H (1980) In: Boachke FL, Topic in Current Chemistry 87. Springer Verlag, New York, p 1
18. Kalyanasundaram K, Thomas JK (1977) J Am Chem Soc 99:2039
19. Turro NJ, Yekta A (1978) J Am Chem Soc 100:5951
20. Binana-Limbelé W, Zana R (1987) Macromolecules 20:1331
21. Chu DY, Thomas JK (1987) Macromolecules 20:2133
22. Thalberg K, Stam JV, Lindblad C, Almgren M, Lindman B (1991) J Phys Chem 95:8975
23. Bohdanecky M, Kovar J (1982) In: Jenkins AD (ed) Viscosity of polymer solutions. Elsevier Scientific Publishing Co., New York, p 177
24. Bock J, Siano DB, Valint Jr. PL, Pace SJ. In: Glass JE (ed) Polymers in aqueous media: performance through association. Advances in Chemistry Series 213, Am Chem Soc, Washington, DC, p 411
25. Agarwal PK, Garner RT, Graessley WW (1987) J Polym Sci: Part B: Polym Phys Ed 25:2095
26. Dondos A, Benoit H (1968) Makromol Chem 118:165
27. Dondos A, Rempp P, Benoit H (1969) Makromol. Chem 130:233

5 Microdomain Composition in Two-Phase Hydrogels

C.G. Varelas, A.J. Dualeh and C.A. Steiner

Hydrogels which form spontaneously from surface-active graft copolymers in solution are characterized by two distinct phases – a continuous phase consisting of water and the hydrophilic backbones of the polymer chains, and a dispersed phase composed of clusters of the hydrophobic side chains. The composition of these clusters, or microdomains, has been investigated using fluorescence probe techniques, and was found to depend on the composition of the solvent. The microdomain composition in turn influences the bulk network properties, since the microdomains act as linkage points in the network. In addition, hydrophobic solutes permeating the network are absorbed into the microdomains, permitting higher solute loading into these hydrogels than in equivalent volumes of water.

5.1 Introduction

We are exploring the phenomenon of network formation in surface-active graft copolymers [1–6]. Under well-defined conditions these polymers phase separate such that the grafted side chains cluster together to exclude solvent, much like the process of micellization in aqueous surfactant solutions. One result of this process is the formation of polymer networks in which the clusters act as temporary crosslinks, imparting elasticity to the material over a finite range of applied shear. More interesting, however, is the fact that on a microscopic level the clusters can act as reservoirs or sinks for solutes permeating the network. Solutes partition into the reservoirs in proportion to their relative solubilities in the bulk solvent and the interior of the reservoirs, or, depending on the solute structure, in proportion to their affinity for the surface of the cluster.

These properties open the possibility for some interesting applications for these networks. They may find use as a new stationary phase for separation processes, where a mixture of surface-active solutes is passed through the network and the solutes selectively adsorb on the interface between the clusters and the bulk network. They may also be used in controlled release technology. In this mode the network would be impregnated initially with solute such that both the clusters and the bulk (continuous) phase of the material are saturated. As the solute diffuses out of the bulk to the delivery site, it will be continually

Macromolecular Complexes
in Chemistry and Biology
Eds: Dubin/Bock/Davis/Schulz/Thies

restored into the bulk from the reservoirs, maintaining a nearly constant level of solute in the bulk and hence a constant driving force to the surroundings, i.e. zero order release kinetics. Alternatively, in wastewater treatment, the clusters would be able to absorb and retain oily contaminants which could then be easily removed with the gel.

We have successfully made networks of this type using hydrophobically modified hydroxyethyl cellulose (HMHEC), a nonionic cellulose ether, in two types of aqueous solvents, surfactant/water solutions and ethanol/water solutions. We have uncovered relationships between the solvent composition and the extent of cluster formation as well as the cluster composition. And we have found that we can indeed obtain higher loading of hydrophobic solutes than would be expected in equivalent volumes of water.

5.2 Background

5.2.1 Microphase Separation in Polymers

To date most of the work published on microphase separation in polymers has focused on block copolymers. The microphases are composed of segments of both types of blocks, as expected from steric constraints on the polymer, but have higher concentrations of the insoluble segments closer to the center of the aggregates, as expected from solution thermodynamics. Structural properties of the polymer such as molecular weight, uniformity, chain flexibility, and stereoregularity of the blocks, as well as the polymer/solvent interaction parameter for each of the blocks in the solvent of interest, all contribute to the tendency for aggregate formation and the structure of the resulting aggregates. The properties of block copolymer solutions have been characterized experimentally using small angle neutron scattering [7], quasielastic light scattering [8], gel permeation chromatography [9], electron microscopy [9, 10], fluorescence spectroscopy [11], and small angle X-ray scattering [11]. Intrinsic viscosity studies have also been used to document the occurrence of hydrophobic interactions and microphase separation [12].

Krause [13] reviewed empirical evidence of the activity of microdomains in block copolymer membranes such as those used for reverse osmosis and ion exchange processes. Depending on whether they are dispersed or continuous, the microphases may either sequester compatible solutes or serve as channels for their transport through the membrane. Partitioning of solutes into the microdomains was found to be a complex function of solute structure and microdomain structure and composition and could not easily be predicted quantitatively.

Even in the solid state, microphase separation resulting in a dispersed-phase structure has been shown to occur in block copolymers [14]. Thus the dispersed-phase structure is clearly the preferred configuration for these polymers in a network. Networks prepared as cast films from block copolymer solutions have been characterized using such techniques as differential scanning calorimetry (DSC) [15–17] and electron microscopy [17, 18].

5.2.2 Importance of Graft Copolymers

Relatively little work has been reported on microphase separation in graft copolymers. Graft copolymers represent an important class of network-forming surface-active polymers because they can be produced easily from synthetic and naturally-occurring homopolymers such as poly(vinyl alcohol), cellulose derivatives, etc. [19] or by copolymerizing vinyl monomers with and without hydrocarbon substituents [20–24] giving well-defined product polymers. Interest in surface-active graft copolymers has grown in recent years with the synthesis of new polymers of this type and the recognition that these materials may interact among themselves and with other components of a solution, such as surfactants and dispersed particles, in a unique and potentially useful manner.

Besides having unique solution properties, our work has shown that surface-active graft copolymers also have the ability to form stiff, water-swollen hydrogels at polymer levels as low as 0.6% by weight [3, 4]. These gels may be either charged or nonionic, depending on the structure of the starting polymer and the composition of the solution from which the gels form. Some interesting characteristics of these gels are described below.

5.3 Results

5.3.1 Hydrogels from Surfactant Solutions

HMHEC forms hydrogels when solubilized in solutions containing sodium dodecyl sulfate (SDS) below the critical micelle concentration (cmc) of the surfactant. The surfactant molecules aggregate along with the hydrocarbon side chains to form clusters resembling mixed micelles. When these clusters incorporate side chains from more than one polymer molecule, as happens when the ratio of side chains to surfactant in the system is high, the clusters behave like noncovalent "crosslinks" and the resulting network exhibits a high ($\sim 10^3$ dyne/cm^2) constant storage modulus (G') over a range of applied shear rates [1, 2].

The structure of the microdomains which form in surfactant solutions is shown schematically in Fig. 5.1. Side chains and surfactant molecules combine to form aggregates resembling mixed micelles dispersed throughout the bulk aqueous network. The composition of these microdomains has been investigated using a fluorescence probe technique adapted from the surfactant literature [2]. Results are summarized in Table 5.1 and Fig. 5.2. The total aggregation number, N_T, of the clusters is equal to the sum of the number of surfactant molecules and alkyl side chains (N_s and N_a, respectively) in each

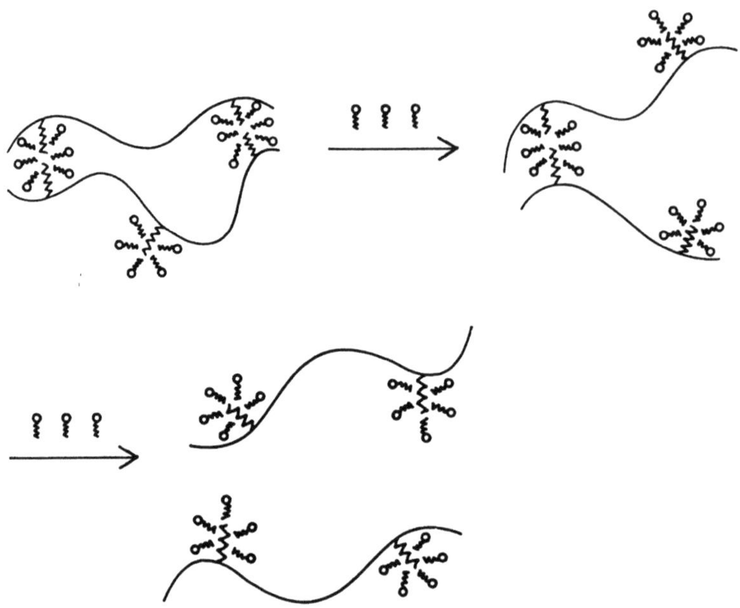

Fig. 5.1. Schematic of network structure for gels made from SDS solutions

Table 5.1. Effect of solution composition on microdomain composition (Ref. 2).

SDS (mM)	Side chain conc. (mM)	N_s	N_a	N_T
4.8	0.105	48.7	8.3	57.1
	0.32	50.7	8.8	59.5
5.6	0.105	58.3	6.9	65.2
	0.32	57.8	8.1	65.8
6.4	0.105	68.4	5.8	74.2
	0.32	65.3	7.0	72.3
7.2	0.157	73.0	4.7	77.7
	0.32	77.0	7.0	84.0

aggregate. As the SDS concentration in the starting solution is increased, N_s goes up, N_a goes down, and the net effect is an increase in N_T. On the other hand, at constant SDS concentration the value of N_a increases with polymer concentration.

The structure of the networks formed in these systems is profoundly affected by the composition of the microdomains. As more side chains become incorporated into each aggregate, the length of the polymer segments between aggregates goes down, and the storage modulus of the network increases, as shown in Fig. 5.3 [2].

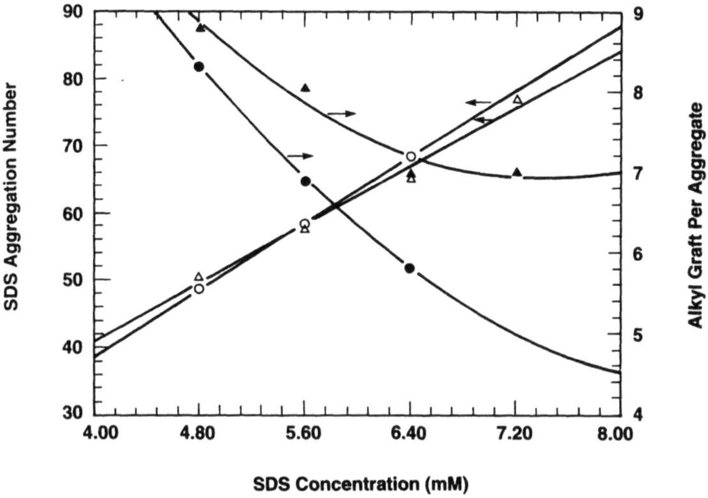

Fig. 5.2. Effect of SDS concentration on microdomain composition. △, ▲ – 0.4% (w/w) polymer, ○, ● – 0.13% (w/w) polymer (Ref. 2).

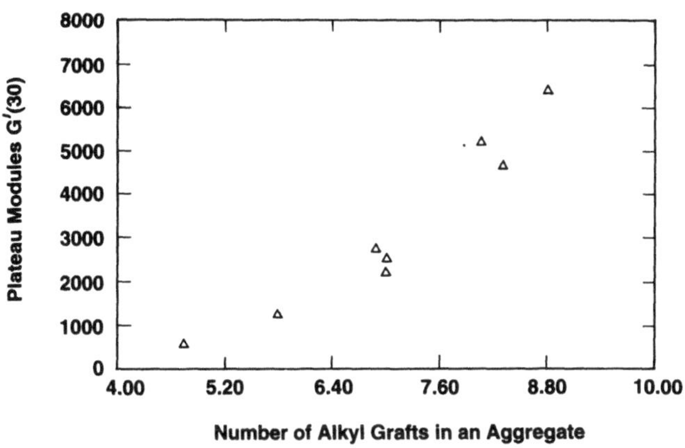

Fig. 5.3. Effect of the number of alkyl side chains per microdomain on the storage modulus, G', of the networks (G' in dynes/cm^2, evaluated at 30 rad/sec)(Ref. 2).

5.3.2 Hydrogels from Ethanol/Water Solutions

HMHEC hydrogels may also be made by solubilizing high concentrations of the polymer in mixtures of ethanol and water and allowing the excess polymer to precipitate out in gel form. The ethanol serves to solubilize the polymer in a relatively extended configuration by disrupting hydrophobic interactions which render the polymer insoluble in pure water. At high enough concentrations ($> 0.6\%$ by weight HMHEC in 50/50 (v/v) EtOH/water, e.g.), however, the

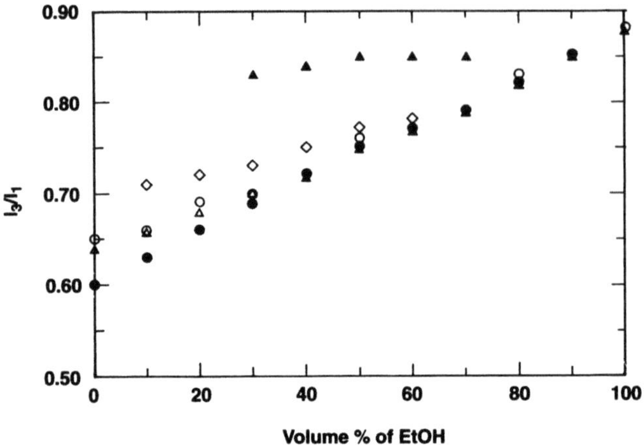

Fig. 5.4. Effect of solvent composition on hydrophobicity of networks made from ethanol/water solutions. ● EtOH/water, △ 0.1% HEC, ○ 0.1% HEC + Brij 30, ◇ 0.1% HMHEC solution, ▲ HMHEC gel: (Ref. 4)

chains begin to interact, with the hydrophobic side chains clustering together selectively, once again acting as linkage points in an aqueous network.

We have employed a fluorescence probe technique, namely, pyrene fluorescence, to demonstrate that these linkage points are in fact hydrophobic. The ratio of the heights of the third and first peaks, I_3/I_1, of the pyrene emission spectrum increases with the hydrophobicity of the pyrene's environment [25]. Since this probe is only sparingly soluble in water, it is expected to partition into any available hydrophobic regions in the network. Pyrene was incorporated into the gels from the initial solutions. Figure 5.4 shows the relationship between I_3/I_1 and the composition of the solvent for a series of gels and control solutions. The peak height ratio in the solvent alone increases linearly with ethanol concentration from 0.6 in pure water to 0.88 in pure ethanol. HMHEC in solution at a level of 0.1% (w/w) has a greater effect on the solution hydrophobicity than the same amount of hydroxyethyl cellulose (HEC), the unsubstituted control polymer, and this effect is felt at higher ethanol levels than is the case for HEC. However, this effect is not due solely to the presence of the hydrophobic side chains on the HMHEC, since control solutions containing HEC and Brij 30 surfactant, whose chemical composition is similar to that of the side chains, do not differ significantly from solutions with HEC alone. This is true for HEC solutions containing Brij 30 at levels up to 10 times higher than the hydrophobe level on the equivalent amount of HMHEC (but still below the critical micelle concentration (cmc) of the surfactant in water). Thus we have evidence of the presence of hydrophobic regions even in dilute HMHEC solutions.

The I_3/I_1 values for HMHEC gels are significantly higher than those of the corresponding solvents or of any of the control solutions (Fig. 5.4). Thus it is clear that there are distinct hydrophobic regions in the gels which are not present in the solutions. We also see that the existence and composition of these domains are relatively insensitive to the bulk solvent composition. While it is obvious that these regions are made up of side chains from the polymer, it is not possible from these experiments to determine their exact composition. The I_3/I_1 ratio for Brij 30 micelles in water is 0.92, which is higher than the value in our gels (0.84). However, some of the pyrene in our gels remains in the bulk (aqueous) phase due to its solubility in ethanol. Thus, the I_3/I_1 ratio, which reflects the average hydrophobicity of the pyrene environment, will have a significant contribution from the aqueous phase.

5.3.3 Solute Uptake by Gels

The affinity of HMHEC hydrogels for two small solutes, tryptophan and theophylline, was determined by reswelling dried gels in large volumes of saturated solutions of these compounds and analyzing the supernatant spectroscopically for the solutes. The partition coefficients, defined as the concentration of solute in the gel divided by the concentration in saturated water solution, were found to be 1.67 ± 0.06 ($N = 3$) and 1.50 ± 0.10 ($N = 3$) for tryptophan and theophylline, respectively [6]. Recall that our measurements of the fluorescence of pyrene unambiguously demonstrated that the pyrene resides primarily within hydrophobic regions of the gels. Thus, we infer that the same is true of these other organic solutes, although their location in the gels cannot be measured directly.

5.4 Conclusion

The microdomain structure of novel two-phase hydrogels has been investigated using a variety of analytical techniques. It has been shown that these domains consist of side chains from the starting polymer which aggregate in such a way as to exclude water. In the presence of surfactant, these clusters incorporate surfactant monomer as well as side chain. The composition of these domains, specifically the number of side chains per cluster, influences the rheological properties of the networks.

These microdomains are capable of solubilizing organic solutes permeating the gel to a greater extent than pure water, which is the primary component of the continuous phase of the gels. Thus, they may be of value in applications such as separations and wastewater treatment.

5.5 References

1. Dualeh AJ, Steiner CA (1990) Macromolecules 23:251
2. Dualeh AJ, Steiner CA (1991) Macromolecules 24:112
3. Varelas CG, Steiner CA (1990) In: Brannon-Peppas L, Harland RS (eds) Absorbent Polymer Technology. Elsevier, Amsterdam, p 259

4. Varelas CG, Steiner CA (1992) J Polym Sci B: Polym Phys Ed, 30:1233
5. Varelas CG, Dixon DG, Steiner CA, AIChE J, submitted
6. Varelas CG (1991) Structural analysis and applications of polymer networks formed from surface-active copolymers in aqueous solvents. Ph.D. Thesis, Department of Chemical Engineering, The City College of the City University of New York, New York
7. Higgins JS, Dawkins JV, Maghami CG, Shakir SA (1986) Polymer 27:931
8. Tuzar Z, Stepanek P, Konak C, Kratochvil P (1985) J Colloid Interface Sci 105:372
9. Price C (1983) Pure and Appl Chem 55:1563
10. Krause SJ, Haddock TJ (1986) Polym Sci B: Polym Phys 24:1991
11. Ikemi M, Odagiri N, Tanaka S, Shinohara I, Chiba A (1982) Macromolecules 15:281
12. Eliassaf J (1965) Polymer Letters 3:767
13. Krause S (1985) In: Lloyd DR (ed) Materials Science of Synthetic Membranes. Am Chem Soc Symp Ser 269, ACS, Washington, DC, p 351
14. Misra BN, Mehta IK, Khetarpal RC (1984) J Polym Sci: Polym Chem Ed 22:2767
15. Henderson CP, Williams MC (1985) Polymer 26:2021
16. Shibanov YD, Godovsky YK (1985) Colloid Polym Sci 263:202
17. Meyer GC, Widmaier JM (1982) J Polym Sci B: Polym Phys Ed 20:389
18. Jiang M, Cao X, Yu T (1986) Polymer 27:1923
19. de la Cruz M, Sanchez IC (1986) Macromolecules 19:2501
20. Bock J, Valint Jr PL, Pace SJ, Siano DB, Schulz DN, Turner SR (1988) In: Stahl GA, and Schulz DN (eds) Water-soluble polymers for petroleum recovery. Plenum Publishing Corp., New York, p 147
21. Valint Jr PL, Bock J, Schulz DN (1989) In: Glass JE (ed) Polymers in Aqueous Media: Performance Through Association. ACS Advances in Chemistry Series No. 223, American Chemical Society, Washington, DC, ch 21, p 399
22. Bock J, Siano DB, Valint Jr PL, Pace SJ, ibid ch 22, p 411
23. Siano DB, Bock J, Myer P, Valint Jr PL, ibid ch 23, p 425
24. Schulz DN, Kaladas JJ, Maurter JJ, Bock J, Pace SJ, Schulz WW (1987) Polymer 28:2110
25. Kalyanasundaram K, Thomas JK (1977) J Am Chem Soc 99:2039

6 Molecular Association and Polymerization of 1-Alkyl-4-vinylpyridinium Ions

Wilmer K. Fife

Complexation in its various forms plays a key role in the homo- and copolymerization of 1-alkyl-4-vinylpyridinium ions. Intermonomer associations are believed responsible for the enhanced polymerizability of monomers with long alkyl chains (C_n, $n \geq 6$) on nitrogen, the ability of the title monomers to copolymerize with anionic and π-rich monomers, and the strong dependence on concentration for homopolymerization of all these cationic monomers. Hydrophobic interactions between lipophilic monomers, electrostatic attraction between cationic and anionic monomers, and charge-transfer complexation between π-rich and π-deficient monomers have all been observed to control polymer formation. Monomer organization/orientation on polyanion templates, at organic solvent-water interfaces and in ordered multiple-phase systems such as micelles, membranes, vesicles, and microemulsions have been used with limited success in attempts to control the microstructure (e.g. tacticity, monomer sequence) in the related polymers. Interpolymer complexes of poly(1-alkyl-4-vinylpyridinium ions) with natural and synthetic polyanions represent a rich resource for the development of selective electroanalytical methods, for efficient new separation procedures, for manipulation of biomembranes in drug delivery, and numerous other applications.

6.1 Complexation in 1-Alkyl-4-vinylpyridinium Ions and Related Polymers

Homo- and copolymers of 1-alkyl-4-vinylpyridinium ions, **1**, are well-known for their ability to form complexes with a wide variety of materials [1, 2]. For example, the interaction of poly(1-alkyl-4-vinylpyridiniums), **2**, **3**, and **4** with natural and synthetic polyanions has been used to form stable inter-polymer complexes [3–10]. Such complexes have been successfully used to attach redox enzymes or electroactive systems of small molecules to working electrodes. When **2** or **3** is used as a matrix for an anionic enzyme such as glucose oxidase, at pH > pKi the enzyme can be effectively "wired" to an electrode and can contribute to a variety of electroanalytical applications. Heller and associates [11] have utilized such interpolymer complexes to provide a glucose-specific

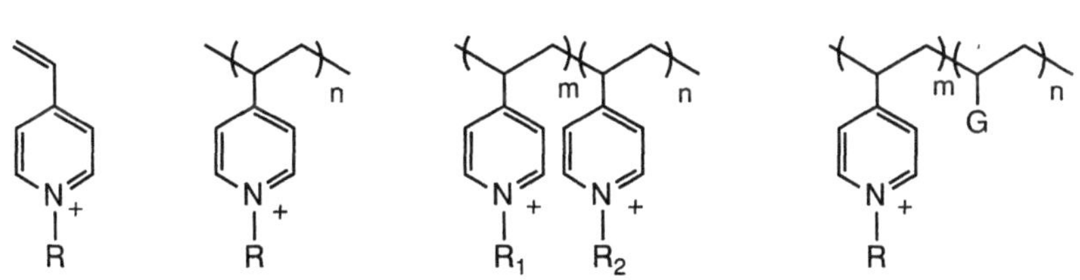

1 2 3 4

Macromolecular Complexes
in Chemistry and Biology
Eds · Dubin/Bock/Davis/Schulz/Thies

enzyme electrode by complexing glucose oxidase and the redox-active ruthenium-bipyridine complex to **4** (R–CH_3, G = pyridine) attached to a glassy carbon electrode, Fig. 6.1. Leidner and coworkers [12] utilized the same electrical properties of polyion-electrode complexes to investigate self-organized electroactive assemblies comprised of a glassy carbon surface, **2** (R = CH_3), dihexadecyl phosphate (DHP) and 2-alkylaminoquinones, Fig 6.2. The electrochemical response of 2-alkylaminoquinone-**2** (R = CH_3) DHP layers prepared from short and long chain 2-alkylaminoanthraquinones was found to depend on the nature of the quinone lipophile (mono- or dialkyl, alkyl chain length), the mol% of quinone within the layer, **2**: DHP ratio, and pH of the contacting aqueous solution. The best performance was observed in basic solutions with low mol% of 2-dimethylaminoanthraquinone.

Dubin and Park [13] have utilized selective complexation of poly(1-alkyl-4-vinylpyridiniums) with mixtures of proteins to develop a convenient new method for isolation of proteins of interest. For example, they were able to cleanly separate bovine serum albumin (BSA) from ribonuclease (RNase) by a selective coacervation technique. The net negatively charged BSA forms a strong ionic complex with **2** and precipitates to leave only the RNase in solution.

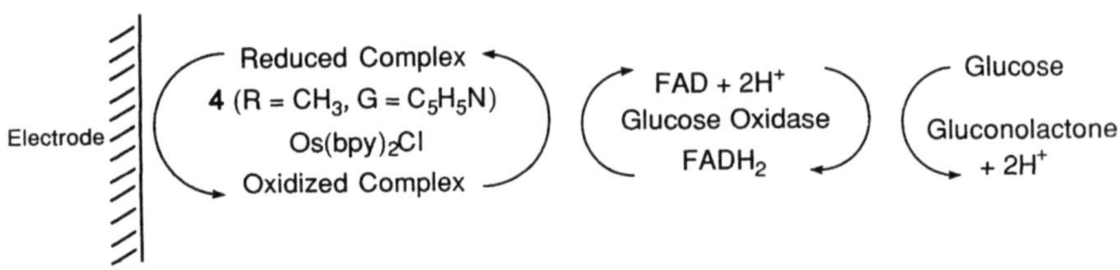

Fig. 6.1. "Electrically wired" glucose oxidase

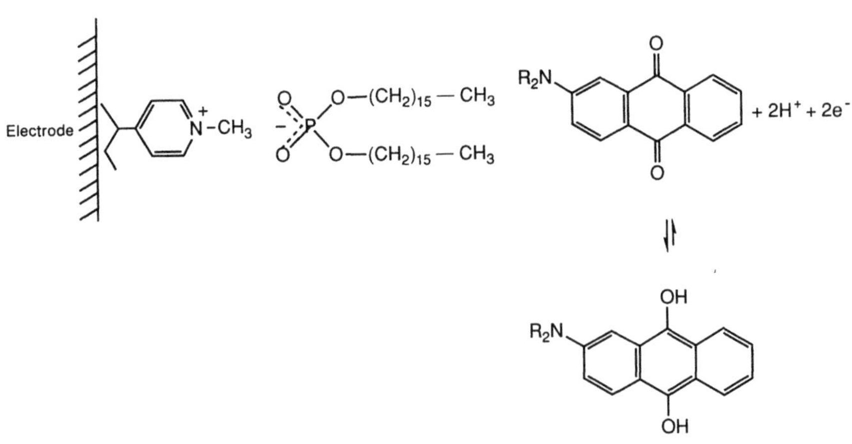

Fig. 6.2. Electroactive lipophile exchange membrane on glassy carbon

Kabanov and associates [14, 15] have recently investigated the effectiveness of **2** in appropriating natural cellular processes for possible use in medical treatment. They have found that interpolymer complexes of **2** and nucleic acids can be used to enhance incorporation into plasmid DNA of fragments that carry the genetic information for a desired protein [15]. They have also found that interpolymer complexes of **2** and the sodium salt of poly(methacrylic acid) coprecipitate with virus-antibody complexes [16]. The presence of the antibody does not affect coprecipitation of the virus as part of a polycation-polyanion complex. This property has been used to develop a sensitive immunoenzymic method for detection of viruses.

Kawabata and others [17–20] have explored the usefulness of macroporous resins that contain **1** (R = benzyl or ethyl) copolymerized with styrene and divinylbenzene in water-purification. They have found them to be particularly effective in removing bacteria (e.g.; *E. Coli*, *Salmonella typhimurium*, and others) from water. They have also successfully used the protein, aspartase, immobilized on a resin much like the water purification material to effect quantitative conversion of ammonium fumarate to aspartate in a continuous flow 'bio-reactor' [21].

6.1.1 *Modes of Complexation Observed in 1-Alkyl-4-vinylpyridinium Ions and Related Polymers*

Complexation of **1** and its polymers with other materials often utilizes the electrostatic forces generated by interaction of a cation or polycation with an anion or polyanion [22]. The structures of **1**, **2**, **3** and **4** also provide other important sources of intermolecular attraction. The pyridinium ring is electron-deficient and represents a potential participant in π-donor-π-acceptor interactions, i.e. charge-transfer complexation [23], with molecules or ions having π-rich electronic systems. Charge-transfer interactions have also been reported between **1** and easily oxidizable anions such as iodide ion [24]. The role of charge-transfer complexes in forming copolymers of 1 and π-rich vinylmonomers will be discussed in Sect. 6.3.2. The vinyl backbone of these polymers provides a hydrocarbon structure that can serve as the locus of hydrophobic interactions that enhance or modify electrostatic and charge-transfer interactions between **1** and other vinyl monomers. The backbone and its stereochemistry also influence the physical and chemical properties of polymers **2**, **3** and **4**. This potentially promising means of controlling polymer-polymer and polymer-small molecule or ion interactions remains largely unexplored.

The *N*-alkyl groups may furnish yet another source of intermonomer and interpolymer association. Considerable evidence has been assembled to establish the surfactant-like properties of these polymers [25]. In fact, copolymers **3** that carry 10–40% *N*-dodecyl groups ($R_1 = C_{12}H_{25}$) and 90–60% *N*-ethyl groups ($R_2 = C_2H_5$) were first prepared by Strauss and associates [26–28] who observed their ability to self-associate in water, both intermolecularly and intramolecularly. Furthermore, these materials, often referred to as polysoaps, were found to have unusual viscosity properties. They exhibited unusually low viscosities under conditions that encourage intramolecular association and formation of internal, micelle-like domains by aggregation of the *N*-alkyl groups. These materials, not surprisingly, solubilize hydrocarbons in water, and locate added lipophilic fluorescent dyes in a hydrocarbon-like environment [29].

6.1.2 Synthetic Routes to Poly(1-Alkyl-4-vinylpyridinium Ions) and Related Copolymers

Polymers **2**, **3** and **4** are available from two synthetic approaches, *N*-alkylation of poly(4-vinylpyridine), **5**, with one or more alkylating agents (Eqs. 6.1.1 and 6.1.2)

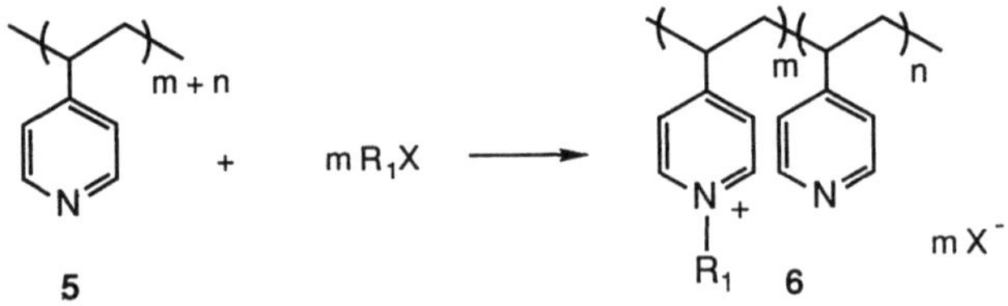

(6.1.1)

$$\mathbf{6} + nR_2X \rightarrow \mathbf{3} + (m + n)X^- \quad (6.1.2)$$

[30] and polymerization of appropriate monomer(s) (Eqs. 6.1.3 and 6.1.4). The former approach has been widely used because of the availability of **5** and a host of effective alkylating agents (e.g. alkyl halides, alkyl sulfonates, and alkyl sulfates). Until recently, the second route, i.e. homo- and copolymerization of appropriate monomer(s), has been restricted primarily to **1** (R = H or CH_3). Attempts in several laboratories [31–33] to synthesize **1** with longer *N*-alkyl groups only furnished the related polymer **2**. Salamone and coworkers [33, 34] correctly analyzed the problem several years ago. They suggested that successful alkylation of 4-vinylpyridine (Eq. 6.1.5) requires a very reactive alkylating agent that rapidly converts 4-vinylpyridine to **1**. This prevents the build up of an appreciable concentration of 4-vinylpyridine in the reaction mixture. They have confirmed the role of 4-vinylpyridine as polymerization initiator during synthesis of **1**. The details of this process will be discussed in Sect. 6.1.3.

$$n\mathbf{1} \rightarrow \mathbf{2} \quad (6.1.3)$$

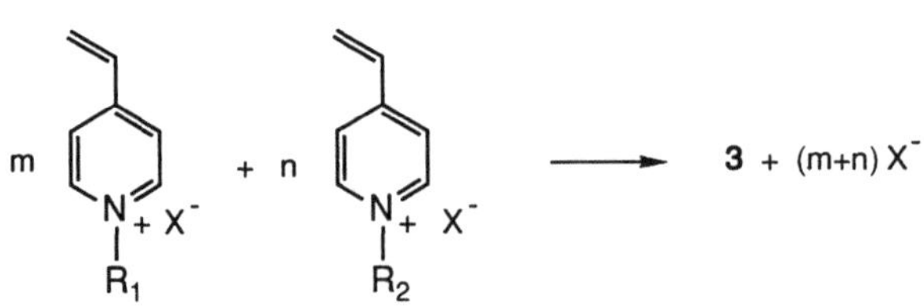

(6.1.4)

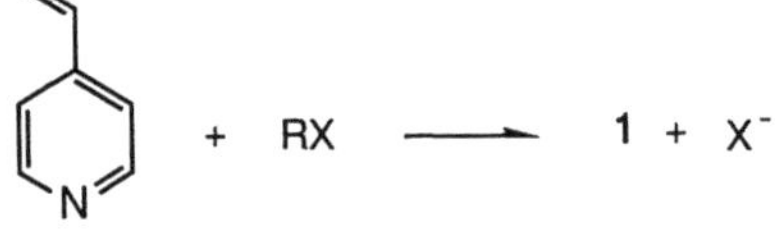

(6.1.5)

Fife and coworkers [35] have found that *n*-alkyl triflates (Eq. 6.1.5, RX $= CH_3(CH_2)_{n-2}CH_2O{-}SO_2CF_3$, $X^- = CF_3SO_3^-$) are sufficiently reactive to alkylate 4-vinylpyridine nearly instantaneously to form a series of **1** (R $= CH_3 \rightarrow C_{12}H_{25}$). The availability of these variants of **1** permits preparation of a wide variety of homo- and copolymers, **2** and **3**, as well as a broader selection of copolymers, **4**.

6.1.3 Pathways for Polymerization of 1-Protium- and 1-Alkyl-4-vinylpyridinium Ions: 1,2-addition vs Ionene Formation

The polymerization of 4-vinylpyridinium salts, **1**, has received extensive attention in the literature [36]. Reports of the polymerization mechanism have often appeared vague and confusing which is undoubtedly a consequence of the complexity of the chemistry. For example, polymerization may follow multiple paths, possibly subject to a variety of factors. Spontaneous polymerization has often been observed and has been attributed to anion (nucleophile), free radical, or photo initiation [36–42]. The polymerization reaction also is highly sensitive to monomer concentration and ionic strength, and polymerization of **1** (R = H) has been observed to follow two different pathways (Path A, Fig. 6.1.3), a typical 1,2-vinyl addition pathway to **2** and a more unusual 1,6-addition to polyionene **7** (Path B, Fig. 6.1.3). Subtle changes in reaction conditions (e.g. monomer concentration) have been reported to cause the change in reaction path and type of polymer produced. Ringsdorf and associates [37, 28] have found that high concentrations of monomer in aqueous solution (conc. > CMC) favor 1,2-addition, whereas lower concentrations (conc. < CMC) lead to **7**. Ionene formation is believed to require both 4-vinylpyridine and **1** (R = H). The mechanism illustrated in Fig. 6.1.4 shows nucleophilic attack by 4-vinylpyridine at the β-vinyl carbon of **1** (R = H) to give a resonance stabilized zwitterionic inter-

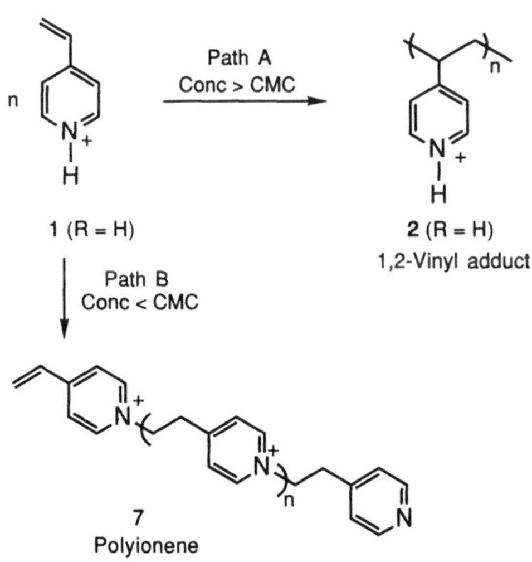

Fig. 6.1.3. Two pathways for polymerization of 4-vinylpyridinium ions

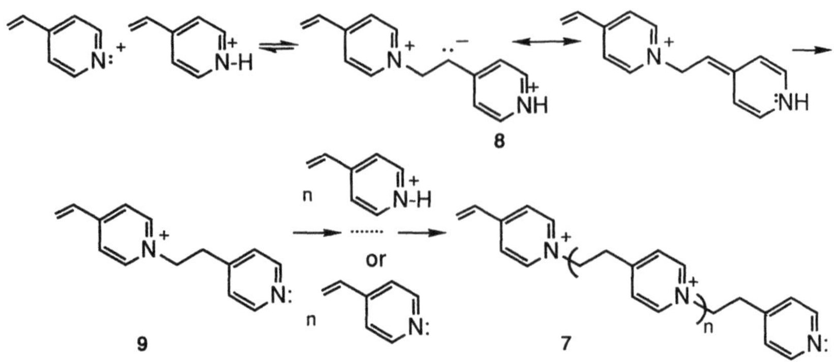

Fig. 6.1.4. Mechanism for polyionene formation

mediate **8**. Intermediate **8** may undergo proton transfer to form *N*-pyridylethyl-4-vinylpyridinium ion, **9**, which can react with another 4-vinylpyridine at its vinylpyridinium end or with **1** (R = H) at nucleophilic nitrogen to extend to polymer chain. The 1-alkyl-4-vinylpyridinium monomers, **1**(R = alkyl), are only able to form the related 1,2-addition polymer **2**. However, the conditions for initiating polymerization of these monomers are quite varied (e.g. radicals, nucleophiles, elevated temperatures ≥ 50 °C, and light) and yet, successful polymerization remains somewhat unpredictable. The so-called "spontaneous" polymerization of **1** to **2**, **3**, or **4** has not been fully explained. Reports from different laboratories have not always agreed to initiating conditions, and inhibitory effects of added acids, bases or radical scavengers [36]. Additional discussion of these matters follows in Sects. 6.2 and 6.3.

6.1.4 Prospects for Control of Stereochemistry and Monomer Sequence During Homo- and/or Copolymerization of 1-Alkyl-4-vinylpyridinium Ions

Paleos [43] was the first investigator to attempt control of microstructure of **2** through orientation of **1** (R = CH_3) at a water-toluene interface. The water-organic solvent interface provides a reaction site, that may orient amphipathic molecules and ions in a particular way. Paleos was interested in obtaining stereoregular **2**, i.e. pure isotactic or syndiotactic forms of **2**, in this fashion but the experimental results were inconclusive [43–45]. Fife and Hu [46] have reinvestigated the possibility for stereo-controlled polymerization of **1** (R = $C_{12}H_{25}$) as a component of aqueous suspensions of microemulsions. This work will be explored in detail in Sect. 6.2.4.

Current work underway in our laboratory is seeking control over monomer sequence in copolymers **3** and **4** by copolymerization of mixtures of 1-alkyl-4-vinylpyridinium ions that contain different alkyl groups under multiple phase conditions. Many of the experiments to-date have been carried out in aqueous medium and have utilized surfactant-like monomers such as **1** (R = $C_{12}H_{25}$) or added non-polymerizable surfactants like sodium dodecylsulfate, 1-dodecylpyridinium chloride or Brij 560 to provide the organic element of a water-organic interface. Polymerization in inverse microemulsions has provided the most exciting results. Successful copolymerization of water soluble cationic monomers (**1**, $R = C_2H_5$) with water insoluble cationic monomers (**1**, $R = C_{12}H_{25}$) has been achieved for the first time in inverse microemulsions that include hexane, water, isoamyl alcohol and the monomers. More details of these experiments along with a brief description of new tandem MS methodology for characterization of copolymers, **2**, will be presented in Sects. 6.3.4 and 6.3.5.

Information currently available provides little or no basis for achieving control over tacticity in homo- and copolymers or monomer sequence in copolymers through monomer pre-organization at water-organic solvent interfaces or in micellar, or microemulsion systems.

6.2 Homopolymerization of 1-Alkyl-4-vinylpyridinium Ions

Polymerization of 1-protium- and 1-alkyl-4-vinylpyridinium ions has been investigated extensively in the laboratories of Kabanov [40], Salamone [36], Ringsdorf [42], Paleos [45] and Fife [35, 46–48]. Unlike the 1-protium analogs, **1** (R = H), which can form polyionenes, **7**, as well as 1,2-addition polymers (see Sect. 1.3), *N*-alkylated monomers **1** (R = alkyl) can only form the typical 1,2-addition polymers. The first report of polymerization of an *N*-alkylvinylpyridinium ion was by Shyluk in 1964 [39]. He observed that 1,2-dimethyl-5-vinylpyridinium ion, **10**, spontaneously polymerized to the related vinyl polymer **11** in aqueous solution at concentrations greater than 25%, (Eq. 6.2.1).

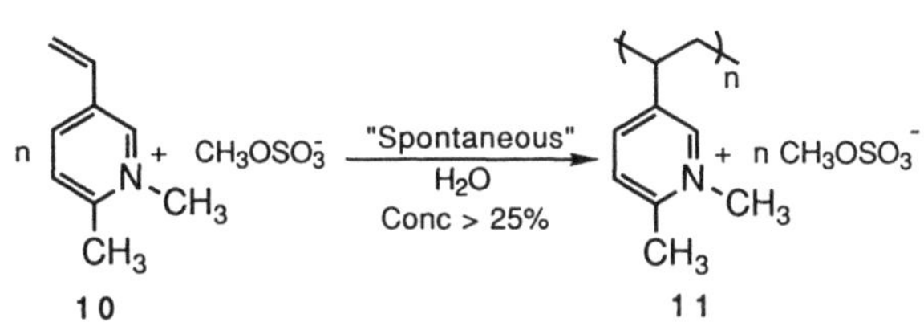

(6.2.1)

The literature on *N*-alkyl-2,3, or 4-vinylpyridinium salts contains many accounts of "spontaneous" polymerization to the 1,2-addition polymer during *N*-alkylation of the vinylpyridine. These attempts to produce other *N*-alkyl analogs had always produced the 1,2-addition polymer instead, (Eq. 6.2.2).

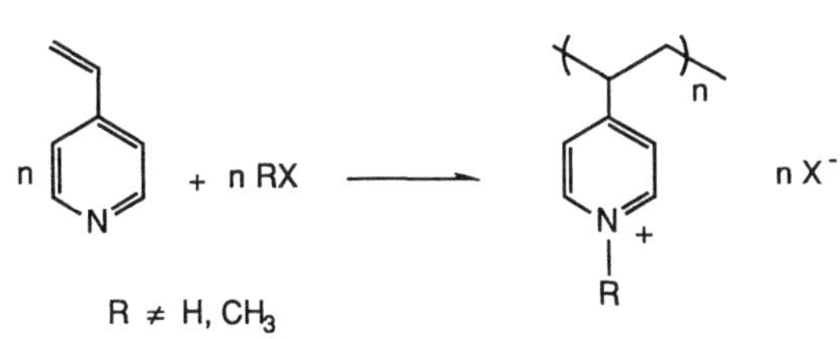

(6.2.2)

Various explanations were advanced to explain the unpredictable behavior of **1**. Kabanov [32] attributed the spontaneous polymerization of **1** to an initiation-propagation sequence that began with nucleophilic attack at the vinyl group by counter-ion X^-, (Eq. 6.2.3). The zwitterionic intermediate **12** was

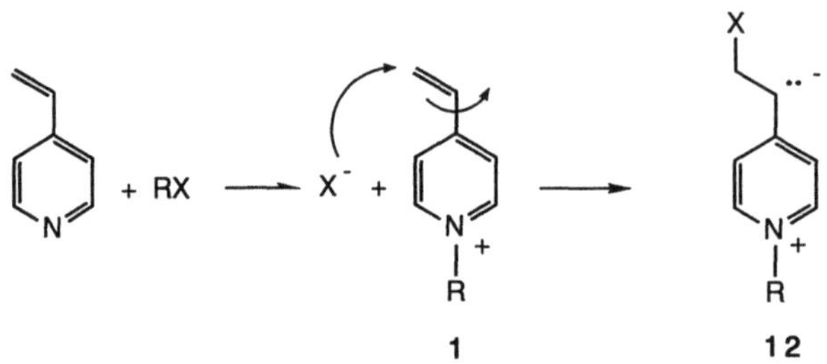

(6.2.3)

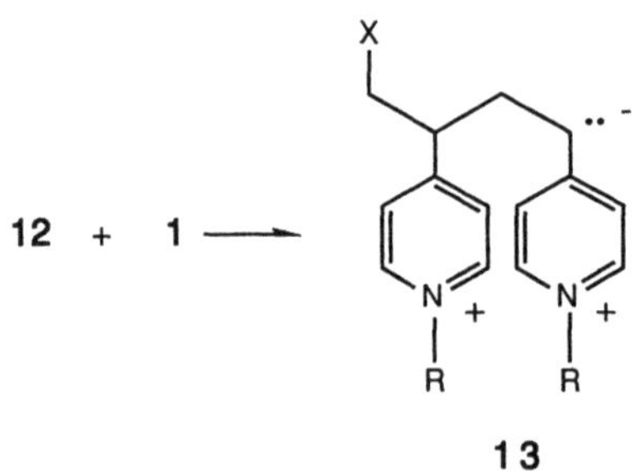

(6.2.4)

believed to react with monomer **1** to form an analogous zwitterion **13** with one additional monomer unit. The 1,2-polymer, of course, would result from repetition of this second step. However, monomers **1** (R = H, CH_3; X^- = Br^- I^-, NO_3^-, ClO_4^-, $CH_3OSO_3^-$, *p*-$CH_3PhSO_3^-$) were prepared as stable intermediates in the laboratories of Salamone [33], Ringsdorf [49] and Paleos [43]. More recently, Fife and coworkers [39] have demonstrated that a wide variety of **1** (e.g. R = H, CH_3, C_2H_5, C_4H_9, C_6H_{13}, $C_{12}H_{25}$; X^- = $CF_3SO_3^-$ are stable to all common halides and many other non-nucleophilic anions. Salamone and associates [33] first suggested that "spontaneous" polymerization of **1** is initiated by nucleophiles. Furthermore, they found 4-vinylpyridine responsible for initiating spontaneous polymerization of **1** during attempted synthesis of it by alkylation of vinylpyridine (Eq. 6.2.5). This explanation has been confirmed in the laboratories of Paleos [43] and Fife [46] who have studied **1** under conditions where it has exhibited long term stability (i.e. in dilute aqueous solution or dissolved in CH_2Cl_2, $CDCl_3$, DMSO-d_6, CH_3CN, CD_3OD) and treated it with reactive nucleophiles such as 4-vinylpyridine, pyridine, triethylamine, CN–, enolate ions, and carbanions to produce 1,2-addition polymer **2**. The simplest mechanism consistent with these observations was first proposed by Salamone and coworkers in 1971 [34]. It is illustrated in Fig. 6.2.1.

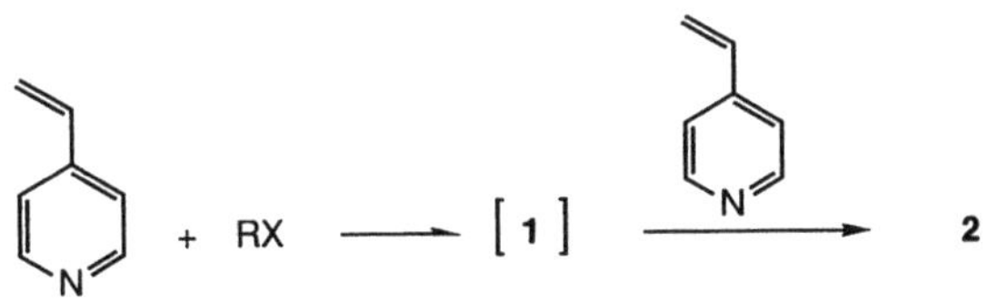

(6.2.5)

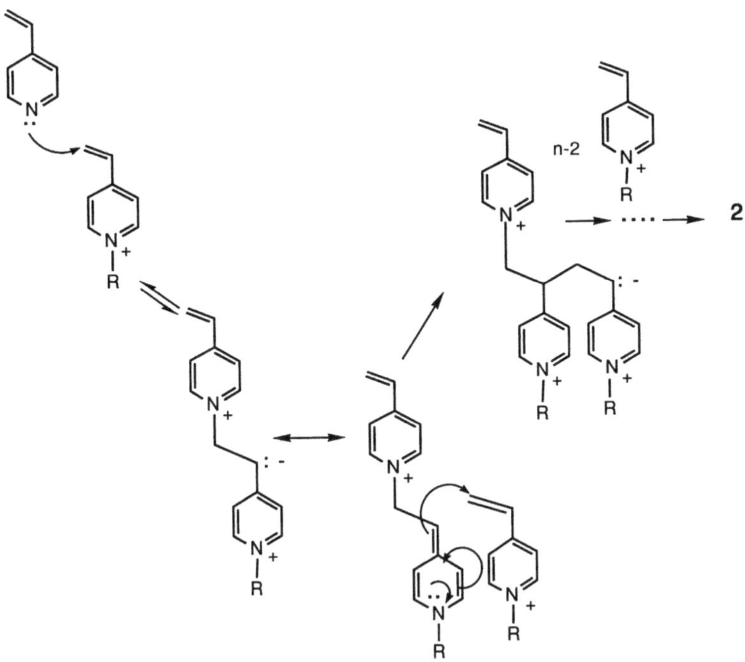

Fig. 6.2.1. Mechanism for nucleophile-initiated addition polymerization

The homopolymerization of **1** (R = alkyl) is discussed in the following sections. Particular attention will be given to studies that have utilized different methods of initiation (thermal-"spontaneous", photochemical, radical, nucleophile) and have explored the influence of aggregation/complexation on polymer formation and structure.

6.2.1 *Polymerization of 1-Alkyl-4-vinylpyridinium Ions: 1,2-Addition and Mode of Initiation*

A multiplicity of methods has been used successfully to initiate polymerization of **1**. These methods have included initiation by nucleophiles (Table 6.2.1), radicals (Table 6.2.2), elevated temperatures [48], and light [50, 51]. However, the details of initiation and the nature of the reactive intermediate(s) remain in

Table 6.2.1. Nucleophile-initiated polymerization of 1-alkyl-4-vinylpyridinium ions[a,b]

1, *N*-Alkyl	X^-	Added nucleophile	1,2-Addition polymer	Reference
CH_3	$CF_3SO_3^-$	F^-, Cl^-, B^-, I^-	No	46, 52
		CN^-	Yes	46, 52
		Pyridine	Yes	34, 43, 46
		4-Vinylpyridine	Yes	46
		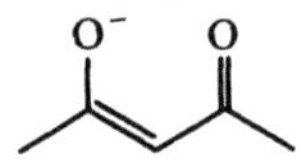	Yes	46
		n-BuLi	Yes	46
		Et_3N	Yes	46
C_2H_5	$CF_3SO_3^-$	4-Vinylpyridine	Yes	46
		CN^-	Yes	46
		n-BuLi	Yes	46
C_6H_{13}	$CF_3SO_3^-$	4-Vinylpyridine	Yes	46
		CN^-	Yes	46
		n-BuLi	Yes	46
$C_{12}H_{25}$	$CF_3SO_3^-$	4-Vinylpyridine	Yes	46
		n-BuLi	Yes	46

[a] Polymeric product was retained by 6000–8000 MWCO dialysis membrane [46].
[b] ^1H NMR and IR spectral data of polymerization product were consistent with literature values for 1,2-addition polymer [46].

question for nearly all of these processes. In the earliest studies polymerization of **1** was observed to accompany alkylation of the vinylpyridine, (Eq. 6.2.2). These results appear to be explained mainly by the nucleophile-initiation mechanism shown in Fig. 6.2.1, in which unalkylated vinylpyridine serves as the initiating nucleophile.

This explanation is supported by the observation that **1** and non-nucleophilic counter-ions form stable, isolatable compounds [45, 46, 52]. Furthermore, polymerization of **1** is reliably initiated by the addition of a wide variety of sufficiently reactive nucleophiles to solutions of **1** in any of several common solvents. Table 6.2.1 contains a listing of these experiments.

Radical initiated polymerization of vinylpyridinium ions has been recognized since 1964 when Shyluk reported 1,2-addition polymerization of 1,2-dimethyl-5-vinylpyridinium methylsulfate initiated by potassium peroxydisulfate [39]. Later, Ringsdorf proposed a radical mechanism for the "spontaneous" polymerization of **1** (R = H) at concentrations greater than 1.0 M in water [37, 42]. The postulated mechanism was supported by experiments in which inhibition of polymerization was demonstrated by radical inhibitors such as dioxygen, copper (II) ions, and 5-butylcatechol. Kinetic analysis of these reactions was also consistent with the proposed radical pathway.

Recent work in our laboratory [47, 51] has confirmed the ability of various radical initiators to promote polymerization of **1** (R = alkyl). Results of representative experiments are listed in Table 6.2.2.

Table 6.2.2. Radical-initiated polymerization of 1-alkyl-4-vinylpyridinium ions[a]

1, *N*-Alkyl	Initiator[b]	Solvent	Temp/Time	Results
CH_3	BP	DMSO	60–80 °C/12 h	No polymer
	AIBN	CH_3CN	55 °C/10 h	1,2-polymer
C_2H_5	AIBN	CH_3CN	55 °C/10 h	1,2-polymer
	NONE	H_2O	50 °C/10 h	No polymer
	$S_2O_8^{-2}/S_2O_5^{-2}$	H_2O	45–50 °C/10 h	1,2-polymer
C_6H_{13}	$S_2O_8^{-2}/S_2O_5^{-2}$	Hexane/H_2O/*i*-AmOH[c]	45–50 °C/10 h	1,2-polymer
$C_{12}H_{25}$	AIBN	Hexane/H_2O/*i*-AmOH[c]	45–50 °C/10 h	1,2-polymer
	$S_2O_8^{-2}/S_2O_5^{-2}$	CH_3CN	55 °C/10 h	1,2-polymer
	$S_2O_8^{-2}/S_2O_5^{-2}$	H_2O	45–50 °C/10 h	1,2-polymer
		Hexane/H_2O/*i*-AmOH[c]	45–50 °C/10 h	1,2-polymer

[a] [1] = 0.15 – 0.20 M, [Initiator] = 0.1–1.0 mole%, Refs [47 and 51].
[b] Polymerization initiator: BP = benzoyl peroxide, AIBN = azobisisobutyronitrile
[c] Hexane/H_2O/*i*-AmOH = 1.3:0.4:0.05 by wt.

Vinylpyridinium ions, **1**, like other vinyl monomers are subject to thermal and light initiated polymerization. A recent study in our laboratory has confirmed that dilute (< 0.20 M) aqueous solutions of **1** (R = CH_3, C_2H_5) and sonicated suspensions of **1** and water are stable at temperatures below 70 °C in normal laboratory light. Suspensions of water-insol monomers, e.g. **1** (R = $C_{12}H_{25}$), polymerize rapidly at temperatures above 80 °C. Dilute aqueous solutions of the water-soluble monomer **1** (R = CH_3 or C_2H_5) resist polymerization at 80–100 °C but polymerize when subjected to UV radiation (300 nm).

In our experience, the single most important requirement for reproducible, predictable behavior of **1** is its purity.

6.2.2 Hydrophobic Effects on Polymerization of 1-Alkyl-4-vinylpyridinium Ions

The hydrophobic interaction between hydrocarbons and water is a powerful ordering force in mixtures of lipophiles and hydrophiles [53, 54]. Domains of lipophiles provide a solubilizing phase for all like substances added to lipophile-water mixtures. These same forces have been successfully used to order water-insoluble organic molecules for enhancement of Diels–Alder reactions, the benzoin condensation and others which require molecular ordering in the transition state [55].

The hydrophobic effect represents a potential ordering force for surfactant-like monomers whose like charges create electrostatic repulsion between monomers and significant barriers to polymerization. For example, polymerization of monomers **1** (R = H or CH_3) is known to be strongly concentration dependent [36]. Presumably, the repulsive force between the cationic monomers contributes to the low reactivity of **1** at concentrations below 1.0 M in water. This analysis suggests the possibility of using hydrophobic forces to compensate for repulsion forces between cationic monomers **1**. Thus, **1** bearing increasingly longer *N*-alkyl groups should exhibit decreasing solubility in water and enhanced susceptibility to polymerization. Fife and coworkers [46, 47, 51] studied the reactivity of a series of **1** with R = CH_3, C_2H_5, C_6H_{13} and $C_{12}H_{25}$ in solution or aqueous suspension with initiation by elevated temperature, nucleophiles, radicals or uv light. The results for initiation by nucleophiles are listed in Table 6.2.1 and the results for other methods of initiation are summarized in Table 6.2.3.

All monomers **1** are quickly and completely polymerized when subjected to sufficiently reactive nucleophiles or radicals. However, the reactivity of **1** toward polymerization initiated thermally or photochemically with 300 nm light is clearly dependent on the length of *N*-alkyl chain. In fact, **1** (R = $C_{12}H_{25}$) is selectively converted to the homopolymer when mixtures of it and **1** (R = C_2H_5) are sonicated at 100 °C in water. The dodecyl derivative appears to quickly polymerize within suspended, insoluble aggregates and the soluble ethyl analog remains unchanged in the water phase.

6.2.3 *Polymerization on Polyanion Matrices*

Polymerization of monomers pre-organized by association with macromolecules is known as matrix or template polymerization. The specific, associative interactions between monomers, the growing chain (daughter) and template (parent) enable this mode of polymerization to occur and may control the structure of the daughter polymer [1,2]. The template is usually a linear polymer and it carries functionality complementary in charge, dipole, or chemical property to that of the daughter. For example, polymerization of **1** may be accomplished via association with polyanions such as poly(4-styrene sulfonate) (Eq. 6.2.6)

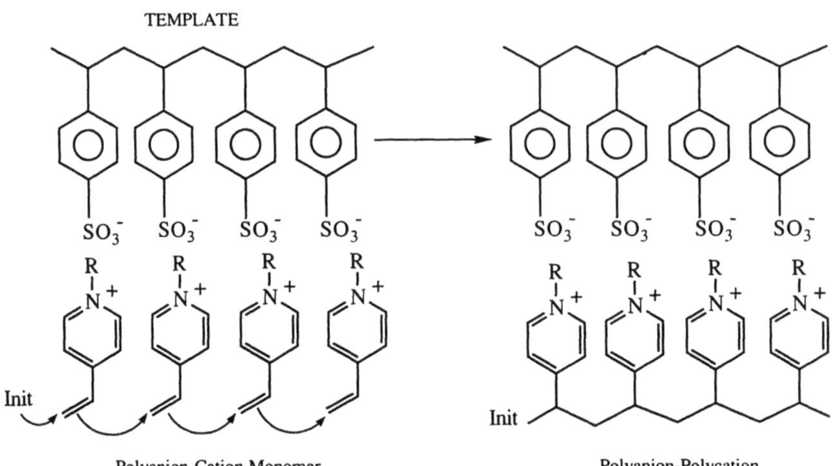

(6.2.6)

polyacrylate or polyphosphates. Templates are known to influence the rate of polymerization of associated monomers, and to control chain length and polydispersity in daughter polymers [1,2].

Table 6.2.3. Effect of *N*-alkyl chain length on polymerizability of **1**

1	Conditions	Solvent	Degas	Polymerization
Thermal Initiation				
R = CH_3 (0.17 M)	− 60 °C, 15 h	CH_3CN	Yes	No
C_2H_5 (0.15 M)	US[a] 100 °C, 13 min	H_2O	No	No
$C_{12}H_{25}$ (0.02 M)	US[a] 100 °C, 13 min	H_2O	No	Yes, 100%
Radical Initiation[b]				
CH_3 (0.17 M)	AIBN (0.1%), 60 °C, 15 h	CH_3CN	Yes	Yes, 100%
C_2H_5 (0.15 M)	AIBN (0.1%, 50% °C, 1 h	CH_3CN	No	Yes, 100%
$C_{12}H_{25}$ (0.02 M)	AIBN (0.1%), 55 °C, 10 h	CH_3CN	No	Yes, 100%
Photo Initiation (300 nm)				
CH_3 (0.17 M)	*hv*, 25 °C, 20 h	CD_3CN	No	No
C_2H_5 (0.15 M)	*hv*, 40 °C, 1.5 h	DMSO-d_6	No	No
$C_{12}H_{25}$ (0 M)	*hv*, 40 °C, 1.5 h	$CDCl_3$	No	Yes, 100%

[a] References 46, 47 and 41.
[b] Sonication with a microprobe

Results from the laboratories of Salamone [52] and Kabanov [56–58] indicate that polyacids promote polymerization of 4-vinylpyridine via protonation to form **1** (R = H) as the reactive monomer. Ionene **7** formation is known to be the usual product when either strong or weak polyacids are used. Monomer concentration does not play the same role in these cases as reported by Ringsdorf and coworkers for monomeric acids [37, 38, 59, 60], where 1,2-addition polymer is obtained at high concentrations.

Apparently electrostatic association of cationic monomers **1** with polyanions does not provide strong enough complexation to control tacticity or other aspects of microstructure in the daughter polymer. This is likely due to the solvation of template, monomers and growing daughter chain by the reaction medium. Conditions that organize solution phase monomers at an ordered, nonsolvated template surface would be more likely to transfer tacticity, or chirality, or to control monomer sequence during polymerization.

6.2.4 Polymerization at Water-Organic Solvent Interfaces

Conformational analysis of isotactic and syndiotactic forms of **1** (R = CH_3) suggests the possibility of control of tacticity by polymerization under conditions that order the monomers prior to polymerization, Fig. 6.2.2. When the backbone of the syndiotactic isomer is arranged in its extended staggered conformation Fig. 6.2.2a, the *N*-methylpyridinium groups are located gauche to one another, Fig. 6.2.2b, but are able to lie widely spaced and relatively flat at an interfacial surface, Fig. 6.2.2c. Similar consideration of the isotactic isomer in its extended staggered conformation is shown in Fig. 6.2.3a. Projection formulas (Fig. 6.2.3b and 6.2.3c clearly predict significant electrostatic repulsion between *N*-methylpyridinium rings stacked/placed face to face approximately 2.5 Å apart. Rotation about carbon-carbon bonds in the backbone will move the

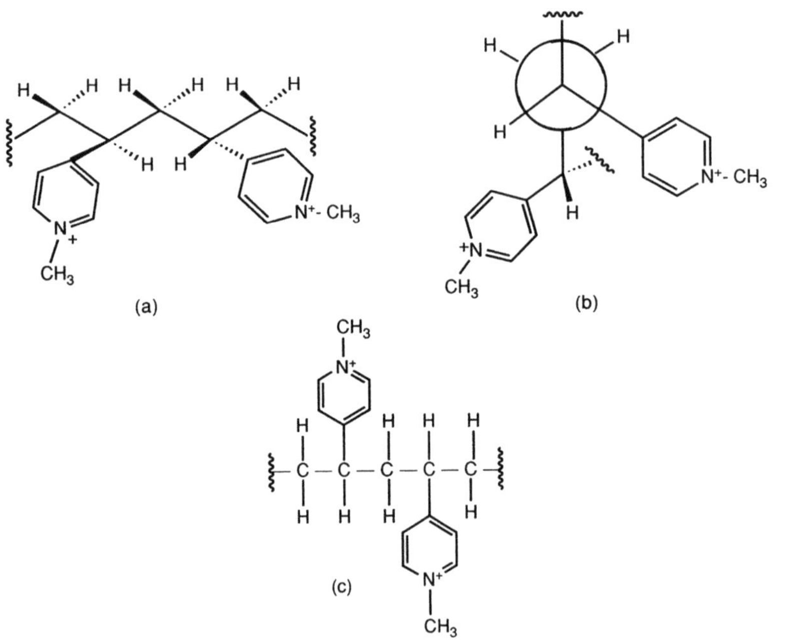

Fig. 6.2.2. Representation of syndiotactic **1**(R = CH_3), (*a*) extended, staggered conformation, (*b*) Newman projection, (*c*) Fischer projection

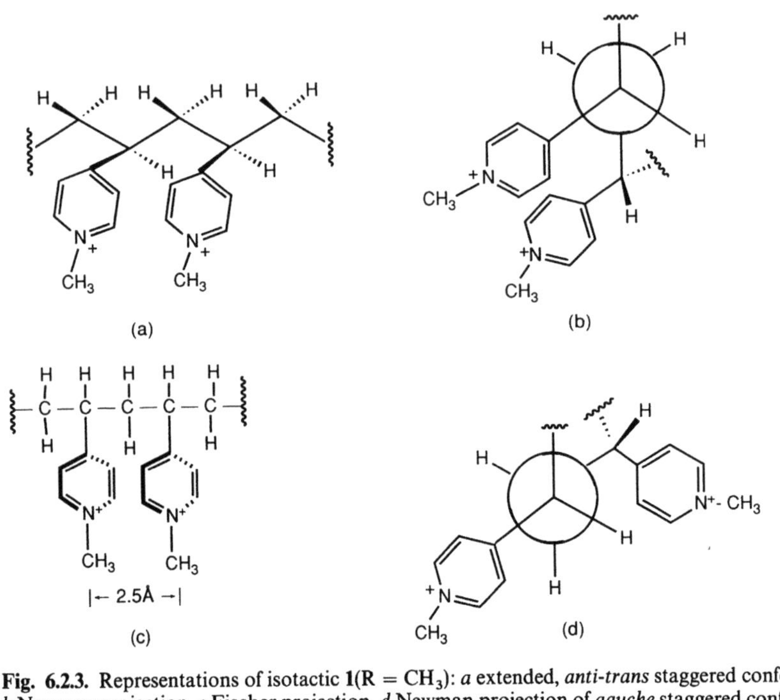

Fig. 6.2.3. Representations of isotactic **1**(R = CH_3): *a* extended, *anti-trans* staggered conformation, *b* Newman projection, *c* Fischer projection, *d* Newman projection of *gauche* staggered conformation

N-methylpyridinium rings farther apart but increase carbon-hydrogen eclipsing. A staggered conformation with backbone linkages gauche to one another and N-methylpyridinium *anti-trans* to N-methyl-4-methylpyridinium (Fig. 6.2.3d appears energetically favored. In this case, the pyridinium rings would not lie along an interfacial surface. They would instead project in different directions from a helical backbone.

Paleos investigated this interesting possibility through polymerization of **1** ($R = CH_3$) initiated by 4-vinylpyridine in toluene-water suspensions, and obtained little if any tacticity control [43, 44]. His data suggest perhaps a slight preponderance of the syndiotactic isomer under the most favored conditions, i.e. vigorous stirring of the reaction mixture to maximize interfacial surface area. We have recently reinvestigated this matter with the lipophilic monomer **1** ($R = C_{12}H_{25}$) and have found no evidence of tacticity control even when polymerization is carried out under inverse microemulsion conditions where interfacial area between aqueous and hydrocarbon (hexane) phases is increased many orders of magnitude over that of the earlier experiments [46, 47]. Apparently ordering of **1** at interfaces between phases that contain immiscible, achiral molecules is not capable of controlling tacticity in polymers.

6.3 Copolymerization of 1-Alkyl-4-vinylpyridinium Ions

Although there are numerous reports of polymerization of 1-protium and 1-methyl-4-vinylpyridinium ions, **1** ($R = H, CH_3$), until very recently successful copolymerization of these monomers had been limited to reaction with two anionic monomers. Salamone and coworkers in 1977 reported the first examples of copolymerization of **1** ($R = H$) with 4-vinylbenzenesulfonate, **20** (Eq. 6.3.1) and 2-acrylamido-2-methylpropanesulfonate, (AMPS), **21**, (Eq. 6.3.2) [61].

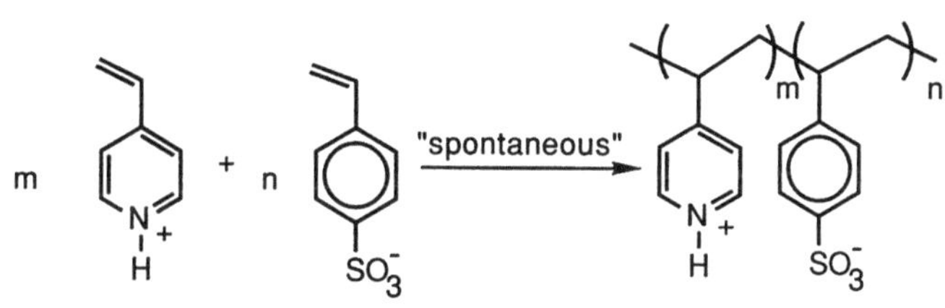

1 **20** **22**

(6.3.1)

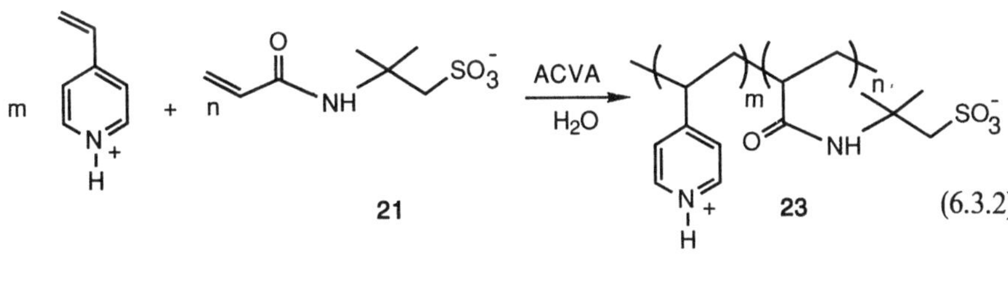

21 **23**

(6.3.2)

More recently, Salamone and associates have successfully copolymerized **1** (R = H, CH_3) with two π-rich monomers, α-methylstyrene and 4-methoxystyrene [62].

Development of a general synthesis of 1-alkyl-4-vinylpyridinium ions in our laboratory permitted investigation of copolymerization of two or more monomers **1** with different *N*-alkyl groups. The results from all of these studies will be discussed in the following sections. It should be noted that intermonomer association and phase-distribution behavior of monomers in multiple-phase media serve as unifying themes for understanding success and failure in these copolymerization experiments.

6.3.1 Copolymerization via Electrostatic Association

All attempts to copolymerize **1** with other vinyl monomers were unsuccessful until Salamone and coworkers obtained a 1,2-addition copolymer **4** of **1** (R = H) and 4-vinylbenzenesulfonate, **20**, obtained by "spontaneous" polymerization of the two monomers as a melt or in concentrated aqueous solution [61]. Vinyl-type copolymers of **1** and **20**, and **1** and **21** were also obtained by initiation with radicals (e.g. from ACVA), light or γ-radiation. These experiments are summarized in Table 6.3.1. Copolymerization of **1** (R = H) with AMPS **21** is complicated when carried out as a "spontaneous" (i.e. absence of an initiator) process in water, methanol, or *N,N*-dimethylformamide (DMF). Two homopolymers derived from **1** (R = H), the 1,2-addition polymer, **2** (R = H), and the polyionene, **7**, are obtained along with poly AMPS. Similar attempts to form copolymer from **1** (R = H) and vinylsulfonate have been unsuccessful. In fact, the latter monomer resisted all conditions of polymerization and **1** (R = H) formed the ionene **7** from a melt or in dilute aqueous solution (< 1.0 m) and yielded **2** (R = H) in concentrated aqueous solution, Table 6.3.1. Recall that this behavior duplicates that reported for **1** (R = H) by Ringsdorf and others. Ringsdorf attributed the change in structure of homopolymer to micellization of **1** (R = H) in concentrated aqueous solution.

Salamone [63] suggested an alternating monomer sequence for copolymers of **1** (R = H, CH_3) and **20** based on ^{13}C NMR data and the observation that copolymerization of **1** with excess **20** stops when all of the latter is consumed. Roitman and Harwood [50] confirmed the essential features of this process, but found the reaction to be photo-initiated and inhibited by *t*-butylcatechol. More recent work by Lai [64] in Salamone's laboratory supports the earlier conclusions of that group.

These results are consistent with a free-radical reaction mechanism. A puzzling aspect of the reaction, however, is the exclusion of added reactive vinyl monomers such as acrylamide or hydroxyethyl methacrylate from the copolymer product. Clearly the reaction is not a normal radical process. Roitman and Harwood dispute the earlier suggestion of an alternating structure for copolymers produced by "spontaneous" polymerization based on ^{13}C NMR spectral analysis. Salamone and coworkers [63] have interpreted data from UV-Vis

Table 6.3.1. Copolymerization of **1**(R = H or CH_3) with anionic and π-rich monomers

1	X^- –	Comonomer	Conditions	Polymeric product(s)	Refs.
R = H	—	$CH_2{=}CHSO_3^-$	137–138 °C	$(7)^{+n}nCH_2{=}CHSO_3^-$	61, 65
H	—	$CH_2{=}CHSO_3^-$	0.9 M/H_2O/25 °C	$(7)^{+n}nCH_2{=}CHSO_3^-$	61, 65
H	—	$CH_2{=}CHSO_3^-$	> 1.0 M/H_2O/25 °C	**2**, (R=H)$nCH_2{=}CHSO_3^-$	61, 65
H	—	$CH_2{=}CHSO_3^-$	ACVA[a]/H_2O/70 °C4 h	**2**, (R=H)$nCH_2{=}CHSO_3^-$	61, 65
H	—	AMPS	H_2O or MeOH or DMF ACVA[a]/H_2O/70 °C/4 h	**7**, **2** polyAMPS[b] vinyl copolymer	61, 65
H	—	AMPS	$h\nu$(uv) or γ	vinyl copolymer	64, 65
H	Cl^-	p-$CH_2{=}CHPhSO_3^-Na^+$	H_2O	1:1 vinyl copolymer[c,d]	64, 65
H	Cl^-	p-$CH_2{=}CHPhSO_3^-Na^+$	MeOH or DMSO	vinyl copolymer	62
CH_3	—	p-$CH_2{=}CHPhSO_3^-$	Dark or $h\nu$ (vis)/48 h	—	63, 65
CH_3	—	p-$CH_2{=}CHPhSO_3^-$	$h\nu$ (uv or γ)	vinyl copolymer	50, 64, 65
CH_3	I^-	p-$CH_2{=}CHPhSO_3^-Ag^+$ or Na^+	H_2O, MeOH or DMSO	1:1 vinyl copolymer[d]	50, 65
CH_3	I^-	p-$CH_2{=}CHPhSO_3^-Na^+$	$h\nu$(vis)/H_2O	1:1 vinyl copolymer[d]	50, 63
CH_3	I^-	p-$CH_2{=}CHPhSO_3^-Na^+$	t-Butylcatechol/H_2O	—	50
CH_3	CH_3OSO_3	α-methylstyrene	MeOH or DMSO	vinyl copolymer	62
CH_3	CH_3OSO_3	p-methoxystyrene	DMSO	vinyl copolymer	62

[a] ACVA = 4,4′-Azobis-4-cyanovaleric acid.
[b] AMPS = 2-Acrylamido-2-methylpropanesulfonate.
[c] A1:1 vinyl copolymer was also reported when excess p-$CH_3PhSO_3^-Na^+$ was used, and added monomers such as vinylsulfonate, butylvinyl ether, AMPS, or stilbene were not incorporated into the polymer.
[d] Structural analysis by ^{13}C NMR.

spectrophotometric analysis of solutions containing **1** and **20** in methanol as indicating charge-transfer (CT) complexation between the monomers. They postulate that CT complexes play an important role in the propagation step of the polymerization process. This suggestion provides an explanation for the reported differences in behavior of **1** (R = H, CH_3) toward various anionic vinyl monomers. However, Roitman and Harwood have observed that copolymers **21** may contain 55–95 mol% **1** (R = CH_3) when it is present in appropriate excess in reaction mixtures of **1** and **20**.

6.3.2 Copolymerization via Charge-Transfer Complexation

Similarly, α-methylstyrene, **24**, and *p*-methoxystyrene, **25** can be copolymerized with **1** (R = H or CH_3) producing copolymers, **26** and **27**, reported to have alternating monomer sequence, **26** (Eqs. 6.3.3 and 6.3.4) [62]. In each instance copolymer equimolar in the two monomers is obtained. Its formation has been attributed to a charge-transfer interaction.

The above examples of successful copolymerization reactions illustrate the importance of coulombic and π donor-π acceptor interactions in copolymerization of 4-vinylpyridinium ions. Thus, electrostatic attraction and repulsion or CT complexation offer potentially useful means for control of composition and sequence of monomers in copolymers.

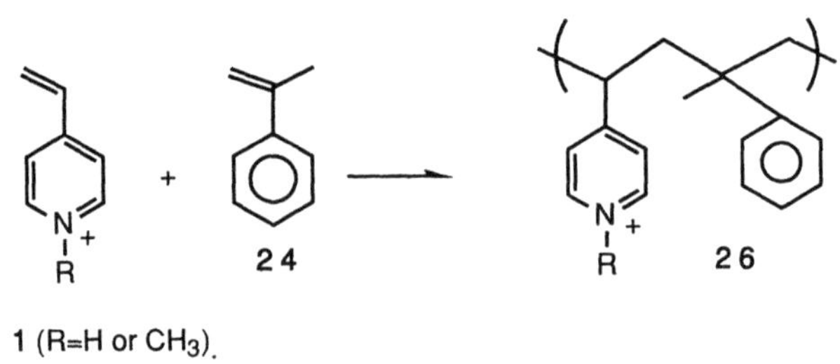

(6.3.3)

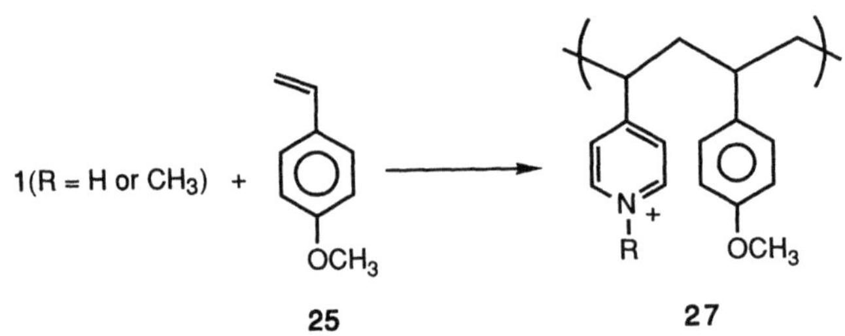

(6.3.4)

6.3.3 Copolymerization of Hydrophobically-Organized Monomers

Hydrophobic forces are responsible for self-organization in many molecular systems [53–55]. These forces are recognized as important determinants of tertiary and quaternary structure in proteins, and are responsible for the assembly of biological membranes and other supramolecular systems. In previous sections, hydrophobic interactions have been seen as responsible for superceding the repulsive interactions of positively-charged monomers **1** (R = H, CH_3) to enable their homopolymerization to **2** in aqueous solutions of high monomer concentration. These same forces may be operative in the copolymerization reactions described in the previous sections (6.3.1 and 6.3.2). Combinations of electrostatic and π-stacking forces acting in concert with hydrophobic forces may well have contributed to these successful copolymerizations. Finally, the greater ease of homopolymerization of **1** (R = $C_{12}H_{25}$) in aqueous suspensions over that of **1** (R = H, CH_3, C_2H_5) provides additional evidence for the reactivity enhancing role of hydrophobic forces in polymerization of **1** [46, 47].

The results from copolymerization of mixtures of two different hydrophobic monomers **1** e.g. **1** (R = C_8H_{17} and $C_{12}H_{25}$) in aqueous suspensions provide additional evidence for the reactivity enhancing role of hydrophobic forces, Table 6.3.2 [46, 47]. Copolymerization of water-soluble monomers, **1** (R = CH_3 or C_2H_5) with water-insoluble monomer **1** (R = $C_{12}H_{25}$) was unsuccessful under conditions that compartmentalized **1** (R = CH_3 or C_2H_5) in an aqueous phase and **1** (R = $C_{12}H_{25}$) in a second "organic" lipophilic phase, i.e. suspended aggregates, micelles (established with added SDS or DPC surfactant, or by the monomer itself). The two monomers polymerized independently of one another to form two homopolymers, and the aggregated lipophilic monomers **1** (R = $C_{12}H_{25}$) polymerized more readily in all cases. The influence of phase distribution of initiator was also examined. Initiators expected to distribute between aqueous and organic phases produced both homopolymers, as did

Table 6.3.2. Copolymerization of **1** (R_1) triflate with **1** (R_2) triflate[a]

1		Monomer ratio **1** (R_2)/**1** (R_1)	Conditions	Products
$R_1 = CH_3$	$R_2 = C_2H_5$	1.0	55 °C AIBN/*n*-BuOH/10 h	88% 1:1 **3**
CH_3	C_2H_5	1.0	55 °C $S_2O_8^{-2}$, $S_2O_5^{-2}$/H_2O/10 h	90% 1:1 **3**
C_2H_5	C_6H_{13}	0.67	125 °C Vac/15 min	[b] 3:2 **3**
C_2H_5	C_6H_{13}	1.5	125 °C Vac/15 min	[b] 2:3 **3**
C_2H_5	$C_{12}H_{25}$	2.3	125 °C Vac/15 min	86% 2:1 **3**
C_2H_5	$C_{12}H_{25}$	2.3	AIBN/CH_3CN/55 °C/10 h	84% ≅ 2:1 **3**
C_2H_5	$C_{12}H_{25}$	0.25	$S_2O_8^{-2}$, $S_2O_5^{-2}$/H_2O/50 °C/10 h	[c]**2**(R = C_2H_5), **2** (R = $C_{12}H_{25}$)
C_2H_5	$C_{12}H_{25}$	0.10	$S_2O_8^{-2}$, $S_2O_5^{-2}$/H_2O/50 °C/10 h	[c]**2** (R = C_2H_5) **2** (R = $C_{12}H_{25}$)
C_2H_5	$C_{12}H_{25}$	0.17	$S_2O_8^{2}$, $S_2O_5^{-2}$/Hexane/H_2O/*i*-AmOH/ 45–50 °C/10 h	60% Copolymer
C_2H_5	$C_{12}H_{25}$	0.17	$S_2O_8^{-2}$, $S_2O_5^{-2}$/Hexane/H_2O/ 45–50 °C/10 h	**2**(R = C_2H_5) + **2**(R = $C_{12}H_5$)
C_8H_{17}	$C_{12}H_{25}$	1.0	$S_2O_8^{-2}$, $S_2O_5^{-2}$/H_2O/50 °C/10 h	76% Copolymer
C_8H_{17}	$C_{12}H_{25}$	1.0	AIBN/*n*-BuOH/50 °C/10 h	82% Copolymer

[a] References 46 and 47.
[b] Product yield was not determined.
[c] Mixture of two homopolymers.

initiators restricted to the aqueous phase like the $S_2O_8^{-2}/S_2O_5^{-2}$ couple. On the other hand, initiators located in the organic phase (e.g. AIBN and lauroyl peroxide) furnished only the lipophilic homopolymer. Successful copolymerization of **1** (R = CH_3) and **1** (R = C_2H_5), both of which are water-soluble was readily accomplished under a variety of conditions (e.g. with thermal, photo at 300 nm, radical and nucleophile initiation). Copolymerization of **1** (R = C_2H_5) with **1** (R = $C_{12}H_{25}$) was also successful in melts at 125 °C and in acetonitrile solution with AIBN initiation. Thus, it appears that compartmentalization of monomers based on hydrophobic-hydrophilic properties represents a poorly appreciated aspect of the chemistry of **1**.

6.3.4 *Copolymerization at Water-Organic Solvent Interfaces*

In the previous section we identified the water-organic component interface as an important barrier to copolymerization of hydrophobic and hydrophilic variants of **1** under a variety of conditions. Recall that such copolymerizations were readily accomplished in solution. Since monomer aggregates or micelles containing monomers suspended in water did not copolymerize with monomer in aqueous solution, we examined the possibility for copolymerization in inverse microemulsions where the weakly associating water-soluble monomer, **1** (R = C_2H_5) would be present at high concentration and the strongly associating hydrophobic monomer, **1** (R = $C_{12}H_{25}$) could be diluted with an organic solvent [47]. A standard formulation of an inverse microemulsion that contained **1** (R = C_2H_5) (0.7 mmol) and **2** (R = $C_{12}H_{25}$) (0.3 mmol) in hexane-water with i-amyl alcohol as co-surfactant produced **2** (m = 0.65, n = 0.35) when initiated by the $S_2O_8^{-2}/S_2O_5^{-2}$ couple. Thus, the copolymer was slightly enriched in hydrophobic monomer during polymerization. This represents the first example of the successful copolymerization of hydrophobic and hydrophilic monomers of similar charge type in an aqueous-organic solvent system, Table 6.3.2. Note that a control experiment also initiated by $S_2O_8^{-2}/S_2O_5^{-2}$ but without the co-surfactant, i-amyl alcohol, was unsuccessful.

6.3.5 *Monomer Sequence Analysis in Copolymers of 1-Alkyl-4-vinylpyridinium Ions*

Unambiguous determination of monomer sequence in copolymers **3** is a requisite for evaluation of various synthetic approaches in controlling their microstructure. Knowledge of microstructure, especially monomer sequence information, is vital to detailed understanding of structure-property relationships. All well established spectroscopic methods, e.g. IR, UV-VIS, ^1H and ^{13}C NMR, are severely limited by the structural similarity of these monomers. All contain the *N*-alkyl-4-vinylpyridinium chromophone. A new method for sequencing synthetic polymers based on pyrolysis/tandem mass spectrometry has been developed by Cooks and associates [66] and applied to analysis of copolymer **3**. Desorption chemical ionization (DCI) and desorption electron ionization (DEI) have been used under conditions that produce mass spectra which display oligomeric ions derived from **3**. Unique and structurally dependent fragmentation processes are monitored by tandem mass spectrometry (MS/MS) experiments. Bernoullian statistical calculations [67] were applied to ion abundances of oligomeric ions recorded in the mass spectra of copolymers **3**($R_1 = C_2H_5$, $R_2 = C_{16}H_{13}$ and $R_1 = C_2H_5$, $R_2 = C_{12}H_{25}$). The results were consistent with randomly distributed monomers. Significantly, randomly ordered monomer sequences were observed in all copolymers investigated whether produced in inverse microemulsions, solutions, or melts. More recent work has shown that pyrolysis/tandem MS cannot be used for analysis of **3** due to rapid randomization through de-alkylation–re-alkylation.

6.4 Conclusion

This review has focused on the importance of molecular association on homo- and copolymerization of 1-alkyl-4-vinylpyridinium ions. Intermonomer association via coulombic interactions of oppositely charged monomers and energetically favorable π-donor-acceptor interactions have been found to play significant roles in copolymerization of 1-alkyl-1-vinylpyridinium ions and anionic or π-donor monomers. Hydrophobic forces appear to enhance reactivity of 1-alkyl-4-vinylpyridinium ions toward homopolymerization and copolymerization of water-insoluble monomers with different *N*-alkyl groups; i.e. **1**($R = R_1$) with **1**($R = R_2$). The repulsive forces between the positively charged water-soluble variants with short chain C_n ($n < 6$), *N*-alkyl groups, appear to impede polymerization as evidenced by the strong concentration dependence of polymerization of these monomers in aqueous solution.

Attempts in Kabanov's laboratory to control macrostructure or tacticity of **2** by polymerization of **1** via pre-association to a polyanion matrix were largely unsuccessful. Similar efforts by the Paleos and Fife groups to order **1**($R = CH_3$ or $C_{12}H_{25}$) at water-organic interfaces to control tacticity in homopolymers and monomer sequence in copolymers also failed.

6.5 References

1. Tsuchida E, Abe K (1982) In: Cantow H-J et al. (eds) Advances in Polymer Science. Springer-Verlag, New York, Vol 45
2. Philipp B, Dantzenberg H, Linow K-J, Kötz J, Dawydoff W (1990) Prog Polym Sci 14:91
3. Tsuchida E, Osada Y, Sanada K (1972) J Polym Sci A-1 10:3397
4. Zezin AB, Lutsenko VV, Kabanov VA (1974) Vysokomol Soyed A16:600
5. Starodubtsev SG, Kabanov VA (1977) Vysokomol soyed A19:1948
6. Izumrudov VA, Kasaikin VA, Ermakova LN, Zezin AB (1978) Vysokomol Soyed A20:400
7. Salamone JC, Paulin S, Watterson AC, Olson AP (1979) Polymer 20:611
8. Otocka EP, Eirich FR (1968) J Polym Sci A-2 6:895
9. Otocka EP, Eirich FR (1968) J Polym Sci A-2 6:913
10. Kabanov VA, Zezin AB (1984) Pure Appl Chem 56:343
11. Heller A (1990) Acc Chem Res 23:128
12. Kasem KK, Fife WK, Zeldin M, Leidner CR (1990) J Electroanal Chem 296:221

13. Park M (1991) MS Thesis, Purdue University
14. Kabanov VA 113, 197702 (1990) Makromol Chem Macromol Symp (Int Symp Mol Des Funct Polym, 1989) 33:279
15. Kabanov VA, Kieselev VI, Chikindas ML, Astaféva IV, Glukhov AI, Gordeev SA, Izumrudov VA, Zezin AB, Levashov AV (1989) Dokl Akad Nauk SSSR 306:226
16. Dzantiev BB, Blintsov AN, Izumrudov VA, Zezin AB, Bobkova AF, Egorov AM, Atabekov IG, Kabanov VA (1990) Dokl Akad Nauk SSSR 311:1482
17. Kawabata N, Hayashi T, Matsumoto T (1983) Appl Environ Microbiol 46:203
18. Kawabata N, Hayashi T, Nishikawa M (1986) Bull Chem Soc Jpn 59:2861
19. Totoki S, Kondo A (1990) Jpn Kokai Tokkyo Koho JP02, 88,505 [90 88,505] Chem Abstr (1990) 113:237168
20. Totoki S, Kondo A (1990) Jpn Kokai Tokkyo Koho JP02, 88,506 [90 88,506] Chem Abstr (1990) 113:237154
21. Kawabata N, Nishimura T, Yoshimura T (1990) Biotechnical Bioeng 35:1000
22. Fuoss RM, Sadek H (1949) Science 110:552
23. Hall H, Padias AB (1990) Acc Chem Res 23:3
24. Kosower EM (1958) J Am Chem Soc 80:3253
25. Menger FM (1986) Top Curr Chem (Biomimetic Bioorg Chem 3) 136:1
26. Strauss UP, Gershfeld NL (1954) J Phys Chem 58:747
27. Strauss UP, Gershfeld NL, Crook EH (1956) J Phys Chem 60:577
28. Strauss UP, Williams BL (1961) J Phys Chem 65:1390
29. Strauss UP (1977) Micellization, Solubilization, Microemulsions In: Mittal KL (ed.) Proc Int Symp (1976) Plenum, New York 2:895
30. Strauss UP, Jackson EG (1951) J Polym Sci 6:649
31. Kargin VA, Kabanov VA, Aliev KV, Razvodovsky (1965) Dokl Akad Nauk SSSR 160:604
32. Kabanov VA, Aliev KV, Patrikeeva TI, Kargina OV, Kargin VA (1967) J Polymer Sci C 16:1079
33. Salamone JC, Snider B, Fitch WL (1970) Macromolecules 3:707
34. Salamone JC, Snider B, Fitch WL (1971) J Polym Sci B, 9:13
35. Fife WK, Ranganathan P, Zeldin M (1990) J Org Chem 55:5610
36. Salamone JC, Mahmud MU, Watterson AC, Olson AP, Ellis EJ (1982) J Polym Sci: Polym Chem Ed 20:1153
37. Martin V, Ringsdorf H, Ritter H, Sutter W (1973) Angew Chem Internat Ed 12:432
38. Mielke I, Ringsdorf H (1972) Makromol Chem 153:307
39. Shyluk WP (1964) J Polym Sci A 2:2191
40. Kargina OV, Mishustina LA, Svergun VI, Lukovkin GM, Yerdakov VP, Kabanov VA (1974) Vysokomol, Soyed A16:1775
41. Salamone JC, Watterson AC, Hsu TD, Tsai CC, Mahmud MU (1978) J Polym Sci Polym Symp 64:229
42. Martin V, Sutter W, Ringsdorf H (1976) Makromol Chem 177:89
43. Paleos CM (1977) J Polym Sci, Polym Lett Ed 15:535
44. Paleos CM, Evangelatos CP, Dais P, Kipouros G (1979) J Polym Sci, Polym Chem Ed 17:1611
45. Paleos CM (1985) Chem Soc Rev 14:45
46. Xin-Y, Hu Y, Zeldin M, Fife WK (1993) Macromolecules 26:4670
47. Xin Y (1991) MS Thesis, Purdue University
48. Ranganathan P, Fife WK, Zeldin M (1990) J Polym Sci: A: Polym Chem 28:2711
49. Mielke I, Ringsdorf H (1970) J Polym Sci C:107
50. Roitman L, Harwood HJ (1984) Polym Prepr (Am Chem Soc Div Poly Chem) 25:152
51. Parish C, Zeldin M, Fife WK (1990) unpublished results
52. Salamone JC, Snider B, Fitch WL (1971) J Polym Sci A-1 9:1493
53. Dill KA (1990) Science 250:297
54. Muller N (1990) Acc Chem Res 23:23
55. Breslow R (1991) Acc Chem Res 24:159
56. Kabanov VA, Aliev KV, Kargina OV, Patrikeeva TI, Kargin VA (1967) J Polym Sci C:1079
57. Kargina OV, Ul'yanova MV, Kabanov VA, Kargin VA (1968) Polym Sci USSR 9:380
58. Kabanov VA, Petrovskaya VA, Kargin VA (1968) Polym Sci USSR 10:1077
59. Mielke I, Ringsdorf H (1971) Makromol Chem 142:319
60. Martin V, Ringsdorf H, Ritter H, Sutter W (1975) Makromol Chem 176:2029
61. Salamone JC, Watterson AC, Hsu TD, Tsai CC, Mahmud MU (1977) J Polym Sci Polym Lett Ed 15:487

62. Salamone JC, Boden MW, Watterson AC (1988) Polym Prepr (Am Chem Soc Div Polym Chem) 29:269
63. Salamone JC, Raheja MK, Anwaruddin Q, Watterson AC (1985) J Polym Sci Polym Lett Ed 23:655
64. Lai KTG (1990) PhD Thesis, Univ of Lowell
65. Salamone JC, Watterson AC, Hsu TD, Tsai CC, Mahmud MU, Wisniewski AW, Israel SC (1978) J Polym Sci: Polym Symp 64:229
66. Majumdar TK, Cooks RG, Fife WK, Zeldin M, Xin Y (1991) 42nd Pittsburgh Conf Anal Chem Appl Spec, Chicago, IL, USA, 302P
67. Majumdar TK, Eberlin MN, Cooks RG, Green MM, Muñoz B, Reidy MP (1991) J Am Soc Mass Spectron 2:130
68. Majumdar TK, Ranasinghe A, Lu L, Cooks RG, Fife WK, Zeldin M (1993) Talanta 40:363

7 Fluorocarbon-Modified Water Soluble Polymers

Y-X. Zhang, F.S. Hwang and T.E. Hogen-Esch

Copolymers of acrylamide (AM) and perfluoroalkyl acrylates and -methacrylates were synthesized in aqueous media containing perfluoroalkyl carboxylates and acetone as surfactants and dispersing agents, respectively and persulfates as initiators. The copolymers, especially those with 2(*N*-ethylperfluoro octane sulfoamido)acrylate (RFA) as comonomers were extremely effective at associating, giving much higher Brookfield viscosities compared to simple laurylacrylates (LA). Also, the molar contents of the RFA copolymers were generally much lower compared with LA copolymers. Addition of salts or increasing the temperature resulted in sharp viscosity increases in many cases. Viscosity decreases were observed with increasing shear, consistent with copolymer association through the perfluorocarbon groups. The Brookfield viscosity of a series of RFA containing copolymers measured at constant copolymer concentration showed a maximum as a function of RFA molar content. This indicated competing intra- and intermolecular hydrophobic association.

Similar results were obtained with hydroxyethylcellulose (HEC) modified with perfluorocarbon chains attached to the HEC backbone through oligoethyleneoxide spacers. Only the HEC derivatives containing two ethyleneoxide spacer units were effective viscosifiers. Again, a sharp maximum in Brookfield viscosity as a function of modifier content is observed, in accordance with earlier results of Landoll in the corresponding HEC alkyl derivatives.

7.1 Introduction

The phenomenon of hydrophobic interactions in aqueous media only recently has been recognized as a significant factor in the self assembly of synthetic macromolecules [1–13].

For instance, the use of hydrophobically associating water-soluble polymers is a new approach to aqueous fluid mobility control over the last few years [4–9]. Much of this research has focussed on the synthesis of copolymers of acrylamide and comonomers containing long hydrocarbon chains such as laurylacrylate or -methacrylate and *N*-alkyl acrylamides [8]. Other approaches to this problem have included the synthesis of cellulose-based water-soluble polymers substituted with long hydrocarbon chains [7].

In all of these cases, the extensive intermolecular association occurring through clustering of the hydrocarbon groups was shown to result in appreciable viscosity enhancements. This was shown to be consistent with several observations including shear thinning (pseudoplastic) behavior, on addition of substances known to compete with hydrophobic interactions such as organic solvent surfactants, etc. Nevertheless, the nature of these hydrophobic

Macromolecular Complexes
in Chemistry and Biology
Eds Dubin/Bock/Davis/Schulz/Thies

interactions has not been fully appreciated due to several factors. For instance, the comonomer sequence in most of these copolymers is not known partially due to the difficulties in spectroscopic observation of very low concentrations of comonomer units. Furthermore, the exact mechanism of the copolymerizations has not yet been fully clarified. Thus, the copolymerizations involving acrylamide are often carried out in the presence of surfactants in order to "dissolve" the hydrophobic comonomers in the aqueous media. As a result, there is the possibility of a dual polymerization mechanism in which the acrylamide polymerizes in an aqueous medium while the hydrophobic comonomer is incorporated into the chain through a type of micellar polymerization process. The resulting polymer would be expected to have a structure in which the comonomer is incorporated in a blocky fashion. Indeed, such a structure has been invoked in several instances [9].

Partially as a result of this, the nature of the hydrophobic interactions is not fully clear. Thus, the number of hydrophobic moieties involved in hydrophobic aggregates causing intermolecular association is not known. However, it appears that these hydrophobic aggregates are not simply micelles containing a large number of hydrophobic groups.

We have recently synthesized new types of water-soluble copolymers containing fluorocarbon moieties [10–13]. This type of copolymer would appear to have advantages over the corresponding hydrocarbon analogs. First, fluorocarbons appear to possess greater apparent hydrophobic character [14, 15]. This is reflected in critical micelle concentrations of perfluorocarbon

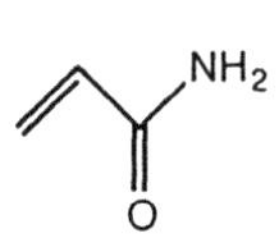

1

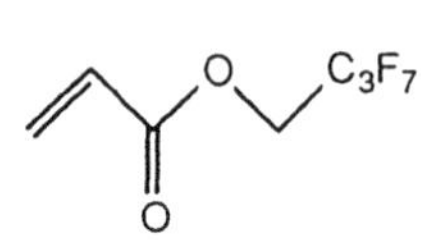

2

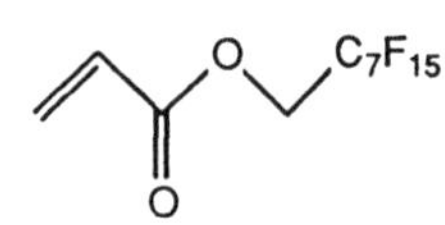

3

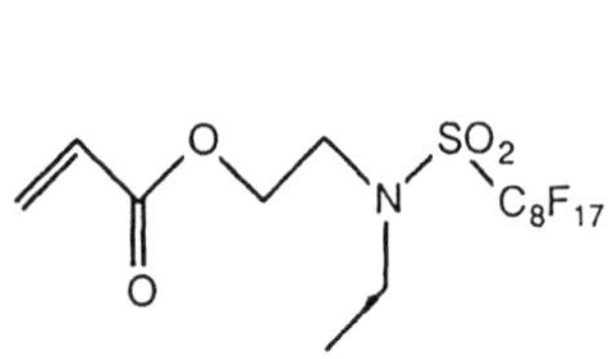

4

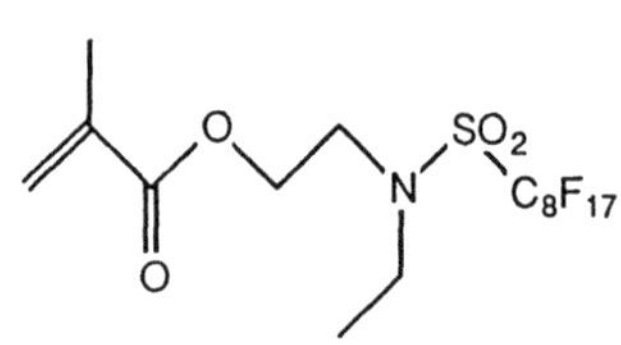

5

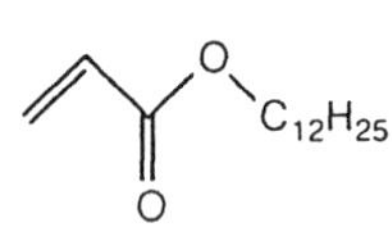

6

surfactants that are typically one order of magnitude lower compared to the hydrocarbon analogs [3]. As a result, intermolecular hydrophobic interactions of perfluoro-containing water-soluble polymers are expected to be stronger resulting in more extensive association. Second, the perfluoro comonomer unit is more readily detected at low concentration either by NMR or by elemental analysis compared to the hydrocarbon comonomers [16]. As a result, the structure of the copolymer may be analyzed more readily.

In the following, we wish to report the synthesis, characterization and properties of several copolymers of acrylamide (**1**) and perfluorocarbon comonomers (**2**–**5**), and of perfluorocarbon derivatives of hydroxyethyl cellulose. The laurylacrylate (**6**)-acrylamide copolymer was also synthesized for comparison purposes.

7.2 Experimental

7.2.1 Reagents and Solvents

Acrylamide (ultra pure electrophoresis grade form Polysciences, Inc.) was used without further purification. Laurylacrylate (**6**) (Monomer-Polymer Labs) was purified by washing with 1 M NaOH followed by vacuum distillation. The 1,1-dihydroperfluorobutyl acrylate (**2**) and 1,1-dihydroperfluorooctyl acrylate (**3**) were purchased from Monomer–Polymer Laboratories and used as received. The fluorine containing surfactants and monomers, 2-(*N*-ethylperfluorooctane-sulfoamido)ethylacrylate (**4**) and -methacrylate (**5**) technical grade were donated by the 3M Company. The monomers were recrystallized from MeOH. All of the solvents and salts were of reagent grade or ACS grade quality. The AMPS monomer (sodium salt Aldrich) was used as received.

Hydroxyethyl cellulose (HEC), (dry powder) with a hydroxyethyl molar substitution level (M_s) of 2.0 and a MW = 90 000–105 000 was obtained from "Polysciences". The 1,1-dihydroperfluoro alcohols ($CF_3(CF_2)_nCH_2OH$, $n = 2$ or 6), epichlorohydrin, tosylchloride and sodium hydride were commercial materials supplied by Aldrich Chemical Company and were used as received. Epichlorohydrin was distilled over CaH_2 before use.

Solvents, such as isopropyl alcohol (IPA), dioxane (DOX), triethylamine (TEA), anhydrous diethyl ether (DEE) and acetone were obtained from the "Mallinckrodt Company" and were used as received.

7.2.2 Synthesis of Modifiers

7.2.2.1 1,1-Dihydroperfluoroalkyl *p*-Toluenesulfonates (**7** and **8**) [17]

Into a round bottom flask, 0.087 mole of $CF_3(CF_2)_nCH_2OH$ ($n = 2$ or 6), 0.092 mole of *p*-toluenesulfonyl chloride and 40 ml of deionized water were added. The mixture was heated with stirring to 50 °C and 0.11 mole of sodium hydroxide dissolved in 20 ml water was added dropwise at a rate designed to keep

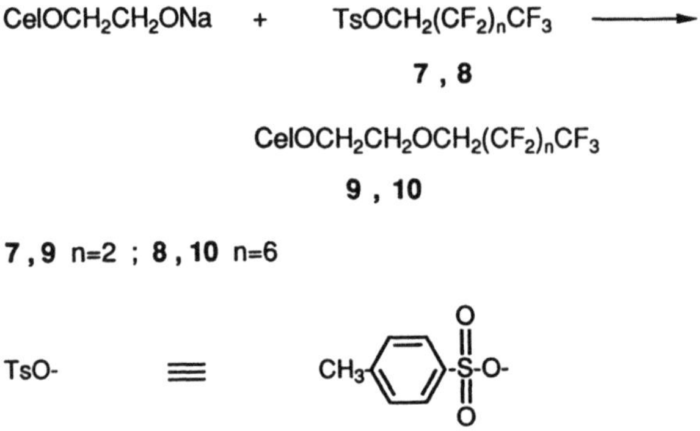

(7.1)

Scheme 7.1

the reaction temperature below 65 °C. The mixture was then stirred for 20 h at 50 °C. The reaction mixture was cooled and the organic layer was extracted by means of petroleum ether. The petroleum ether solution was washed repeatedly with concentrated aqueous ammonia and then deionized water to remove any unreacted tosylchloride. This solution was concentrated by removing the solvent on a steam-bath. The product crystallized at refrigerator temperature. After recrystallization from petroleum ether (bp = 35–60 °C), the structure of the product was confirmed by IR and NMR. Proton NMR ($CDCl_3$) shows the characteristic peaks of *p*-toluenesulfonate at $\delta = 2.45$ (s, 3H, $-CH_3$), 4.43 (t, 2H, $-CH_2R_f$), 7.35 and 7.78 (d, 2H, aromatic protons). The 1,1-dihydroperfluorobutyl *p*-toluenesulfonate, (PFBOTs) (**7**), was obtained in 83% yield (mp = 27–28 °C) and 1,1-dihydroperfluorooctyl *p*-toluene-sulfonate, (PFOOTs) (**8**), in 65% yield (mp = 57–59 °C).

7.2.2.2 1,1-Dihydroperfluorooctyl Glycidyl Ethers (PFOGE) (**12**) [18]

In a typical experiment, 3.5 g of sodium hydride dispersion in mineral oil (0.0875 mole) was treated with three portions (each of 40 ml) of anhydrous diethyl ether to remove mineral oil. To the slurry of NaH and 150 ml fresh anhydrous diethyl ether, 25 g (0.0625 mole) of 1,1-dihydroperfluoro octanol, $CF_3(CF_2)_6CH_2OH$ dissolved in 80 ml diethyl ether was added with stirring at room temperature over a period of 20 min. Epichlorohydrin (7.0 g, 0.076 mole) was then added into the mixture that was stirred overnight. After removal of 160 ml diethyl ether, an additional quantity of 50 g of epichlorohydrin was added. Upon removal of the last traces of ether, the reaction mixture was refluxed for 4 h with stirring. An equal volume of water was then added to the cooled reaction mixture. The organic layer was separated. The aqueous layer was extracted three times with 50 ml diethyl ether. The combined ether solution was dried over $MgSO_4$. Reduced pressure distillation provided a center fraction (b.p. 73–75 °C) at 1.5 mmHg with 53% yield. Elemental analysis (Galbraith, Knoxville, TN) confirmed the structure: Calc. for $C_{11}F_7O_2$: C, 28.96%; F,

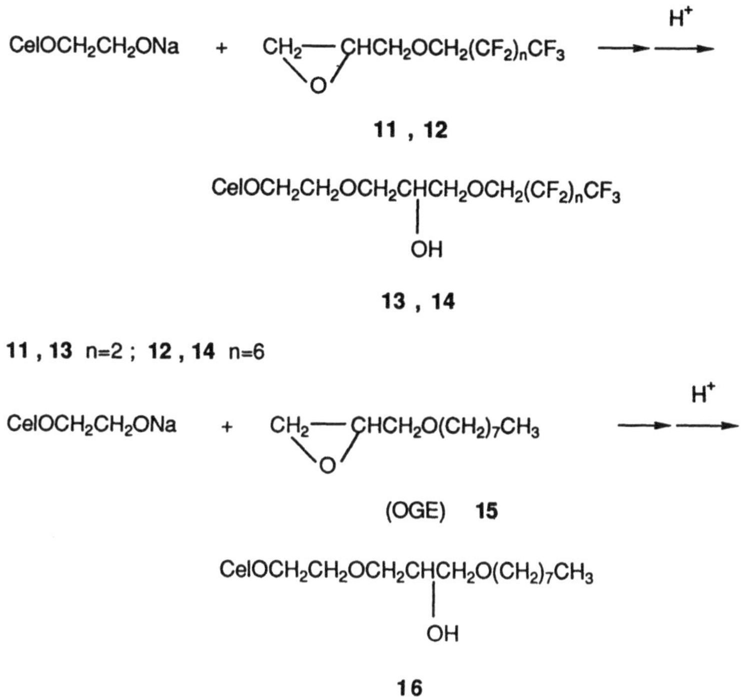

(7.2)

(7.3)

Scheme 7.2

62.47% H, 1.55%. Found: C, 28.74% F, 62.35% H, 1.56%. Proton NMR ($CDCl_3$) shows the characteristic peaks of epoxide ring at δ: 2.62 (q, 1H), 2.80 (t, 1H), 3.17 (m, 1H), and 3.53 (q, 1H, $-OCH_a-$), 3.97 (q, 1H, $-OCH_b$), 3.92–4.16 ppm (m, 2H, $-OCH_2R_f$).

The 1,1-dihydroperfluorobutylglycidylether, (PFBGE) (**11**), was prepared using a similar procedure. Boiling point was 75–78 °C at 40 mmHg and the yield was 60%. Proton and fluorine NMR show the same characteristic peaks as those of **12**.

7.2.2.3 Polymerization

The polyacrylamide copolymers were prepared by dispersion of 2.0 g of acrylamide and varying amounts of comonomers in a 50 ml capped round bottomed flask containing 20.0 ml of deionized water, 4.0 ml acetone and 0.12–0.25 wt% surfactant (sodiumlauryl sulfate or potassium perfluoro-octanoate). After purging the solution with argon for 30 min, 0.50 ml of a solution of sodium metabisulfite (19.0 mg in 10.0 ml deionized water) and 0.50 ml of a solution of ammonium persulfate (22.8 mg in 10.0 ml deionized water) were injected into the rapidly stirred contents of the flask. Each polymerization was carried out for 24 h at 50 ± 1.0 °C. Gels of various consistencies were obtained at periods of 2–24 h. The homopolymer was prepared in a similar fashion except that no comonomer was present during polymerization.

7.2.2.4 Modification of Hydroxyethyl Cellulose (HEC) [7, 19, 20]

Hydroxyethyl cellulose (8.0 g) was added to 100 ml isopropyl alcohol and was then sparged with nitrogen for 1 h. Sodium hydroxide (2.0 g) dissolved in 4 ml H_2O was then added dropwise over a period of 10 min. The mixture was stirred for another hour with a slow flow of nitrogen gas. The mixture was then heated to 50 °C and a certain amount of modifier dissolved in 10 ml degassed isopropyl alcohol was added dropwise over one hour. The reaction mixture was kept at 60 °C for another 3 h. The cooled reaction mixture was neutralized with acetic acid and then filtered. The solid product was washed with four portions each of 100 ml (1:4 water/acetone) solvent and then pure acetone repeatedly. The product was dried in vacuum at 50 °C for 24 h.

To the dry powder product was added a suitable amount of deionized water and the mixture was then transferred into Spectra/por 1 (MWCO 6-8000) dialysis tubing and dialyzed against deionized water for seven days. The polymer was then precipitated out from the resulting solution in a large amount of acetone and dried in vacuum at 50 °C for 24 h. Fluorine NMR analysis shows that unbound modifier can be removed quantitatively in this way from HEC derivatives.

7.2.2.5 Polymer Dissolution and Viscosity Measurements

The copolymer solutions were prepared by two methods: in the first method, the contents of the flask were diluted with several volumes of deionized water and allowed to stand for periods varying from 24 h to several weeks in order to dissolve the gels. The solution was then dialysed with deionized water for three or four days. The concentration of the solution was determined by precipitation in acetone. In the second method, the contents of the flask were dissolved as described above and were then precipitated into acetone and dried in a vacuum oven at 40 °C for 24 h. The solid polymer was then dissolved in deionized water for subsequent measurements. For semiconcentrated polymer solutions, (1.0 or 0.5 wt%), a Wells Brookfield viscometer was used. The Brookfield viscosity (cp) and shear rate were obtained by direct readout and rotational speed (rpm), using the range table supplied by the manufacturer.

Solutions of HEC derivatives were prepared using the dry powder products at different concentrations. The suspensions were stirred gently for at least three days in order to obtain homogeneous solutions prior to viscometry measurements. Brookfield viscometry measurements were carried out at various shear rates using appropriate spindles. The apparent intrinsic viscosity of modified and unmodified HEC were also measured by using "Ubbelohde" viscometry.

7.2.2.5 Determination of Modifier Content

Using the dry polymer, a 2.0 wt% aqueous solution was prepared with sodium triflate (CF_3SO_3Na) as internal standard and subjected to ^{19}F NMR analysis. Since $CFCl_3$ is insoluble in aqueous polymer solutions, we used $CFCl_3$ (dissolved in $CDCl_3$) as external reference. From the ratio of the peak area ($NaSO_3CF_3$ at -80.8 ppm and $-(CF_2)_nCF_3$ at -82.8 ppm), the weight percent of fluorine in the polymer was calculated and thus the molar ratio of modifier over anhydroglucose unit (AGU). The results are tabulated in Table 7.9.

7.3 Results and Discussion

7.3.1 Polyacrylamide Copolymers

The synthesis of the polyacrylamide copolymers containing comonomers **2–6** was carried out as reported in the Experimental Section. Tables 7.1–7.5 show the Brookfield viscosities of the copolymers as a function of comonomer composition. Table 7.1 also shows the results of a homopolymer prepared under conditions identical to the preparation of the copolymers. Comparison of Tables 7.1–7.5 shows that comonomers **4** and **5** appear to be most effective in promoting associative behavior in the copolymers. Thus, the viscosities of 0.5 wt% solutions of the copolymers of **4** and **5** are about two orders of magnitude higher than that of hydrocarbon comonomer **6**. Furthermore, comparison of Tables 7.1 and 7.5 shows that the optimal comonomer weight fractions are 10 times smaller for the copolymer containing **4** than for that containing **6**. Thus, sample AL-5 having 4.80 wt% of **6** contains about 10 times

Table 7.1. Synthesis and properties of RFA (**4**) containing polyacrylamides[d]

	Mole % comonomer	Weight % comonomer	Conversion[e] %	Brookfield[a], [b] viscosity (Centipoise)
PAM	0	0	99.9	50
RFA-22	0.006	0.05	90.3	480
21	0.012	0.10	90.7	1760
18	0.023	0.20	88.5	7000
17	0.045	0.40	95.2	10800
12	0.070	0.62	95.0	12000
10	0.14	1.25	96.3	6200
9	0.28	2.40	96.1	300
8	0.56	4.80	101.4	50
7	1.12	9.10	100.0	50
AFT-13[c]	0.14	1.25	78.9	2500

[a] Polymer conc = 0.50 gr/dl.
[b] Shear rate = 0.40 s^{-1}.
[c] Terpolymer AM/AMPS/RFA containing 16% mole NaAMPS.
[d] Emulsifier was $CF_3(CF_2)_6COOK$ (obtained from 3M Company) at 12 wt%.
[e] Determined gravimetrically

Table 7.2. Synthesis and properties of RFM (**5**) containing copolymers[d]

	Mole % comonomer	Weight % comonomer	Conversion	Brookfield[a, b] viscosity (Centipoise)
RFM-24	0.022	0.20	101.0	120
23	0.044	0.40	100.0	2220
15	0.069	0.62	110.7	6000
16	0.14	1.25	83.1	6800
19	0.28	2.40	74.3	10 800
20	0.55	4.80	75.9	9000
32	1.11	9.10	71.8	200
33	1.66	13.0	67.8	120

Footnotes: See Table 7.1

Table 7.3. Synthesis and properties of RF4 (**2**) containing copolymers[b]

Comonomer g	Comonomer/Content[c] mole%	wt%	Yield g	Conv.[d] %	Viscosity[a] at conc. 0.50%	Viscosity[a] at conc. 1.00%
0.30	4.22	13.0	1.87	81.3	30	250
0.20	2.81	9.10	1.77	80.5	50	350
0.10	1.41	4.76	1.91	90.8	40	320
0.050	0.70	2.44	1.75	85.4	35	180
0.027	0.38	1.34	1.92	94.7	35	170
0.011	0.16	0.57	1.82	90.4	35	160

[a] Brookfield viscosity was measured at shear rate 0.40 s^{-1}.
[b] For preparation, see Experimental Section.
[c] Calculated values.
[d] Determined gravimetrically

more comonomer than sample RFA-17 (0.40 wt%) that has a viscosity that is about 100 times greater. Comparisons on the basis of mole fractions favor the copolymers of **4** and **5** even more. Thus, the comonomer molar content of the most strongly viscosifying copolymers of **4** and **5** is about 100 times less than that of the corresponding copolymers of **6**. Although less strongly viscosifying, the copolymer of **3** is still significantly more strongly viscosifying than that of laurylacrylate although **3** contains a shorter hydrophobic chain (8 vs 12 carbons).

The Brookfield viscosities of the copolymers of **2** and **6** are roughly comparable but in this case, the hydrophobic group of the comonomer **2** only contains 4 C atoms.

The viscosity-concentration profile of some of the copolymers of **4** is shown in Fig. 7.1 along with that of **6**. Shown is also a terpolymer of acrylamide **4** and AMPS. Figure 7.1 illustrates the rather dramatic differences with the laurylacrylamide copolymers and with the homopolymer.

As shown above, the level at which the fluorocarbon-containing groups appears to be effective in the copolymers is much less than that for the lauryl

Table 7.4. Synthesis and properties of RF-8 (3) containing copolymers[b]

Comonomer g	Comonomer/Content[c] mole%	wt%	Yield g	Conv.[c] %	Viscosity[a] at conc. 0.50%
0.30	2.36	13.0	1.98	86.3	120
0.20	1.57	9.10	2.03	92.5	200
0.134	1.06	6.29	1.99	93.3	600
0.065	0.51	3.14	1.91	92.5	1400
0.034	0.27	1.69	1.92	94.5	650
0.013	0.10	0.66	1.88	93.4	170
0.008	0.06	0.40	1.77	88.1	120
0.004	0.03	0.20	1.89	94.2	100

[a] Brookfield viscosity was measured at shear rate 0.40 s^{-1}.
[b] For preparation, see Experimental Section.
[c] Determined gravimetrically

Table 7.5. Synthesis and properties of Laurylacrylate (6) containing polyacrylamides[a]

Sample	Comonomer mole%	wt%	Conversion[b] %	Brookfield Centipoise (1.0%)	Viscosity[c] (0.50%)
AL-1[d]	0.74	2.44	100.4	84	10
-2[d]	1.64	4.80	99.5	120	30
-3[d]	2.99	9.13	100.4	170	40
-4[d]	4.52	13.20	92.3	44	10
Al-7[e]	0.75	2.40	86.4	400	50
-5[e]	1.50	4.80	86.7	420	100
-6[e]	3.00	9.10	83.4	760	120
-8[e]	4.50	13.00	76.6	250	40
PAM	0	0	99.9	450	50

[a] Polymerization conditions: 2 g AM in 20 ml deionized water.
[b] Based on weight of precipitated polymer.
[c] Shear rate is 0.40^{-1}.
[d] Sodium laurylsulfate (25 wt %).
[e] Emulsifier was $CF_3(CF_2)_6COOK$ (obtained from the 3 M Co) at 12 wt %

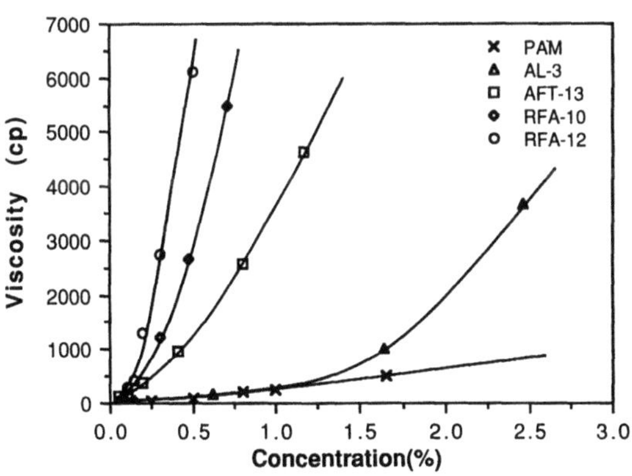

Fig. 7.1. Dependence of Brookfield viscosity on concentration for several acrylamide homo- and copolymers

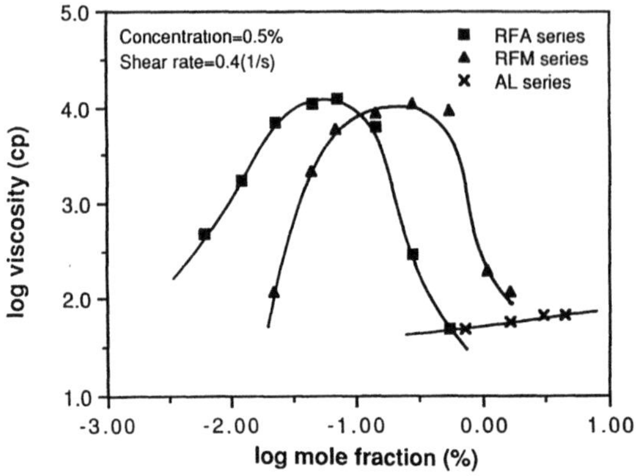

Fig. 7.2. Dependence of Brookfield viscosity on comonomer content of monomer feed for 0.5 weight percent copolymer solutions

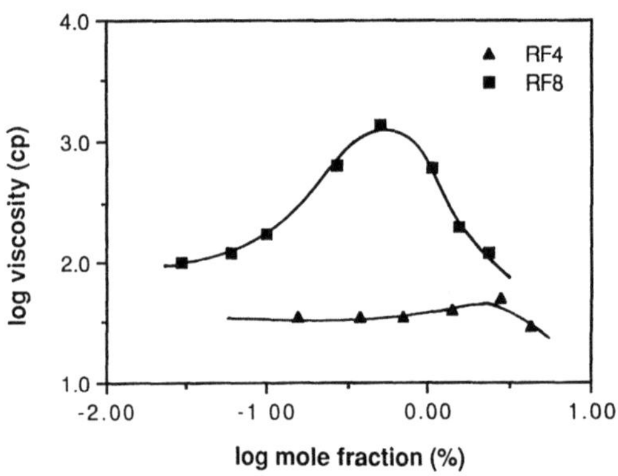

Fig. 7.3. Dependence of Brookfield viscosity on comonomer content of 0.5 weight percent polymer solutions of RF4 and RF8 comonomers

acrylate. For instance, even assuming complete incorporation of the comonomer RFA, appreciable viscosity increases over the PAM homopolymer are observed at comonomer mole fractions as low as 6×10^{-5}. Thus, appreciable intermolecular association probably occurs at comonomer levels corresponding to less than one in 10000 monomer units. This remarkable result may indicate that the number of hydrophobic groups per aggregate is quite low [21]. Because of the extremely low comonomer content, it would appear that the occurrence of comonomer blocks in the copolymer is unlikely, especially for the lower comonomer feed ratios.

The data in the Tables are also plotted in Figs. 7.2 and 7.3 and show the occurrence of maxima of the Brookfield viscosity of 0.5 wt% solutions at constant (0.40 sec^{-1}) shear rate as a function of comonomer content.

The maxima in the Brookfield viscosity versus comonomer content plots for the RFA and RFM-containing copolymers are of special interest. Although it is conceivable that variations in molecular weight may have affected the results to some extent, it seems plausible that the maxima are primarily due to a variation in comonomer content. Viscosity maxima in water-soluble cellulose derivatives

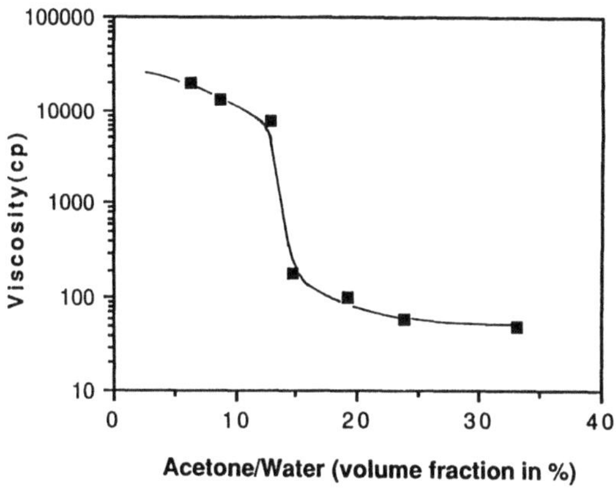

Fig. 7.4. Effect of acetone content of reaction medium on Brookfield viscosity of RFA copolymer

as a function of hydrophobe content have been observed by others [7, 8]. In the present case, the maxima may reflect relatively increased intramolecular association at higher comonomer content. This may reflect a greater degree of cooperative association for intra- as compared to intermolecular association.

In order to overcome the lack of solubility of the hydrophobic comonomers, acetone was used as the cosolvent during polymerization. As shown in Fig. 7.4, the viscosity of the copolymer solutions was adversely effected by the concentration of acetone in the reaction medium presumably because acetone may act as a chain transfer agent.

7.3.2 Polymer Characterization

The reduced viscosities of a number of associating polyacrylamide copolymers were determined as a function of concentration by capillary viscometry at 30 °C and are shown in Table 7.6. In some cases, no reasonable intercepts could be obtained. Although the higher values of $[\eta]$ may indicate much higher molecular weights, we think that they reflect residual intermolecular association persisting at a very low concentration of copolymer. Dynamic light scattering measurements on some of these samples appear to confirm residual association at concentration around 100 ppm [22]. Thus, there is a distinct possibility that residual hydrophobic association of some of RFA and FRM copolymers occurs even at the lowest concentrations. In order to gain insight into this question, one sample, RFA-18, was studied in detail in several media including H_2O, H_2O/NaCl, DMF/H_2O (50/50, V/V) and DMSO/H_2O (50/50, V/V) (Table 7). There is a considerable decrease in apparent intrinsic viscosity of RFA-18 in going from water ($[\eta] = 8.08$ dl/gr) to DMSO/H_2O ($[\eta] = 5.70$ dl/gr) or to DMF/H_2O ($[\eta] = 2.05$ dl/gr). The intrinsic viscosity of the homopolymer on the other hand, is changed very little in going from water to DMF/water. The very large drop in intrinsic viscosity of RFA-18 in going from water to DMF/water at very low concentrations is thus most likely caused by

Table 7.6. Synthesis and properties of associating polyacrylamide copolymers

	Mole%[a]	Brookfield[b]	$[\eta]^d$ Viscosity (cp)	K_H^e dl/gr
RFA-22	0.006	480	8.66	—
-18	0.023	7000	8.08	1.03
-17	0.070	10800[d]	10.13	2.12
-10	0.140	6200	9.71	0.91
-9	0.280	300	6.48	0.69
-13[f]	0.140	2500	4.81	3.47
RFM-19	0.280	10800[c]	11.07	2.34
-24	0.022	120		
-32	1.11	200		
AL-3	3.00	40[d]	4.37	1.08
PAM	—	10	5.22	0.81

[a] Mole % comonomer.
[b] At 0.50 wt % copolymer in water at 25 °C. Shear rate is $0.40\ s^{-1}$.
[c] Maximum values in a series prepared with varying comonomer fractions, see Table 7.2.
[d] Apparent intrinsic viscosity in 2% NaCl at 30 °C.
[e] Huggins constant.
[f] AMPS Terpolymer (Table 7.1)

Table 7.7. Intrinsic viscosity and Huggins constant of RFA-18 and PAM in aqueous solution

	Solvent/H_2O	Intrinsic viscosity[b] dl/g	Huggins constant (K_H)
RFA-18	DMSO/H_2O[a]	5.70	1.39
-18	DMF/H_2O[a]	2.05	6.47
-18	H_2O	8.08	1.03
-18	2% NaCl	8.28	1.23
PAM	DMF/H_2O[a]	5.22	0.81
PAM	H_2O	5.07	0.22

[a] Fifty percent by volume.
[b] Measured at 30.0 °C

dissociation of the aggregated macromolecules, presumably into unassociated polymers. The high value of the Huggins constant (K_H) in the DMF/H_2O system nonetheless, suggests substantial aggregation at higher concentration.

7.3.3 *Dependence of Viscosity on Shear, on Temperature and on Addition of Salts and Other Additives*

7.3.3.1 Evidence for Hydrophobic Association

The pronounced intermolecular hydrophobic interaction occurring between the fluorocarbon chains of RFA and RFM copolymers is supported by the

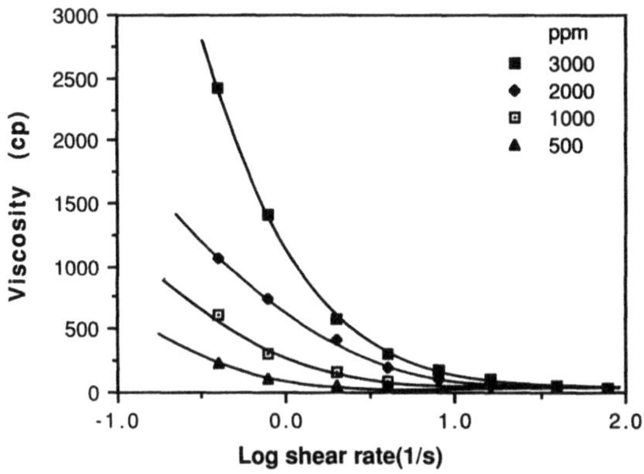

Fig. 7.5. Dependence of Brookfield viscosity on shear rate of RFA-12 copolymer

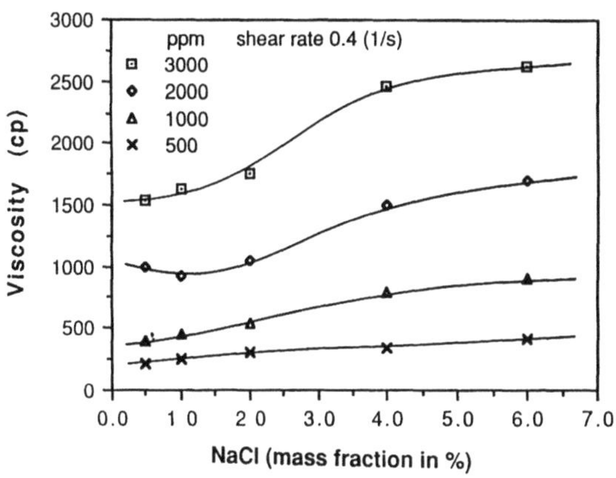

Fig. 7.6. Effect of NaCl concentration on Brookfield viscosity of copolymer RFA-18 at various concentrations

viscosity-concentration profiles, the relationships between comonomer structure and copolymer viscosity and by the effects of comonomer content.

The solutions of all of these copolymers are also quite pseudoplastic as shown in Fig. 7.5. The very sharp decrease in viscosity with shear rate is again consistent with the existence of network-like loose structures held together by the relatively weak hydrophobic interactions that would be expected to be disrupted readily upon application of shear. Of special interest is the fact that upon removal of shear the viscosity of the polymer solutions completely recover their original low shear value. This result again is consistent with reversible cleavage of the hydrophobic "bonds" holding the network together. We did not observe indications of dilatant behavior in any of the systems investigated.

Additional evidence for intermolecular hydrophobic association is provided by the effects on the Brookfield viscosity by addition of NaCl.Thus, the viscosity of the copolymer solution increases with increasing NaCl concentration, especially at lower shear rate ($0.4\ s^{-1}$) (Fig. 7.6). At polymer concentrations above 2000 ppm, the viscosity increases are especially pronounced between 1.0 and 5.0

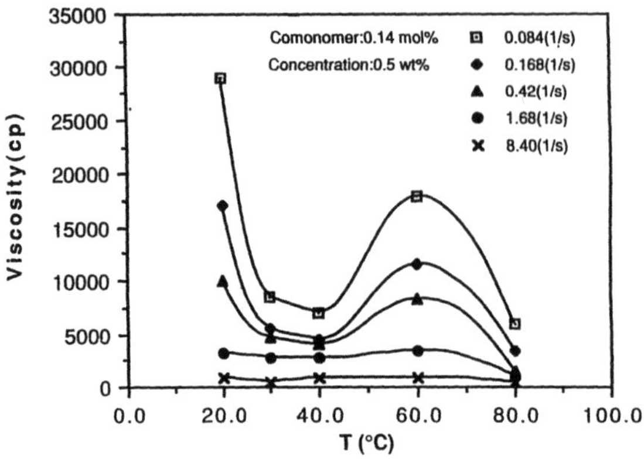

Fig. 7.7. Effect of temperature on Brookfield viscosity of typical RFM copolymer containing 14 mole% of **5**

weight % NaCl. These increases appear to correlate well with the known effects of NaCl on hydrophobic association [23].

The effect of temperature on the Brookfield viscosity is shown in Fig. 7.7. Especially at low shear, the results are quite dramatic. From 20 to 40 °C, there is a viscosity decrease followed by an increase and a maximum at about 60 °C. Other samples show a similar behavior with local viscosity maxima occurring between 60 and 80 °C. The viscosity increases observed upon heating are again consistent with an entropy driven increase in hydrophobic bonding [1, 2]. At very high temperatures (> 80°) the viscosity decreases with increasing temperature again dominate presumably as a result of changes in the hydration of the hydrophobic groups and/or in the structure of water itself.

For the PAM homopolymers, the viscosity versus temperature behavior is completely different and shows a monotonous decline in viscosity with temperature. This strongly indicates that the viscosity-temperature profile of the RFA copolymer is related to hydrophobic association.

The pseudoplasticity, the salt tolerance and the relatively favorable viscosity-temperature profiles may make these copolymers of potential interest in oil recovery or drilling applications.

7.3.3.2 Effects of Additives

Consistent with the postulated hydrophobic association in aqueous media, addition of organic solvents such as DMSO, DMF and acetone at constant polymer concentration causes a drop in Brookfield viscosity. This behavior is illustrated in Fig. 7.8 for the addition of acetone. There is no effect until about 5% by volume and then there is a dramatic 3–30 fold decrease at an acetone concentration of 15%. Similar effects are observed for the addition of urea and ionic and nonionic surfactants (Fig. 7.9). In the latter case however, the viscosity vs additive profile is more complicated. There is a sharp initial decrease followed by an increase. The nature of this increase at higher surfactant concentration is unclear at present. The sharp decrease is consistent with association of the

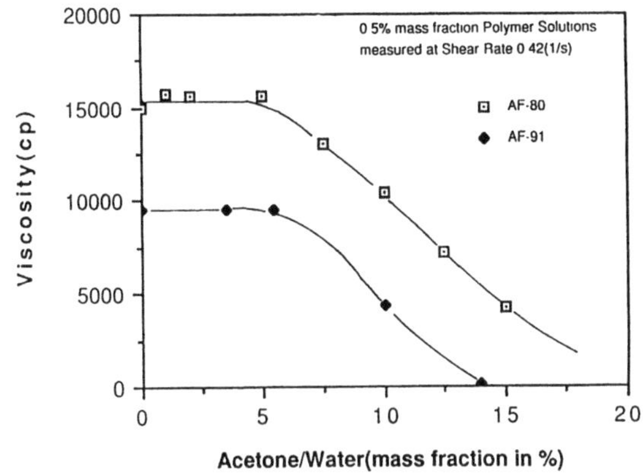

Fig. 7.8. Effect of addition of acetone on Brookfield viscosity of typical RFA copolymer

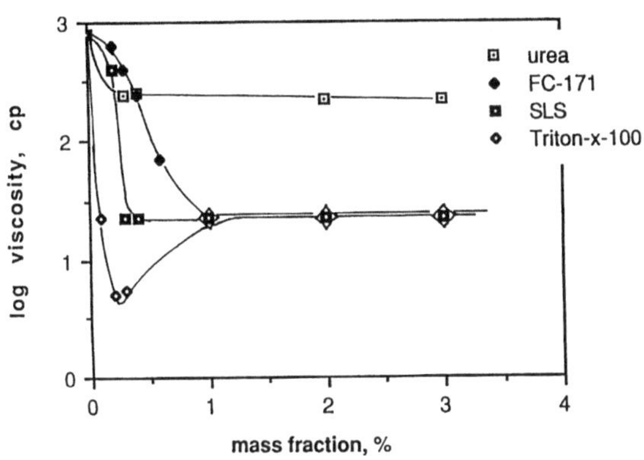

Fig. 7.9. Effect of addition of urea and surfactants on Brookfield viscosity of RFA-17 copolymer at 25 wt% and 40 sec^{-1}.

surfactant with the hydrophobic groups of the copolymer thus competing with intermolecular polymer-polymer association. However, the systems are sufficiently complex to warrant more extensive investigations.

7.3.3.3 Fluorocarbon-Containing HEC Derivatives

The synthesis of the 1,1-dihydroperfluoroalkyl derivatives of HEC is shown in Eqs. (7.1), (7.2) and (7.3) in Schemes 7.1 and 7.2.

The 1,1-dihydroperfluorobutylhydroxyethyl cellulose **9** is prepared by reaction of a HEC-NAOH slurry with 1,1-dihydroperfluorobutyltosylate (PFBOTs) **7** at 50 °C in various solvents including dioxane (DOX), isopropylalcohol (IPA) and triethylamine (TEA). The incorporation of perfluorocarbon modifier ([modifier]/[AGU]) into the substrate polymer was evaluated from fluorine elemental analysis in the case of **9**-1 and **9**-3 and was found to be 14.2 and 3.6 mole%, respectively (Table 7.8). However, with the exception of very few samples (**9**-6 and **9**-9) the products were generally insoluble

Table 7.8. Synthesis of HEC derivatives **9**[a] and **10**[b] in various solvents at 50 °C

Sample No.	Modifer mole %[h] added	Solvent	Product solubility (H_2O)	Modifer mole %[h] ncorporated[c]	Brookfield viscosity[g] (cp)
9-1[a]	100	IPA[d]	Insol.	14.2	
9-2[a]	36	DOX[e]	Insol.		
9-3[a]	25	IPA[d]	Insol.	3.6	
9-4[a]	18	TEA[f]	Insol./Sol.		
9-5[a]	10	DOX[e]	Insol./Sol.		
9-6[a]	2.5	TEA[f]	Sol.		13.4
9-7[a]	2.5	IPA[d]	Cloudy		
9-8[a]	1.0	IPA[d]	Cloudy		
9-9[a]	0.25	IPA[d]	Sol.		< 5
10-1[b]	0.50	IPA[d]	Cloudy		

[a] Modified by PFBOTs (**7**).
[b] Modified by PFOOTs (**8**).
[c] Calculated from fluorine elemental analysis.
[d] Isopropanol.
[e] Dioxane.
[f] Triethylamine.
[g] At 25 °C, shear rate is 0.40 s^{-1}.
[h] With respect to AGU

especially when the mole percent of the PFBOTs with respect to anhydroglucose units (AGU) was large (> 25%). Below 25 mole percent of modifier more or less soluble products were generally obtained. In the case of polymer **10**-1 modified by **8** (0.5 Mole%) mostly insoluble product was obtained. Therefore, in the case of both modifiers **7** and **8**, only low viscosity filtered solutions were observed. Similar results were obtained in the case of modifier **11** prepared by reaction of the 1,1-dihydroperfluorobutanol with epichlorohydrin under basic conditions. Only partially soluble products were obtained and the viscosities of the filtered solutions were generally low.

However, much better results were obtained in the case of **12** as modifier. The ^1H and ^{19}F NMR spectra of modifier **12** are shown in Figs. 7.10 and 7.11. The peaks at 2.63, 2.82 and 3.18 ppm in ^1H NMR are characteristic epoxide peaks. The two quartets at 3.53 and 3.98 ppm represent the two nonequivalent protons, H_d and H_e. The multiple peaks at 4.04 ppm are due to the protons next to the fluorocarbon chain. In the ^{13}C NMR spectrum, the carbon alpha to the fluorocarbon chain is split by the adjacent fluorines into the expected triplet centered at 68.6 ppm. The carbons bearing fluorine instead of hydrogen atoms show very small multiple peaks between 105 to 124 ppm. This is due to the strong shielding caused by the fluorine atoms. Modifier **12** also shows a clean ^{19}F NMR spectrum. The sharp downfield peak (− 81 ppm) is due to the $-CF_3$ group and it is this peak that was chosen in the determination of modifier content in modified polymer. Elemental analysis also confirms the proposed structure of PFOGE. Calc. for $C_{11}F_{19}H_7O_2$: C, 28.96; F, 62.47; H, 1.55%. Found: C, 28.74; F, 62.35; H, 1.56%.

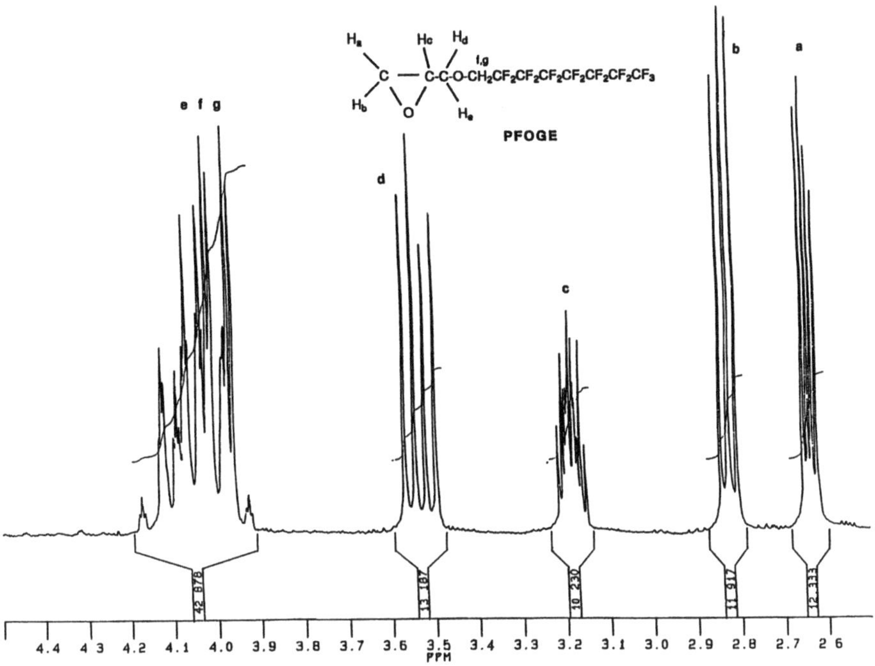

Fig. 7.10. Proton NMR (250 MHz) of modifier PEOGE (**12**)

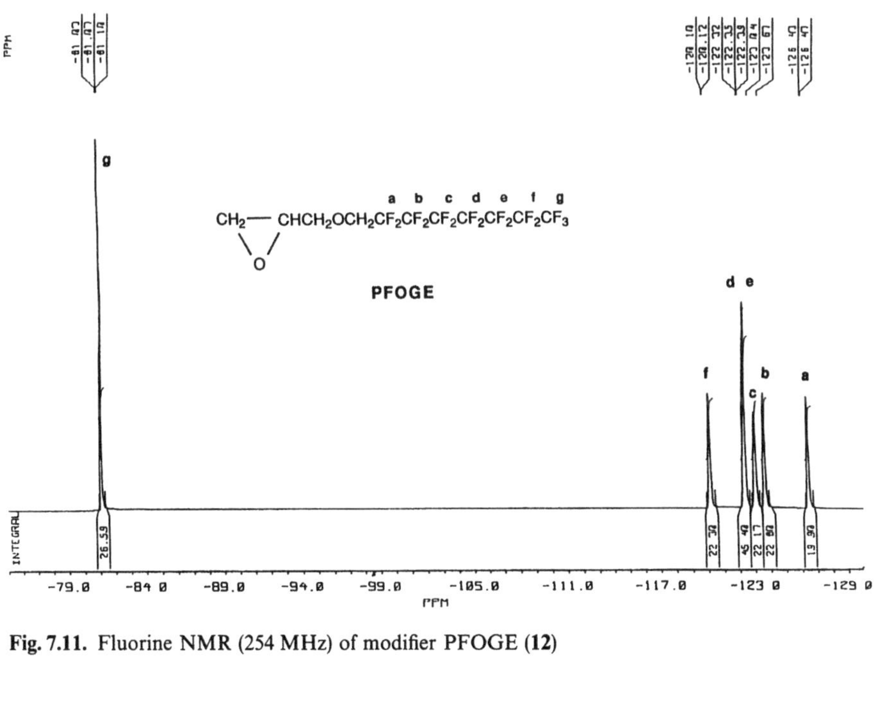

Fig. 7.11. Fluorine NMR (254 MHz) of modifier PFOGE (**12**)

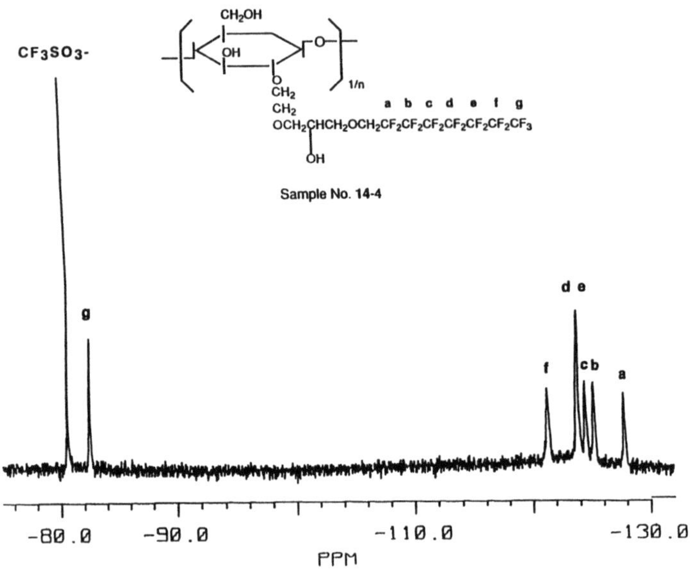

Fig. 7.12. Fluorine NMR (254 MHz) of modified (by PFOGE) HEC (sample no. **14**-4) with CF_3SO_3Na as internal standard

The fluorine NMR spectra of the modified polymers (**14**-4 and **14**-6) are shown in Figs. 7.12 and 7.13. Although the peaks and the chemical shifts of the bound modifier are quite similar to those of the unbound one (**12**), the set of three peaks at 122–124 ppm is shifted slightly upfield with respect to the corresponding peaks of **12** and the peak width is also slightly broadened. The molar fraction of **12** with respect to AGU was kept between 0.12 and 5.0 mole percent. Evidently, the modification reaction is inefficient, since the actual molar content of perfluorocarbon groups with respect to AGU was found to be between 0.04 and 2.16 mol% (Table 7.9). Thus, the grafting efficiency was found to be between 23% and 44%. The results of the Brookfield viscometric measurements at 0.40 sec^{-1} at 25 °C, also shown in Table 7.9, show that there is a substantial viscosity enhancement at a modifier level of between 2 and 4 mole percent. thus, at 3 mole percent of modifier feed, the viscosity, at 2.0 wt% polymer, is increased more than 200 percent compared to the unmodified HEC. Significantly, as the PFOGE (**12**) modifier is increased to 4 and 5 mole percent, the viscosity decreases. Thus, at 5.0 mole percent of added PFOGE, the Brookfield viscosity of the product is decreased to about the value of unmodified HEC. Furthermore, at 10 mole percent of added PFOGE, the polymer becomes partially insoluble and the viscosity plunges to a very low value. A plot of the Brookfield viscosities of 2.0 wt% solutions of samples **14**-1 to **14**-7 against incorporated PFOGE shows a clear maximum at about 3 feed mole% (Fig. 7.14). Similar maxima have been observed in HEC derivatives prepared with hydrocarbon epoxides. Although the increases in viscosity are relatively modest

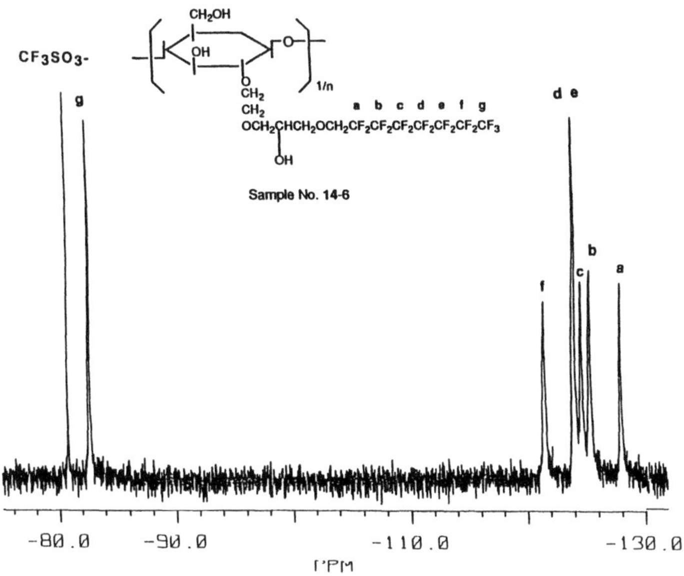

Fig. 7.13. Fluorine NMR (254 MHz) of modified (by PFOGE) HEC (sample no. **14**-6) with CF_3SO_3Na as internal standard

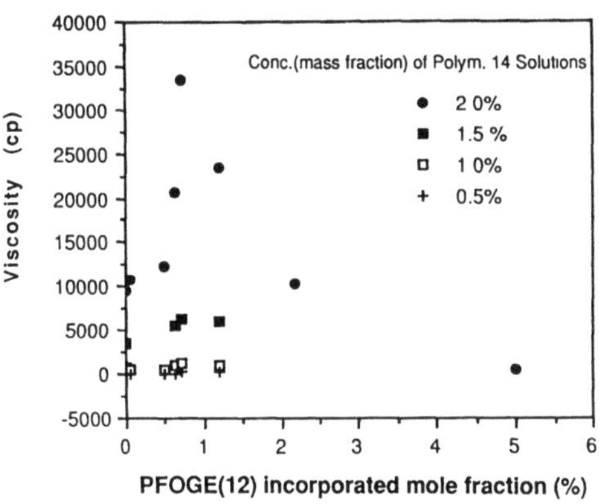

Fig. 7.14. Brookfield viscosities of modified (by various amounts of PFOGE) HEC's (**14**) at different concentrations

compared with that of the polyacrylamide copolymers previously studied, it seems clear that the increases are due to similar factors invoked for the polyacrylamide copolymers. Starting with very low mole fractions of perfluorocarbon groups, hydrophobic association is largely intermolecular to form network-like structures and this increases as the mole fraction of perfluorocarbon groups increases, resulting in an increase of the apparent hydrodynamic size and thus of the viscosity of the polymer in aqueous solution. However, at higher perfluorocarbon content, intramolecular hydrophobic

Table 7.9. Synthesis[a] and Brookfield viscosities (cp)[b] of HEC derivatives **13**[c], **14**[d] and **16**[e]

Sample No.	Modifier mole %[h] added	Polymer Concentrations (Wt%) 0.5	1.0	1.5	2.0	Incorporated[f] mole %[h]	wt. %[i]	Grafting[g] efficiency (%)
HEC	0	55	695	3600	9600	0	0	
14-1[d]	0.12	50	570	—	10800	0.042	0.058	35
14-2[d]	1.13	75	565	—	12400	0.490	0.697	43
14-3[d]	2.0	145	1010	5520	20560	0.628	0.857	31
14-4[d]	3.0	175	1200	6240	33400	0.695	0.946	23
14-5[d]	4.0	160	920	6080	23320	1.195	1.611	30
14-6[d]	5.0	—	—	—	10400	2.162	2.856	43
14-7[d]	10.0	—	—	—	650[j]			
13-1[c]	1.0	—	—	—	70[j]			
13-2[c]	3.0	—	—	—	80[j]			
16-1[e]	3.0	—	—	—	5520			

[a] At 50 °C, isopropanol as solvent.
[b] At 25 °C, shear rate is 0.40 s^{-1}.
[c] Modified by PEBGE (**11**).
[d] Modified by PFOGE (**12**).
[e] Modified by OGE (**15**).
[f] Calculated from ^{19}F NMR analysis.
[g] Obtained from [Incorporated modifier]/[modifier].
[h] With respect to AGU.
[j] Wt% of modifier in polymer.
[i] Solutions only partially soluble: viscosities are of filtered solutions

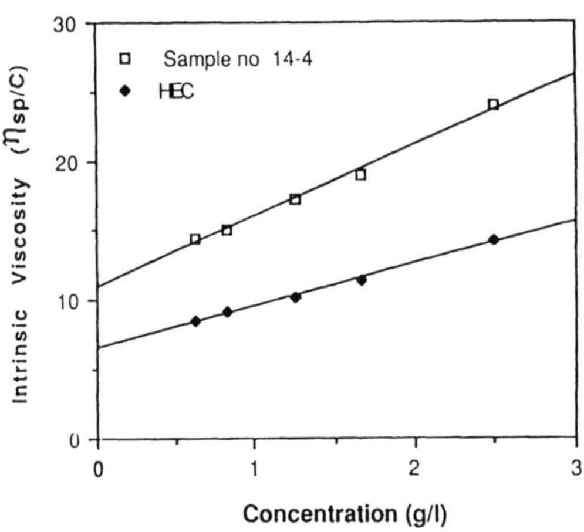

Fig. 7.15. Intrinsic viscosities of unmodified HEC (◆) and modified HEC (sample No. **14-4**, □)

association of the perfluorocarbons becomes more prevalent and this of course tends to compete with intermolecular association. Thus, at higher perfluorocarbon content, intramolecular association prevails causing a "collapse" of the polymer coils resulting in lower viscosity. Similar effects have been observed by Landoll [7, 20] for analogous hydrocarbon derivatives of HEC and a similar explanation was used to describe the results.

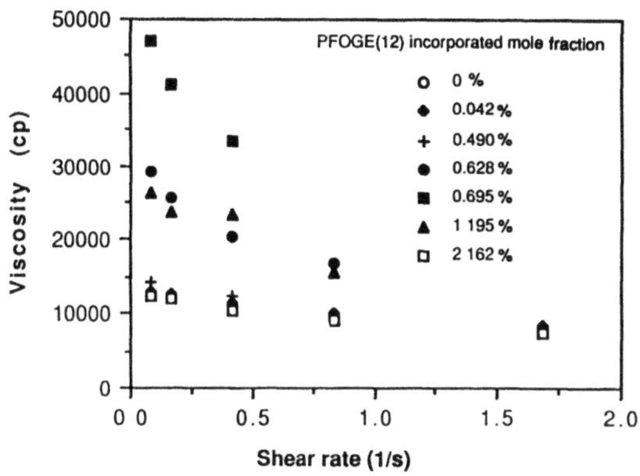

Fig. 7.16. Brookfield viscosities of HEC's modified by various amounts of PFOGE at different shear rates

Figure 7.15 shows that the apparent intrinsic viscosity of modified polymer (**14**-4) is higher (10.9 dl/g) than that of unmodified HEC (6.6 dl/g). This fact is consistent with the existence of hydrophobic association formed by the bound fluorocarbon chains in this system. Apparently, intermolecular association of sample **14-4** still prevails at the lowest concentration (~ 0.06 gr/dl).

The corresponding hydrocarbon equivalent of **12**, OGE (**15**) was also used as modifier at 3.0 mole percent modifier feed with respect to AGU's. As was the case for the polyacrylamide copolymers, the viscosity was found to be significantly (about 6 times) lower than for **12** as modifier at the same modifier level (Table 7.9).

As in the case of the polyacrylamide copolymers, the aqueous solutions of **14** were found to be highly pseudoplastic (see Fig. 7.16). The explanation for the sharp decreases of viscosity with increasing shear is the shear induced disruption of the weakly associating hydrophobic groups.

7.4 Conclusions

The above results, at least qualitatively, confirm similar behavior for hydrophobically modified HEC and polyacrylamides. As in the case of the polyacrylamide copolymers, the perfluorocarbon containing HEC polymers are generally more strongly viscosifying compared to the hydrocarbon analogs. In the case of HEC derivative **14**, there is clearly a viscosity maximum with regard to perfluorocarbon content. Similar observations were made by Landoll for alkyl derivatives of HEC [4]. As in the case of the polyacrylamide copolymers, the solutions of perfluorocarbon modified HEC were found to be highly pseudoplastic.

Although the increases in viscosity for polymer **14** are quite modest compared to the polyacrylamide copolymers of **4**, they are comparable to that observed for the polyacrylamide copolymers of **2** and **3**. The presence of the sulfonamide group in the case of comonomers **4** and **5** appears to be unusually effective in enhancing perfluorocarbon association. Studies aimed at exploring these effects are in progress.

Acknowledgements. This work was supported by the U.S. Department of Energy, Division of Materials Sciences, Office of Basic Energy Sciences (Grant No. DE-FG03-89ER45407). We appreciate a gift of the RFA and RFM comonomers and fluorine-containing surfactants from the 3M Company. We wish to thank Mrs. A-H. Da, Ms. M. Yassini and Mr. X. Xie for their help in carrying out some of the measurements.

7.5 References

1. Tanford C (1980) The hydrophobic effect, 2nd ed. Wiley, New York
2. Ben-Naim A (1980) Hydrophobic interactions. Plenum Press, New York
3. Myers D (1988) Surfactant science and technology. VCH Publishers, New York, Chapter 3
4. Evani S, Rose GD (1987) Polym Mat Eng, Preprints, 57:477
5. Evani S (1984) U.S. Patent No. 4432881
6. Constien VG, King MT (1985) U.S. Patent No. 4541935
7. Landoll LM (1982) J Polym Sci, Chem. 20:443; Sau AC (1987) Polym Mat Sci and Eng, Preprints, 57:497–501
8. For hydrocarbon containing polyacrylamides, see: Schulz DN, Kaladas JJ, Maurer JJ, Bock J, Pace SJ, Schulz WW (1987) Polym 28:2110; McCormick CL, Hoyle CE, Clark D (1987) Proceedings Polym Mat Sci & Eng. Vol. 57:643; Middleton JC, Cummins D, McCormick CL (1989) Polym Preprts. 30(2):348; Bock J, Pace SJ, Valint PL (1989) Polym Preprts (30)2:386; Flynn CE, Goodwin JW (1989) Proceedings Polym Mat Sci & Eng. 61:522; Turner SR, Schulz DN, Siano DB, Bock J (1986) Proceedings Mat Sci and Eng Vol 55(2):355; Turner SR, Siano DB, Bock J, U.S. Patent No. 4520182; Turner SR, Schulz DN, Siano DB, Bock J (1986) Proceedings Mat Sci and Eng 55(2):355; Valint Jr PL, Bock J, Ogletree J, Lushma S, Pace SJ (1990) Polym Preprts 31(2):67
9. a. Glass JE (1989) "Polymers in Aqueous Media" ACS Advances in Chemistry, Series 223, Washington, DC; b. Branham KD, Middleton JC, McCormick CL (1991) Polym Preprts. 32(1):106; c. Thibeault JC, Sperry PR, Schaller EJ (1984) (ed) In: Glass JE Water-Soluble Polymers. Adv Chem Ser. 213 Chapter 20; d. Ezzell A, McCormick CL (1987) In: McCormick, Butler (eds) Water-Soluble Polymers ACS Symp Ser 467 Chapter 8
10. Zhang YX, Da AH, Hogen-Esch TE, Butler GB (1989) Polym Preprts 30(2):338
11. Hogen-Esch TE, Yassini M, Zhang YX, Huang F, Amis EJ, Seery T (1990) Polym Preprts. 31(2):460
12. Zhang YX, Da AH, Butler GB, Hogen-Esch TE (1990) J Polym Sci Part C Polym Ltrs Vol 28:213–218
13. Zhang YX, Da AH, Butler GB, Hogen-Esch TE (1992) J Polym Sci Part A (Chem Ed 30:1383)
14. Schonfeld VP, Selibt H (1976) Z Chem 16(12):497
15. Mukerjee P, Handa T (1981) J Phys Chem 85:2298
16. "Electrochemical fluorine analysis methods appropriate for the determination of the comonomer content of the copolymers (Table 1 and 2) are currently being developed", Zhang YX, Hogen-Esch TE, To be published
17. Tiers VD, Brown HA, Reid TS (1953) J Am Chem Sci 75:5974
18. Bondakian MM, Raes MC, Urs SV (Feb. 1987) Development of solvent-resistant sealants. Report #NOW 66-03-3d, Naval Acr Systems Command
19. Brodo GL, Stanley JP, Partain EM, U.S. Patent No. 4579942
20. Landoll LM, U.S. Patent No. 4228277
21. Jiang XK (1988) Acc Chem Res 362
22. Seery T, Amis EJ, To be published
23. Saeto S (1969) J Polym Sci Chem Ed 7:1789

Part II

Polyelectrolyte Complexes

8 Static Light Scattering of Polyelectrolyte Complex Solutions

H. Dautzenberg, J. Koetz, K.-J. Linow, B. Philipp, and G. Rother

The study of highly diluted polyelectrolyte complex (PEC) solutions by physico-chemical methods allows the assessment of the influence of the molecular characteristics of the polyelectrolyte components and of the external parameters of the systems investigated. Especially static light scattering (SLS) has been proved to provide instructive quantitative data about the course of PEC formation and the structural parameters of the generated particles. After a brief survey of the characterization of PEC solutions by SLS an improved algorithm of light scattering curve analysis for higher aggregated structures is described. Its efficiency will be demonstrated by the results of SLS studies on the complex formation between anionically and cationically modified poly(acrylamide) of high molecular weight in comparison with the information obtained by other methods.

8.1 Introduction

Polyanion-polycation-complexes are known for a long time on an empirical basis from the mutual precipitation of proteins. Already at the end of the previous century Kossel [1] recognized the electrostatic interaction between the oppositely charged polyions as the driving force for precipitation. Willstaetter [2] also introduced the term symplexes for polyelectrolyte complexes.

In contrast to the simple procedure of PEC formation, a deeper understanding of the processes, governing the formation of super molecular structures in these systems, is still a challenge to the scientists at present, due to the great variety of different species of polyelectrolytes, the influence of a lot of external parameters and the interfering interactions of non-Coulombic forces.

Starting with the pioneering work of Michaels [3] in preparing well-defined polysalts with a 1:1 stoichiometry a continuously increasing number of papers were published, which were reviewed by several authors in the last years [4–7]. Successful contributions to elucidate the mechanism of PEC formation have been achieved by the work of Tsuchida [8, 9], Kabanov and Zezin [10–12] about sequential water soluble PECs in diluted solutions of polyelectrolyte components with significant different molecular weights in nonstoichiometric systems. In contrast, our work was focused on the investigation of PEC formation between polyanions and polycations both of high molecular weight. In highly diluted solutions (polymer concentration lower than 10^{-3} g/ml) the PEC formation leads to aggregates with dimensions on a colloidal level, building optically homogeneous and stable dispersions, also called quasi-soluble

Macromolecular Complexes
in Chemistry and Biology
Eds. Dubin/Bock/Davis/Schulz/Thies

PECs by us. This offers the possibility to study the course of PEC formation as well as the structure of PEC particles by methods, which are normally used for the characterization of polymer solutions. In this paper we will especially discuss the influence of the charge density on the formation and stability of complexes between such high molecular weight polyelectrolytes, stressing methodical aspects and results of SLS.

8.2 Survey of Static Light Scattering Studies on PEC Solutions

Due to the high sensitivity of the scattering intensity to the presence of supermolecular structures in polymer solutions, static light scattering is a very helpful tool to study the course of PEC formation and to characterize the structure of the generated particles. Nevertheless, in comparison to such methods as potentiometry, conductometry, turbidity and viscometry, SLS studies of PEC formation are rather scarce.

The main field of application of SLS on PEC solutions was the characterization of nonstoichiometric water soluble complexes, consisting of one or a few high molecular polyelectrolyte molecules, whose charged groups are sequentially blocked by polycounterions of significant lower degree of polymerization as minor component. These PECs may be considered as a new class of block copolymers. Most of the SLS studies on such systems were done by Kabanov, Zezin, Kharenko and Izumrudov (for a review see [11]). However, SLS was employed in its traditional manner of polymer characterization, i.e. the concentration and angular dependence of the scattering intensity were measured and analyzed by Zimm plots, yielding the particle mass M_w, the radius of gyration R_g and the second virial coefficient A_2. This procedure was successful due to the low level of aggregation of these PECs and their stability. The SLS studies provided a deciding contribution to recognize the molecular structure of the nonstoichiometric water soluble complexes and to investigate the conditions of their existence in dependence on their composition as well as the ionic strength and pH-value of the medium for a variety of combinations between synthetic polyelectrolytes with weak and strong ionic groups [13–18]. But also the complex formation between biopolymers and synthetic polyelectrolytes (bovine serum albumin/poly(4-vinyl-*N*-ethyl pyridinium bromide)) was investigated [19]. In [20] it was shown by SLS that the substitution of salt bonds between poly(acrylic acid) and poly(ethyleneimine) by amidic ones did not cause significant changes of M_w, i.e. only an intracomplex reaction took place.

Dubin et al. studied the formation of polyelectrolyte-micelle complexes between poly(dimethyldiallylammonium chloride) and anionic/nonionic mixed micelles by static and dynamic light scattering [21–23]. Dynamic light scattering yields the hydrodynamic radius of the particles investigated (very useful information), especially in combination with SLS, allowing an assessment of the particle architecture.

Djadoun [24] investigated the association between cationically modified poly(styrene) and anionically modified poly(methylmethacrylate) of low charge density in ethyl benzoate (isorefractive for PMMA) by low angle light scattering. The aggregation behavior of poly(methacrylic acid)/poly(ethylene oxide) complexes after the formation of the polymer complexes in dependence on polymer concentration, pH-value and temperature was studied by Tsuchida et al. [25, 26], measuring the scattering intensity at scattering angle 30°. After reaching an equilibrium state the angular dependence was also registered and analyzed by the dissymmetry factor I_{45}/I_{135}, assuming the model of monodisperse spheres. The spherical shape could be proved by scanning electron micrographs. The same technique of scattering curve analysis was used by Hara and Nakajima [27], who studied the complex formation between Heparin and partially aminoacetalized poly(vinyl alcohol) in dependence on the composition and the ionic strength of the medium also for highly aggregated complex particles. They compared the experimental scattering curves with theoretical ones for monodisperse spheres, coils and rods, concluding the best fit for spheres. Moreover, they recognized that neither the size nor the molecular weight of the complex particles can be obtained from the Zimm plot (measured angular range $30° \leq \Theta \leq 150°$). However, the dissymmetry method yields only rough information about the particle size, because the entire information of the scattering curves is not used.

PEC formation between two components of high molecular weight leads generally to higher degrees of aggregation and in appropriately diluted systems to particle sizes in the range of a few hundred nm, yielding a strong angular dependence. Therefore, we derived some years ago an improved algorithm of scattering curve analysis [28, 29], which allows one to assess the structure type and to determine the polydispersity of the scattering particle system, the particle mass, and the size and the internal polymer packing density. This algorithm will be briefly described in the following section.

Using this kind of light scattering data analysis, we studied the general features of PEC formation of synthetic polyelectrolytes in dependence on polymer concentration, degree of conversion (titration), ionic strength of the medium, charge density of the polyelectrolyte components and the strength of the ionic groups [7, 30–35]. The main results may be summarized as follows:

- Increasing dilution of the polymer concentration leads to a decrease of the level of aggregation.
- The growth of PEC particles started immediately after the dosage of an oppositely charged PEC solution to the given one. In most systems investigated, the structural parameters do not change significantly with rising degree of conversion, i.e. the generation of new particles is the dominating process. At a certain ratio of the polyelectrolyte components flocculation occurs.
- The response of PEC particles, formed in pure water, to the addition of low molecular salts may be very different. Aggregation, flocculation, swelling and

dissolution were observed, caused by the screening and salt-out effect of the electrolyte.
- The charge density and the going together of the charge distances of the components play an important role for the degree of swelling of PEC particles and their salt stability.
- Between highly charged polyelectrolytes with strong ionic groups a 1:1 stoichiometry of the PEC is favored. For polyelectrolytes with weak ionic groups the 1:1 stoichiometry is more an exception than the rule.
- All processes of PEC formation, structural changes and disaggregation are relatively slow, i.e. they take place in a time range from a few seconds to hours.

To demonstrate the high efficiency of SLS studies using the improved scattering curve analysis, the results obtained on PEC formation between anionically and cationically modified poly(acrylamide) are given in this paper in more detail, also considering the information from other methods.

8.3 Interpretation of Light Scattering Experiments

For the quantitative analysis of the scattering data the concentration and the refractive index increment of the scattering entities must be known, which can be calculated after the determination of the PEC stoichiometry from the corresponding values of the components.

Light scattering of highly diluted particle systems provides in its traditional manner of interpretation average data about the particle mass and the radius of gyration, requiring an extrapolation to the scattering angle zero.

Characterizing the complex particle systems we are confronted with two problems:

- Polydisperse systems of large particles yield scattering curves with strong curvature in the small angle range. Therefore, the extrapolation to zero scattering angle may be very erroneous.
- The weight average particle mass M_w and the z-average of the square of the radius of gyration represent different average values of the particle size; thus, not allowing to distinguish between effects of polydispersity and swelling.

In principle the scattering curves contain information about the structure type and the polydispersity of the scattering system, which may be obtained by a detailed analysis of the shape of the experimental scattering curves. However, the shape of the scattering curves is determined by four different parameters: particle size, particle shape, internal structure of the particles and the polydispersity of the scattering system. This impedes an unambiguous interpretation of the light scattering data. In many cases the problem can be solved by a comparison of the experimental scattering curves with theoretically calculated ones for different structure models, using a fitting procedure. This kind of scattering curve analysis offers substantial advantages:

- The structure type and the polydispersity of the system may be assessed.
- The extrapolation to zero angle can be carried out in a reliable way, using the known shape of the appropriate theoretical curve.

Recently a simple fitting algorithm was derived by us [28, 29], based on following ideas:

- A special logarithmic distribution function separates the influence of the size (e.g. radius a_m for spheres, radius of gyration s_m for coils) and polydispersity parameter σ on the scattering curve.
- In the Rayleigh–Debye-approximation the argument of the scattering function is the product of the size parameter and the length of the scattering vector h ($h = 4\pi\,\lambda^{-1}\,\sin\,\theta/2$). Therefore the size dependence can be represented by one scattering curve.
- A double logarithmic plot reduces the multiplicative combination of the argument to an additive one so that the fitting procedure can be realized by axis parallel shifts.

The results of model calculations can be represented in a scaled manner for all kind of plots, which are proposed in the literature and are given by the expression [36]

$$R(h)\,h^{\beta}\,K^{-1}\,c^{-1}\,,$$

where $R(h)$ is the Rayleigh ratio of the scattering intensity, h is the length of the scattering vector, K is a constant, which contains the optical parameters of the system and c is the mass concentration of the scattering entities. $\beta = 0$ corresponds to a Debye plot, $\beta = 1$ to a Holtzer plot and $\beta = 2$ to a Kratky plot.

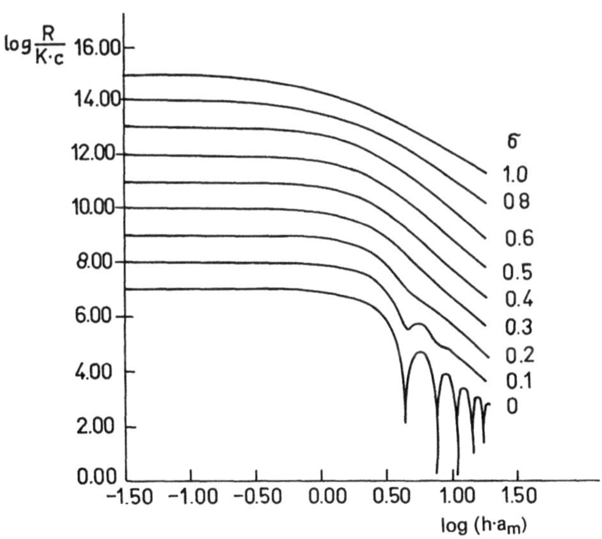

Fig. 8.1. Master scattering curves of polydisperse systems of spheres (Debye plot). Each curve is shifted by 1 unit upwards from the previous one with increasing values of σ.

For a given structure type we obtain one set of master curves in dependence on the polydispersity. Model calculations were carried out for polydisperse systems of spheres and Gaussian coils, which cover the scattering behavior, observed on PEC particle systems. Figure 1 gives the master representation of the Debye plot for polydisperse spheres, showing significant deviations for different polydispersities.

The assignment of an experimental scattering curve to a section of an appropriate theoretical one allows the assessment of the structure type and the polydispersity of the scattering system. From the position (X_E, Y_E) of the experimental curve on the theoretical curve section (X_M, Y_M) the weight average of the particle mass M_w and the size parameter can be calculated. For spheres from these parameters the average value of the polymer packing density ρ may be evaluated too. This quantity is the reciprocal of the degree of swelling and has proved to be an instructive structural parameter.

8.4 Experimental

8.4.1 Materials

For systematic investigations anionically (AX) and cationically (CX) modified poly(acrylamide), containing varying amounts of acrylic acid or ether units

$$\sim CH_2\text{–}\underset{\displaystyle COO^-}{\underset{|}{CH}} \sim \quad \text{or} \quad \sim CH_2\text{–}\underset{\displaystyle O{=}C\text{–}ORNR_3^+}{\underset{|}{CH}} \sim$$

in the polymer chains were employed. X indicates the average number of acrylamide units between two charged groups. Characteristics of samples investigated are given in Table 8.1. The molecular mass of the samples was estimated by viscosimetric measurements, using the [η]-M-relations of (acrylamide sodium acrylate) copolymers [37]. The attempt of M_w-determination by light scattering failed due to a nonseparable small amount of aggregates or gel particles in the samples.

Table 8.1. Characteristics of samples employed in complex formation

Polymer	$M_w\ 10^{-6}$ [g/mol]	l_0 [Å]	dn/dc in H_2O [ml/g]
C0	1.1	2.5	0.133
C4	2.9	12.6	0.147
C8	0.9	22.7	0.178
A0	5.3	2.5	0.175
A4	2.2	12.6	0.189
A8	2.2	22.7	0.192

M_w –weight average molecular mass.
l_0 –average distance between two charges.
dn/dc –refractive index increment

8.4.2 Methods of Investigation

To obtain more comprehensive information about the formation, structure and stability of the polyelectrolyte complexes, we used a combination of several methods of investigation, which are listed in Table 8.2.

The titration methods were employed simultaneously in a concentration range of initial phase separation (10^{-3}–10^{-2} g/ml), where a clearly pronounced maximum in the turbidity indicates the endpoint of titration, allowing the assessment of the PEC stoichiometry.

Viscosimetric data yield information about the coil density, i.e., the polymer mass per unit volume of the polymer particles. PEC formation often leads to a decrease of coil density in comparison to the dissolved components and therefore it can be studied by this method. However, the main contribution of the viscosity is given by the dissolved excess component.

For the systems investigated, light scattering experiments are much more sensitive to PEC formation. Especially in salt free solutions the scattering of the components is suppressed by the polyelectrolyte effect to a great extent. PEC formation leads to a drastic increase of the scattering intensity, even at the very start of the process.

Light scattering investigations were carried out with the instrument Sofica 42000 (Wippler and Scheibling, Strasbourg, France), equipped with a 5 mW-He–Ne-Laser as light source (accuracy of measurement better than 1%). PEC formation could be realized directly in the scattering cell by dosage of the polycounterion solution via an immersing capillary into the excess component solution under gentle stirring. First the scattering intensity at the scattering angle 45° was continuously registered and then the dosage was interrupted and the complete scattering curves were measured after reaching the equilibrium. All solutions have been dustfree purified by filtration (0.45 μm pore size membrane filters). The addition of salt solution was performed in an analogous way.

To evaluate the binding strength of the PEC the binding of acridine orange on the polyanions and its release by complex formation with a polycation was investigated by UV/VIS-spectroscopy.

Additionally electrophoresis was employed to obtain information about the surface charges. Measurements were carried out under conditions of isotachophoresis, but missing a terminating electrolyte, due to the extremely low

Table 8.2. Methods of investigation

Method	Information
Titration:	course of phase separation, stoichiometric ratio of ionic groups at the endpoint
Conductometry	
Turbidimetry	
Viscosimetry	hydrodynamic volume of symplex particles
Light scattering	average values of particle mass, size and polymer packing density
UV/VIS-spectroscopy	dye binding on polyelectrolytes and release by complexation
Electrophoresis	surface charges

mobility of the PEC particles. The migration of the PEC particles could be observed with a UV-detector.

8.5 Results and Discussion

The results of titration studies are demonstrated in Fig. 8.2. While for the system A0/C0 with high charge density the break point in the conductometric curve and the flocculation point of the turbidimetric curve are clearly expressed, these effects flatten with decreasing charge density, not allowing a determination of

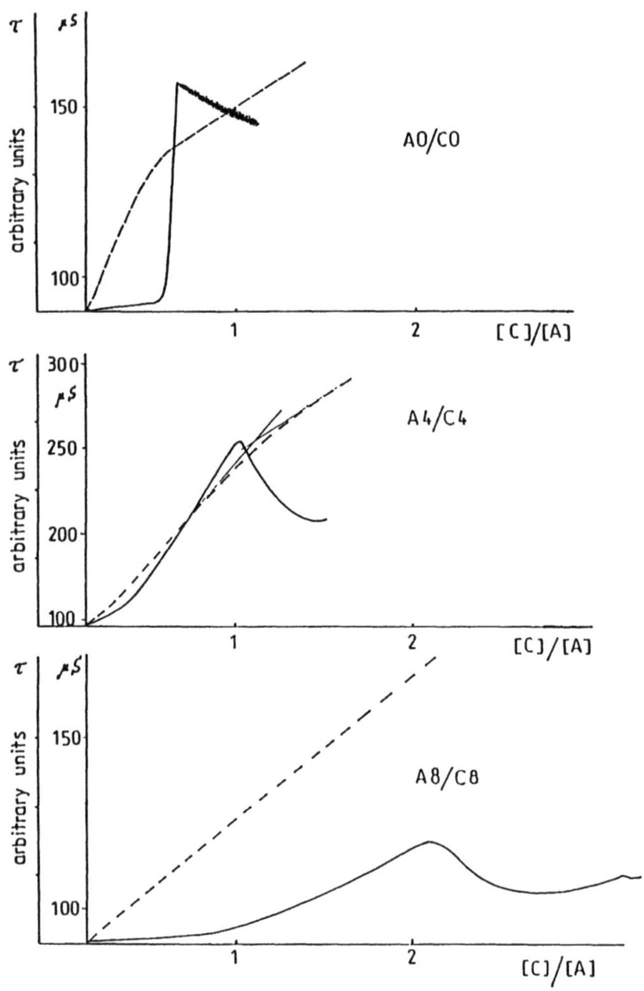

Fig. 8.2. Conductometric (- - -) and turbidimetric (———) titration curves for systems of different charge density

the electrochemical endpoint for the system A8/C8. The stoichiometric ratio of charged groups [C]/[A], determined from the concentration ratio of the components at the titration endpoints (Table 8.3) shows that the 1:1 stoichiometry is more an exception than the rule. With decreasing charge density the composition of the PEC particles changes from an excess of anionic groups to an excess of cationic ones. The stoichiometric ratio for the different combinations of components may be understood on the basis of a mass action law, assuming a different degree of dissociation between polyions and low molecular counterions for the polyanions and the polycations [38], but the physical reason for the varying mass action constant is still in question.

The changing course of the phase separation phenomena with alteration in charge density suggest the formation of different PEC structures. This could be qualitatively confirmed by the viscosimetric data of Fig. 8.3, which showed a

Table 8.3. Stoichiometric ratio of cationic to anionic groups at the titration endpoints (polyanion as excess component)

	Polycation		
Polyanion	C0	C4	C8
A0	0.58	0.72	1.11
A4	1.29	1.19	2.14
A8	0.95	1.60	2.11

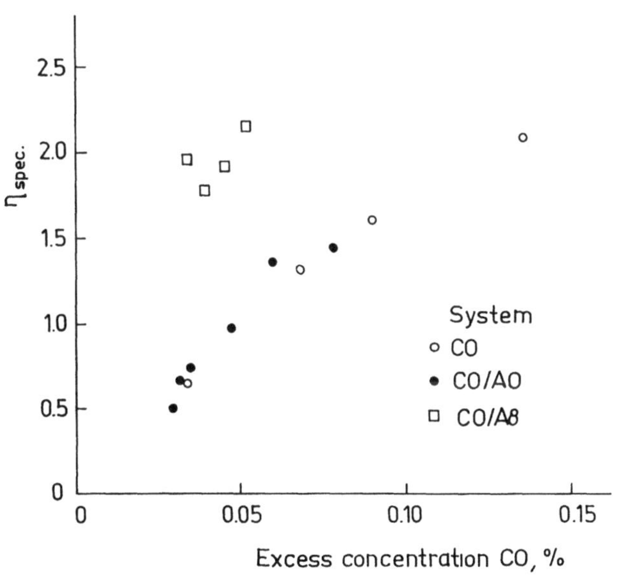

Fig. 8.3. Specific viscosity as a function of the excess concentration of the polycation C0 for the system C0/A0, C0/A8 and the PEC free system

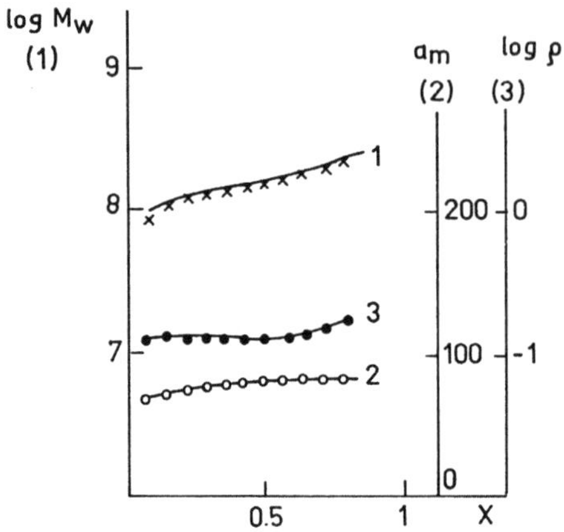

Fig. 8.4. Structural parameters M_w (*1*), a_m (*2*) and ρ (*3*) as a function of the degree of conversion X for the PEC C0/A0 ($c_C = c_A = 1 \times 10^{-3}$ g/ml)

negligible contribution of the PEC particles C0/A0, but a remarkable one for the system C8/A8, indicating an increasing degree of swelling of the PEC particles with decreasing charge density.

Light scattering studies yield more detailed information about the structure of the PEC particles. The scattering curves were analyzed by the master curve interpretation. The dependence of the structural parameters of the system C0/A0 on the degree of conversion is given in Fig. 8.4. Mass and size as well as the degree of swelling change only slightly with increasing conversion. This means that the generation of new particles is the dominant process occurring during PEC formation. The particle size is already fixed from the very beginning of the reaction, and no significant growth occurs later on.

However, an increasing polymer concentration causes a higher degree of aggregation, but the polymer packing density ρ remains nearly constant. The increasing degree of aggregation may be explained by two competing processes of charge compensation via conformational changes of the polyelectrolyte chains or additional incorporation of polycounterions.

For the system C4/A4 we obtained equivalent results (the main difference consists in the higher degree of swelling of the PEC particles), while the system C8/A8 shows a completely different behavior. Here, in the first steps of titration the mass of the PEC particles increases slightly, but the size decreases drastically with increasing degree of conversion, yielding a steep rise in the compactness of the PEC particles. At a higher degree of conversion the level of aggregation increases, but the degree of swelling remains constant. These results suggest an alternation of the PEC stoichiometry and structure with increasing degree of conversion (for details see [7]).

Additional information about the PEC structure and stability could be obtained by investigation of the influence of the ionic strength of the medium. Figure 8.5 shows the scattering curves of the PEC particles C0/A0, formed in saltfree solution, at different contents of subsequently added NaCl. The crosses are the experimental points, while the full lines represents the sections of

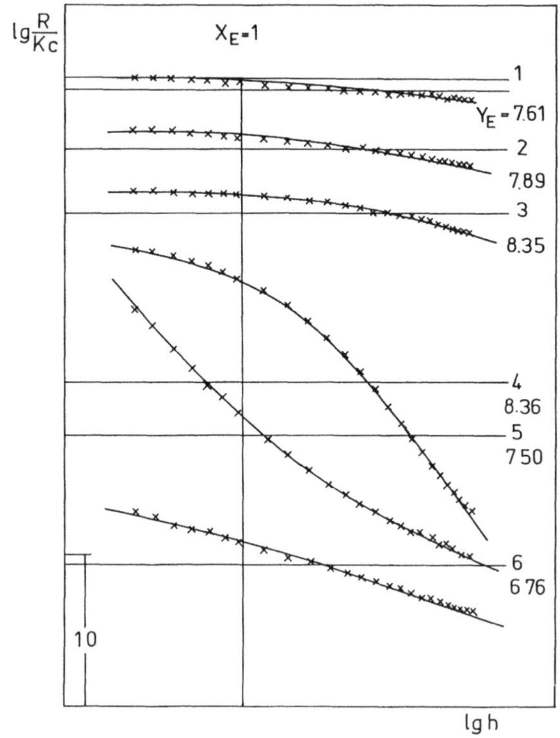

Fig. 8.5. Master scattering curves of PEC particles C0/A0 as a function of subsequently added NaCl (Debye plot): × experimental point, — sections of appropriate theoretical scattering curves. Conditions of PEC formation: $c_C = 2 \times 10^{-4}$ g/ml, $c_A = 1 \times 10^{-4}$ g/ml, $X = 32\%$; for salt content and structural parameters see Table 8.4

appropriate theoretical master curves. The shape of the scattering curves changes significantly with the ionic strength of the medium. The scattering curves 1–4 could be fitted quite well by monomodal polydisperse systems of spheres. Curve 5 corresponds to a bimodal system of spheres and the last curve reveals the scattering behavior of Gaussian coils. The scattering curve 5 was interpreted in detail by an iterative procedure of curve separation. Figures 8.6 and 8.7 show in master representation of the Debye plot and as Guinier plot the experimental curve 5, the scattering curves of the two components and the sum of these two curves, which is nearly identical with the experimental one.

The quantitative analysis of all scattering curves yields the structural parameters, given in Table 8.4. Up to a salt content of about 0.3 mol/1 an additional aggregation occurs, accompanied by a slight increase of the polymer packing density ρ. At a NaCl-concentration of 0.4 mol/1 a drastic increase of the PEC particle mass and an initial decrease of the density ρ were observed. After the next step of salt dosage two components exist in the system, a component of very large particles and a second disaggregated one. The degree of

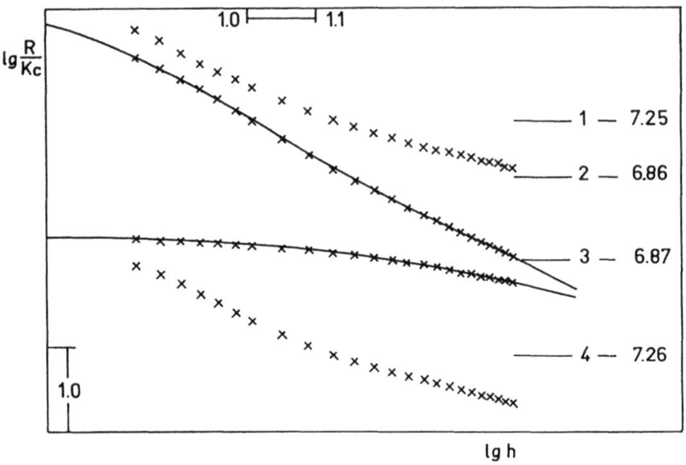

Fig. 8.6. Results of an iterative curve separation procedure of a bimodal system of C0/A0-PEC in the stage of salt-caused disaggregation (curve 5 in Fig. 8.5). Master scattering curves of Debye plot: (*1*) experimental points, (*2*) theoretical scattering curve of the component of large particles, (*3*) theoretical scattering curve of the disaggregated particles, (*4*) sum of the scattering intensities corresponding to curve 2 and 3

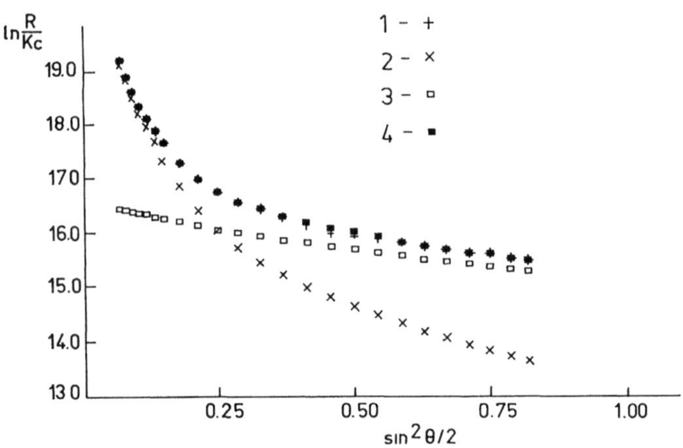

Fig. 8.7. Scattering curves of Fig. 8.6 as Guinier plot

swelling increases significantly. In the last stage investigated, the particle mass corresponds to the small component in the previous system, but the particle size is clearly increased and therefore also the degree of swelling. Obviously, a process of dissolution occurs at a critical salt concentration. This could also be observed for the systems C4/A4 and C8/A8. The changes of the structural parameters of the PEC C4/A4 as a function of the salt content are demonstrated in Fig. 8.8.

Table 8.4. Salt-dependence of the structural parameters of PEC particles C0/A0 at subsequent addition of NaCl. ($c_C = 2.0 \times 10^{-4}$ g/ml, $c_A = 1.0 \times 10^{-4}$ g/ml, conversion: 32%)

No.	c_{NaCl} mol/l	σ_{sp}	M_w 10^{-7} g/mol	a_m nm	ρ g/ml	Structure type
1	0	0.3	4.9	52	0.14	spheres
2	0.065	0.3	10.7	62	0.18	spheres
3	0.286	0.3	33.0	69	0.40	spheres
4	0.40	0.25	263	195	0.14	spheres
5 bimodal						
$x_1 = 0.39$	0.45	0.3	439	431	0.022	spheres
$x_2 = 0.61$	0.45	0.3	2.6	78	0.022	spheres
6*	0.50	1.5	2.6	158	0.0026	coils

* σ_M and s_m are given. ρ was estimated considering the particle volume as a sphere of the radius s_m.
x_1, x_2 – mass ratio of the two components, calculated under the assumption of equal polymer packing density

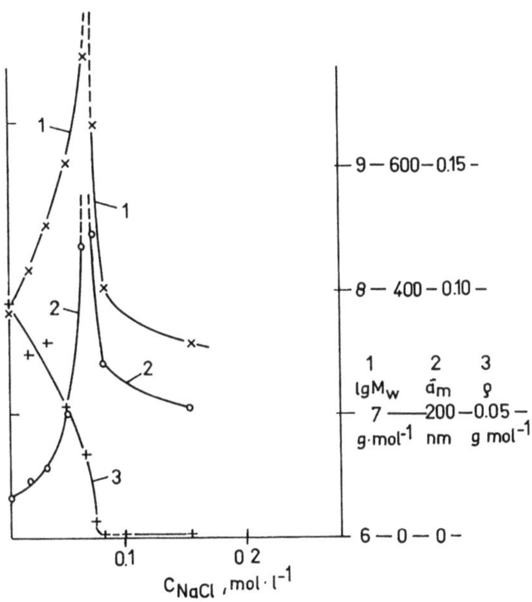

Fig. 8.8. Dependence of the structural parameters of PEC C4/A4 on NaCl-content ($X = 40\%$, $c_C = 5 \times 10^{-5}$ g/ml, $c_A = 3.3 \times 10^{-5}$ g/ml)

The critical salt concentration depends strongly on the charge density of the components. The quantitative results on the compactness and the salt stability of the PEC particles (Table 8.5) show that with decreasing charge density the degree of swelling of the PEC particles increases and the binding strength between the poly-counterions decreases. PEC formation between components of different charge density could be detected in a limited range of conversion only. The dissolution of PEC structures may be explained by the shielding effect of the low molecular salt.

Table 8.5. Aggregate packing density ρ and critical NaCl concentration

Combination	ρ g/ml	$c_{NaCl\,crit.}$ mol/l
C0/A0	0.13–0.17	0.45
C4/A4	0.07–0.13	0.065–0.09
C8/A8	0.02–0.08	0–0.02

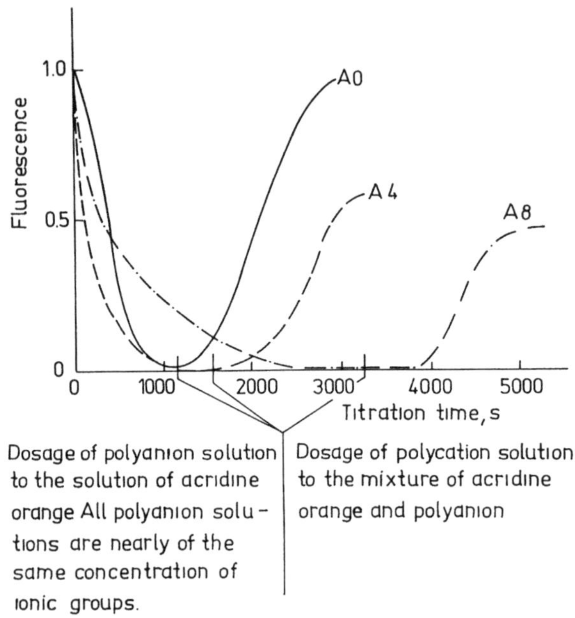

Fig. 8.9. Change in fluorescence intensity of acridine orange due to binding by A0, A4 or A8 and release via complex formation with C0

To obtain more information about the binding strength, we investigated the binding and release of ionic dyes in polyanion-polycation systems by spectrophotometry. Figure 8.9 depicts the quenching of the fluorescence of acridine orange by binding on the different polyanions employed and the recovery of fluorescence due to dye release via PEC formation with the polycation C0. With decreasing charge density an increasing amount of anionic sites per dye molecule is required for quenching, indicating the cooperative character of this process. The release of dye by PEC formation is increasingly incomplete with decreasing charge density of the polyanions, confirming the results of light scattering.

Electrophoretic studies led to the conclusion that PEC particles behave as polyampholytes. By variation of the pH-value the direction of migration in the

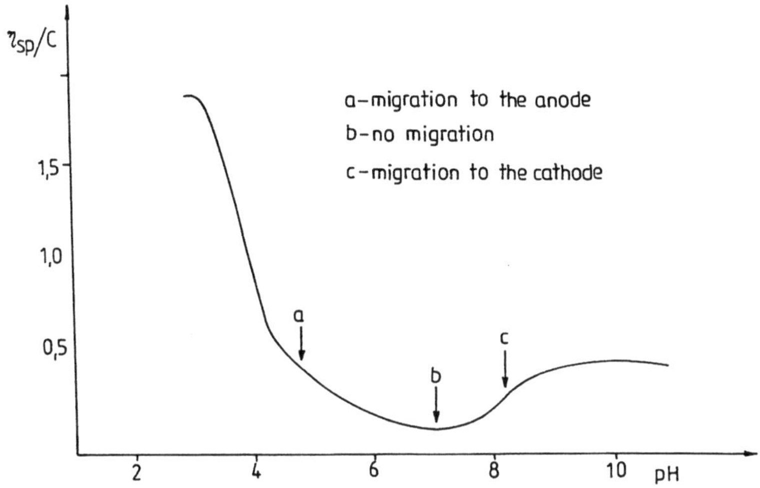

Fig. 8.10. pH-dependence of η_{sp}/c if the A0/C0 PEC particles; similar to electrophoretic migration

electric field can be changed; in the neutral range no migration occurs. The pH-dependence of the viscosity of the PEC system A0/C0 shows the typical curve of polyampholytes, with a minimum near pH = 7 (Fig. 8.10)

8.6 Conclusion

Static light scattering studies in combination with several other analytical methods allow detailed information to be obtained about the course of polyelectrolyte complex formation and the characterization of the structure of the particles formed. In the range of higher polymer concentration (10^{-3}–10^{-2} g/ml) normally phase separation occurs. In highly diluted systems quasi-soluble PEC particles on a colloidal level are built, even between two oppositely charged polyelectrolytes of high molecular weight. Compactness and stability of PEC particles depend strongly on the charge density of the polyions. Larger differences in the average distances between two charged groups of the components impede PEC formation.

8.7 References

1. Kossel A (1896) J Physiol Chem 22
2. Willstaetter R, Rohdewald M (1934) Hoppe-Seiler's Z Physiol Chem 225:103
3. Michaels AS, Miekka RG (1961) J Phys Chem 65:1765
4. Bekturov EA, Bimendinna LA (1981) Adv Polymer Sci 41:100
5. Tsuchida E, Abe K (1982) Adv Polym Sci 45:1
6. Smid J, Fish D (1988) Encycl Polym Sci and Eng 11:720
7. Philipp B, Dautzenberg H, Linow K-J, Koetz J, Dawydoff W (1989) Prog Polym Sci 14:91

8. Tsuchida E, Osada Y, Samada K (1972) J Polym Sci A-1 10:3397
9. Tsuchida E, Osada Y, Ohno H (1980) J Macromol Sci B17:683
10. Kabanov VA, Zezin AB, Mustafaev MI, Kasaikin VA (1980) Polym Aminos Amm Salts (IUPAC 1979, Gent) 173 Pergamon Press Ltd.
11. Zezin AB, Kabanov VA (1982) Uspechi Chimii 9:1447
12. Kabanov VA, Zezin AB (1984) Pure appl Chem 56:343
13. Kharenko OA, Kharenko AV, Kalyuzhnaya RI, Izumrudov VA, Kasaikin VA, Zezin AB, Kabanov VA (1979) Vysokomolek soed A21:2719
14. Kharenko OA, Kharenko AV, Kasaikin VA, Zezin AB, Kabanov VA (1979) Vysokomolek soed A21:2726
15. Kharenko OA, Izumrudov VA, Kharenko AV, Kasaikin VA, Zezin AB, Kabanov VA (1980) Vysokomolek soed A22:218
16. Izumrudov VA, Kharenko OA, Kharenko AV, Gulyaeva ZhG, Kasaikin VA, Zezin AB, Kabanov VA (1980) Vysokomolek soed A22:692
17. Kharenko AV, Neverova YeA, Kalyuzhnaya RI, Zezin AB, Kabanov VA (1981) Vysokomolek soed A23:2083
18. Rogacheva VB, Ryzhikov SV, Shchors TV, Zezin AB, Kabanov VA (1980) Vysokomolek soed A26:2417
19. Izumrudov VA, Kasaikin VA, Yermakova LN, Mustafaeva MJ, Zezin AB, Kabanov VA (1981) Vysokomolek soed A23:1365
20. Grishina NV, Rogacheva VB, Lopatina LJ, Zezin AB, Kabanov VA (1985) Vysokomolek soed A27:1154
21. Dubin PL, Davis DD (1984) Macromolecules 17:1294
22. Dubin PL, Rigsbee DR, Gan LM, Fallon MA (1988) Macromolecules 21:2555
23. Dubin PL, Thé SS, Gan LM, Chew CH (1990) Macromolecules 23:2500
24. Djadoun S (1982) Polymer Bulletin 7:607
25. Tsuchida E, Osada Y, Ohno H (1980) J Macromol Sci-Phys B17:683
26. Ohno H, Matsuda H, Tsuchida E (1981) Makromol Chem 182:2267
27. Hara M, Nakajima A (1989) J. Polym Sci, Part B: Polymer Physics 27:1043
28. Dautzenberg H, Rother G (1988) J. Polym Sci, Part B: Polymer Physics 26:353
29. Dautzenberg H, Rother G (1991) J Appl Polym Sci, :Appl. Polym Symp 48: 351
30. Dautzenberg H, Linow K-J, Philipp B (1981) Acta Polymerica 32:225
31. Dautzenberg H, Linow K-J, Philipp B (1982) Acta Polymerica 33:619
32. Dautzenberg H, Linow K-J, Philipp B (1982) Plaste und Kautschuk 29:631
33. Dautzenberg H, Dautzenberg Ho (1985) Acta Polymerica 36:102
34. Dautzenberg H, Rother G, Linow K-J, Philipp B (1988) Acta Polymerica 39:157
35. Dautzenberg H, Linow K-J, Rother G (1990) Acta Polymerica 41:98
36. Kerker M (1969) The Scattering of Light and other Electromagnetic Radiation. Academic Press, New York
37. Klein J, Conrad KD (1978) Makromol Chem 179:1635
38. Haronska P, Koetz J, Philipp B (1989) Acta Polymerica 40:110

9 Interaction Between Oppositely Charged Low Ionic Density Polyelectrolytes: Complex Formation or Simple Mixture?

Dominique Frugier and Roland Audebert

The notion of a polyelectrolyte complex is well established for the mixture of two homopolymers one anionic, the other cationic. The complex formation is generally associated with a phase separation (often a precipitation) although water soluble polymeric complexes are known. Only few data are available when the charge density of the polyelectrolytes is progressively decreased. This situation is exemplified with a series of mixtures of polyacrylamide derivatives, one family of copolymers has various contents in anionic units, the other is similar but with cationic units. We found that even for a few percent in ionic units, the mixtures water/copolymer anionic/copolymer cationic phase separates: a polymer rich phase (including the two types of copolymers) is in equilibrium with a very diluted polymer solution.

In a classical triangular diagram, the diphasic area, which does not depend on the polymer molecular weight, strongly increases with the ionic content of the macromolecules but vanishes if the ionic strength of the medium rises.

Various techniques (conductimetry, gel permeation chromatography, viscometry) are used to study the polyanionic/polycationic affinity both in dilute and semidilute regimes. It appears that at least in semidiluted solutions there is no specific interactions between the two types of polymers. Although a phase separation occurs, there is no evidence for the formation of a complex.

In fact this kind of behavior is very general between species of poor entropy of mixing in a common solvent. It was described between attractive polymers in organic medium and between polyanionic polymer and cationic surfactants in aqueous medium. The general shape of the phase diagram can be predicted from thermodynamic calculations but various difficulties, which are discussed, occur to take into account the attractive effects.

9.1 Introduction

More than twenty years before the idea of the macromolecule was definitively accepted, Beijerinck [1] observed that by mixing the solutions of two water soluble polymers an inhomogeneous medium was obtained. It was promptly recognized that this incompatibility is the general rule [2–4]. A typical phase diagram for the system polymer A/water/polymer C is pictured in Fig. 9.1; when the interactions between the units A and C of the two homopolymers poly A and poly C are not too large, especially when no specific interaction between A and C is involved. Under these conditions the two phases in equilibrium are liquid. On the other hand, strong interactions may occur between the units A and C, if for instance A = anionic and C = cationic. Then a diphasic system is generally also obtained but most of the polymeric species are recovered in the same phase [5–6], frequently solid. Both the chemical analysis of the polymeric rich phase and classical experiments where the evolution of a physical property (for

Macromolecular Complexes
in Chemistry and Biology
Eds. Dubin/Bock/Davis/Schulz/Thies

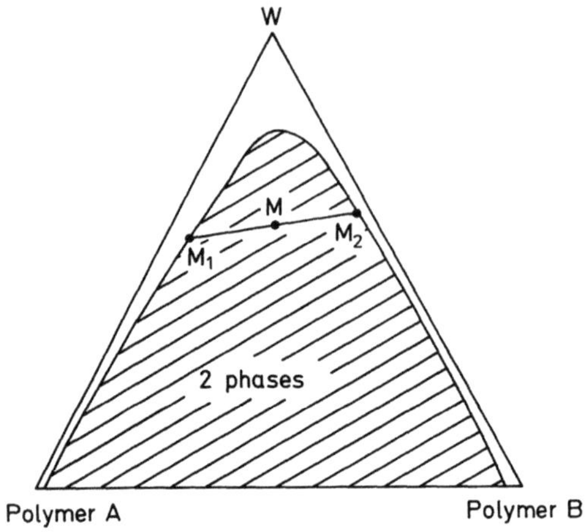

Fig. 9.1. A typical phase diagram for dilute and semidilute solutions of polymers A and C in their common solvent water (W). The medium is homogeneous only for very diluted systems. A typical tie-line is M_1M_2

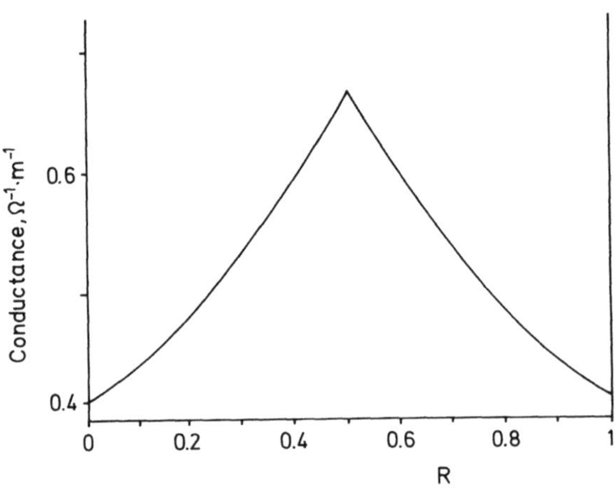

Fig. 9.2. Variation of the conductance of a 10^{-3} M solution of polystyrene sulfonate by titration by polytrimethylbenzylammonium chloride according to Michaels and Miekka [7]. An equivalent point appears when the ratio *R*: number of ammonium groups/total number of ammonium and sulfonate groups is 0.5

instance conductivity, see Fig. 9.2) of a poly A solution when poly B is added, leads to the formation of a kind of stoichiometric system called a "polymeric complex". For the example depicted in Fig. 9.2 (mixture of sodium polystyrene sulfonate and polytrimethyl benzyl ammonium chloride) a 1:1 stoichiometry appears [7].

Between these two limiting cases only few studies were devoted to the behavior of aqueous mixtures of oppositely but moderately charged polyelectrolytes [8–15].

One of the most efficient ways to progressively change the relative affinity of two macromolecular chains is to use series of two types of copolymers for which one pair of homopolymers has no specific interactions and the other pair of

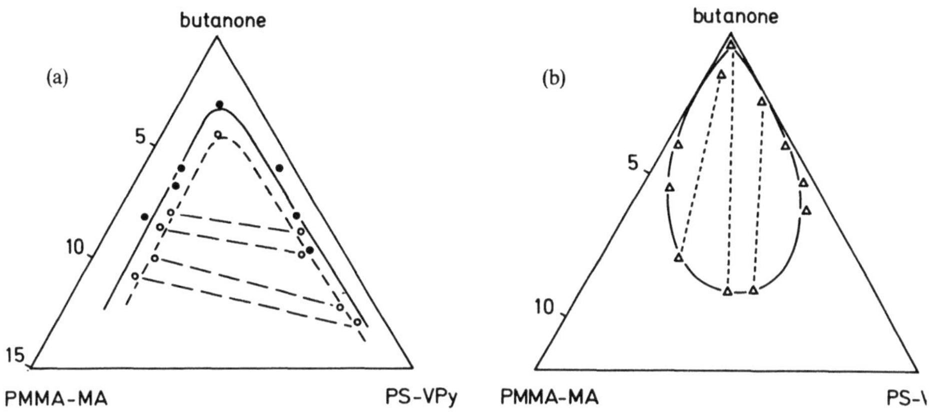

Fig. 9.3 a, b. Phase diagram in butanone of **a**: copoly(methyl methacrylate–methacrylic acid) and **b**: copoly(styrene–vinylpyridine) according to Morawetz et al. [16]. The contents in methacrylic acid and vinylpyridine (percent of monomer units in the chains) are respectively:
a. 0%; 0% (homopolymers) —●—
a. 2.24%; 1.62% ----○----
b. 8.52%; 8.85% —△—

homopolymers are, for instance, polyanionic and polycationic macromolecules. A typical example of this kind of situation is given in the paper of Morawetz et al. [16]. They studied the compatibility of polymethylmethacrylate and polystyrene in butanone. The corresponding phase diagram, limited to total polymer concentrations smaller than 15%, has the characteristic cloud curve of two poorly interacting polymers and an area of compatibility limited to the very diluted polymer solutions (Fig. 9.3a). The affinity between the two macromolecular chains is increased by copolymerizing methylmethacrylate with methacrylic acid and styrene with vinylpyridine since these two new units can be engaged in hydrogen bonding. Even for a low content of interactive monomers (Fig. 9.3a) an expansion of the compatibility area is observed, nevertheless the shape of the tie-lines stay similar. However, when the content of interactive monomers becomes large enough (Fig. 9.3b) a new type of incompatibility area, limited by a closed loop, appears. In this situation, for a composition corresponding to a point inside the loop, a polymer-rich phase is in equilibrium with a very diluted solution.

According to a similar approach we prepared two families of acrylamide copolymers, one anionic the other cationic, with various molecular weights and compositions. The corresponding monomers are given below:

$CH_2{=}CH$
$\quad\quad |$
$\quad\quad C{=}O$
$\quad\quad |$
$\quad\quad NH{-}C(CH_3)_2{-}CH_2{-}SO_3^-\ Na^+$

AMPS

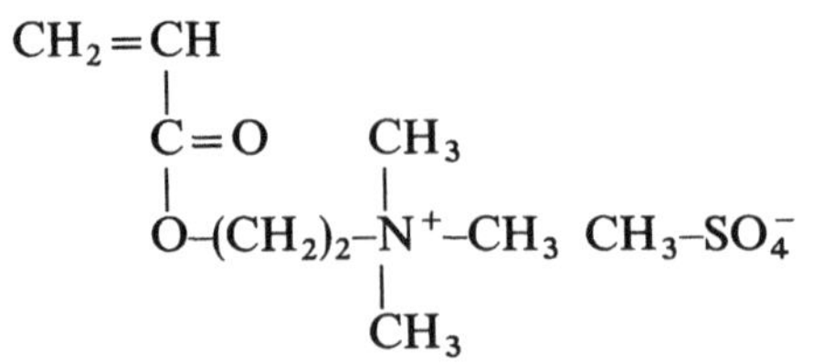

MSA

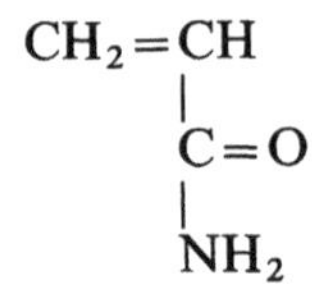

AM

The behavior of the aqueous mixtures of anionic and cationic copolymers was studied versus their molecular weight, their composition (ratio of ionic units) and the ionic strength of the medium.

9.2 Material and Techniques

9.2.1 Polymer Synthesis

Polymers were obtained by radical copolymerization in aqueous media of acrylamide and the corresponding ionic acrylic comonomers (AMPS or MSA). By using a redox system (potassium persulfate, *N,N,N′,N′*,tetramethylene diamine) the polymerization can be performed at a temperature lower than 35° [17–18] which avoids branching; the latter may occur at a reaction temperature higher than 60°. Polymers are recovered from the reaction medium by precipitation in ethanol.

9.2.2 Polymer Characterization

The composition of the copolymers was checked by chemical analysis (ratio of the sulfur and nitrogen contents). As this method is not very accurate the results were confirmed by proton NMR spectroscopy. In all cases the copolymer composition is very close to the corresponding monomer mixture in the reaction bath. The ionic monomer ratio ranges between 0 and 11.5% (Table 9.1). Taking into account the usual difficulties to determine the molecular weight of water soluble copolymers several ways of determination were used.

Table 9.1. Characterization of the copolymer samples; the intrinsic viscosity [η] and mean molecular weight $\bar{M}_w$ were obtained by viscometry (visco), light scattering (DDL) and size exclusion by chromatography (GPC)

Copolymer	$[\eta]^{visco}_{(ml/g)}$	$\bar{M}_p^{visco}$	$\bar{M}_p^{DDL}$	$\bar{M}_p^{GPC}$
P(AM-AMPS) 11.5%	103	1.8×10^5	1.7×10^5	2.7×10^5
P(AM-AMPS) 4%	149	3.0×10^5	3.0×10^5	3.1×10^5
P(AM-AMPS) 3.5%	275	7.2×10^5	6.8×10^5	5.2×10^5
P(AM-AMPS) 1.8%	130	2.5×10^5	2.0×10^5	2.1×10^5
P(AM-MSA) 8.5%	160	3.4×10^5	3.7×10^5	4.8×10^5
P(AM-MSA) 4%	175	3.8×10^5	3.1×10^5	3.4×10^5
P(AM-MSA) 3.5%	427	1.3×10^6	1.5×10^6	9.9×10^5
P(AM-MSA) 1.2%	106	1.9×10^5	2.0×10^5	1.4×10^5

As previously we determined a relation between the intrinsic viscosity $[\eta]$ and the mean molecular weight (in weight, $\bar{M}_w$) for cationic copolymers of acrylamide with similar structure [19], we assumed that, at least in the range of low ionic compositions, the same relation holds and the measurements of intrinsic viscosity leads to $\bar{M}_w$ according to the relation:

$$[\eta]_{\text{ml/g}}^{\text{NaCl 1M, 30°C}} = 1.91 \times 10^{-2}\, \bar{M}_w^{0.71} \,.$$

The absolute value of $\bar{M}_w$ was also established by light scattering thanks to a FICA 42000 apparatus equipped with a laser He–Ne source. Mean values of molecular weight and additionally of distribution were obtained by gel permeation chromatography (Water ALC 201 equipment with Shodex OH-pak columns). The eluent was lithium nitrate (with 400 ppm of sodium azide as bactericide). Our apparatus was equipped with three detectors: a refractometer, a light scattering detector (chromatix CX 100) and a viscometric detector developed in the Laboratory [20–21].

The results are shown in Table 9.1. The $\bar{M}_w$ determined by the three techniques are in reasonable agreement. In all cases the molecular weight distribution is very large. We can mention that we found two samples – one anionic P(AM-AMPS) 4%, the other cationic P(AM-MSA) 4% – with practically similar elution patterns in gel permeation chromatography.

9.2.3 Other Techniques

As these polymers are hygroscopic the effective concentration of their aqueous solution was determined from the total carbon analysis (apparatus Dohrman, Xertex DC 80). The latter was also used for the composition determination to establish the tie lines in the phase diagram. Viscometric determinations were performed with a Couette's type viscometer (Low shear 30 Contraves) at 25 °C. In the range of shear rate and concentration we used, the sample exhibited a Newtonian behavior.

9.3 Phase Diagram

9.3.1 Phase Diagram Representation

The thermodynamic equilibrium between anionic copolymer/cationic copolymer/aqueous medium are given in classical triangular representation. Only the diluted and semidiluted (up to 10% in weight for the total polymer concentration) are studied.

9.3.2 Influence of the Charge Density

For very low contents in ionic units, for instance 1.8% in anionic groups and 1.2% in cationic groups, the copolymers behave in water as pure polyacrylamide

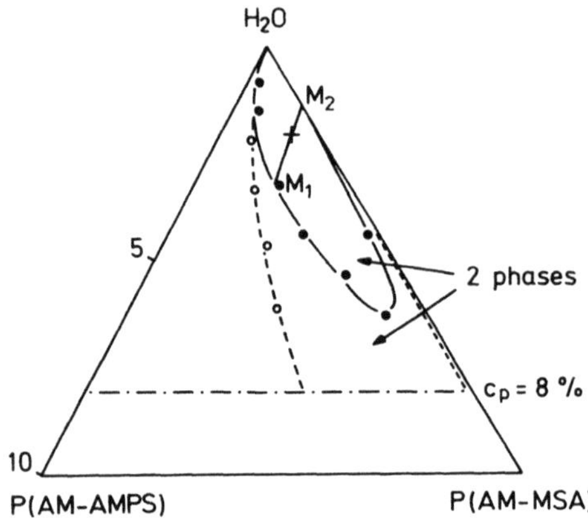

Fig. 9.4. Effect of the charge density on the phase diagram in pure water

P(AM-AMPS) 1.8%	P(AM-MSA) 1.2%	compatible system
P(AM-AMPS) 4%	P(AM-MSA) 4%	—●—
P(AM-AMPS) 11.4%	P(AM-MSA) 8.5%	----O----

would do, and the mixture is homogeneous in the whole range of concentrations studied. However, the increase of the copolymer ionicity promptly induces a phase separation. The cloud curve is given in Fig. 9.4 for 4% in both anionic and cationic groups. This curve is a closed loop. For all the mean compositions corresponding to points inside the loop a coacervation is observed (a tie line is given): a polymer rich phase is in equilibrium with a polymer poor phase (which in some cases can be practically free of polymer). When the ionic content increases the diphasic area promptly increases. For instance, the lower level of the loop is out of the limit of a total polymer concentration of 8% for the polymer system with an anionic content of 11.4% whereas the cationic ratio is 8.5%. The behavior is very similar to the one depicted in Fig. 9.3b.

As our samples have a very large distribution in molecular weight, a segregation according to the molecular weight is possible. So we analyzed by gel permeation chromatography the two phases corresponding to the points M_1, M_2 in Fig. 9.4 and the original pure copolymers. These copolymers correspond to the pair of similar elution patterns (Fig. 9.5). The polymer rich phase has a molecular weight distribution shifted towards high molecular weight and of course the opposite is observed for the supernatant (polymer poor phase). However, the shift is not so large that the separation according to the molecular weight can be considered as the driving force in this kind of phase separation.

9.3.3 *Influence of the Ionic Strength*

The small ions in the system are Na^+ and $CH_3SO_4^-$. If we add an excess of sodium methylsulfate in the system the diphasic area promptly decreases and finally disappears (Fig. 9.6).

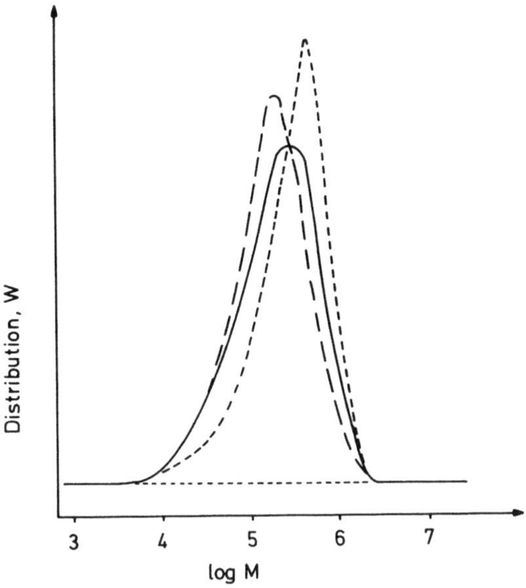

Fig. 9.5. Molecular weight distribution for P(AM-AMPS) 4% and P(AM-MSA) 4%: —— (similar distribution for the two copolymers). Molecular weight distribution for the polymer rich phase: ---- (composition corresponding to the point M_1 in Fig. 9.4). Molecular weight distribution for the polymer poor phase: — — (corresponding to M_2 in Fig. 9.4)

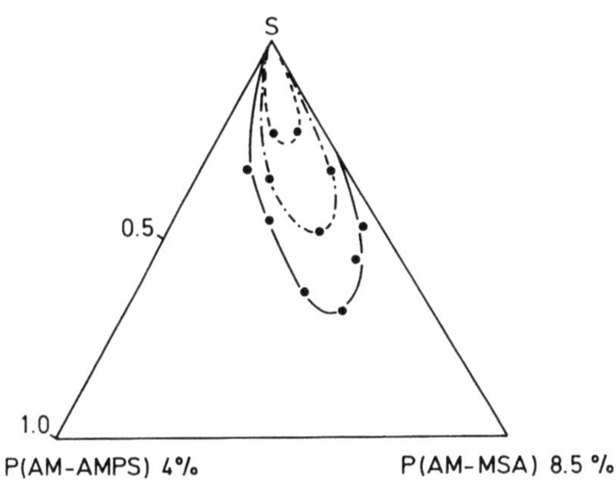

Fig. 9.6. Effect of the ionic strength on the phase diagram P(AM-AMPS) 4% P(AM-MSA) 8.5%: —— pure water; —·— $NaCH_3SO_4$ 10^{-2} M; ----- $NaCH_3SO_4$ 1.35×10^{-2} M

9.3.4 Influence of the Molecular Weight of the Samples

Obtaining two couples of copolymers with similar ionicity but very different molecular weight is difficult. This is nevertheless practically the situation depicted in Fig. 9.7. Taking into account the range of experimental error, we can conclude that the phase diagram is practically unaffected by the changes in molecular weight.

9.3.5 Phase Diagram and Complex Formation

For homopolymers the formation of a polymer complex is very frequently depicted as correlated with a phase separation although the soluble polymer

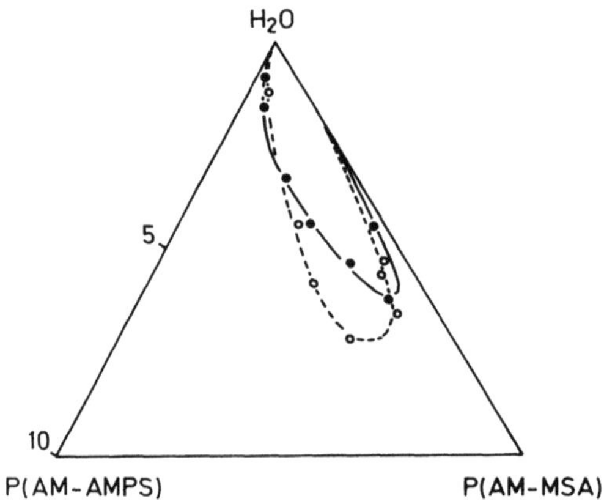

Fig. 9.7. Effect of the molecular weight on the phase diagram P(AM-AMPS) 4%, $\bar{M}_w = 3 \times 10^5$; P(AM-MSA) 4%, $\bar{M}_w = 3.5 \times 10^5$ —— and P(AM-AMPS) 3.5%, $\bar{M}_w = 7 \times 10^5$; P(AM-MSA) 3.5%, $\bar{M}_w = 1.4 \times 10^6$ ----

complexes are known [22–24]. Some cases of copolymer soluble complexes are also mentioned [12], [25–27]. We studied various systems with a composition corresponding to a composition on the border of the cloud curve.

Initial information can be obtained from the gel permeation chromatography results (Fig. 9.5): as the two types of copolymers have the same elution pattern it means that if they are injected simultaneously the two kinds of macromolecules are eluted at the same speed in the column. If an interpolymer complex was formed the hydrodynamic volume of the complex would be larger than the hydrodynamic volume of the original macromolecules and the chromatogram would be largely shifted towards high molecular weights, for each phase. As this is not the case we can deduce that in the experimental conditions of chromatography (diluted regime) no complexation occurs.

Another view of the behavior in dilute medium can be obtained from conductimetric measurements (see for instance Fig. 9.1). We started from a solution of the anionic copolymer P(AM-AMPS) 11.5% with a 10^{-2} M total concentration in monomer unit and we plotted (Fig. 9.8) the evolution of specific conductivity observed by adding a solution of P(AM-AMS) 8.5%. Two borderline cases can be predicted: if a stoichiometry based upon an electrical balance of the charge is expected then an equivalent point would appear (curve 8a). On the contrary, if there is no interaction between the two polymers a regular increase of the specific conductivity, proportional to the amount of cationic copolymer introduced should be noticed (curve 8b). For intermediate situations it is possible to calculate the ratio of anionic sites associated to cationic groups [28] if we speculate that this association leads to the release of the two corresponding small counterions. In this situation an intermediate curve between 8a and 8b, but with an equivalent point would be expected. In practice no equivalent point can be determined (curve 8c). For copolymers with ionic contents smaller than those corresponding to the curve 8c the titration curves are even smoother. Such behavior cannot be explained if a specific interaction between anionic and cationic sites occurs, a non specific long range interaction between these sites would generally be more expected.

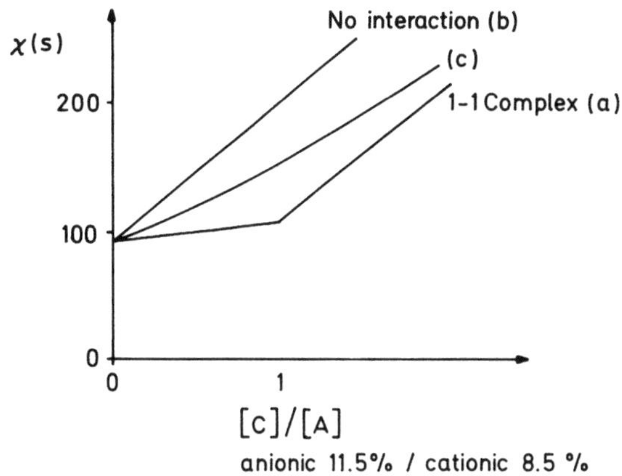

Fig. 9.8. Conductimetric titration curve of a 10^{-2} M (molar units) solution of P(AM-AMPS) 11.5% by P(AM-MSA) 8.5% (χ = specific conductivity, Siemens): *a* Expected curve if a stoichiometric complex based upon electrostatic charge balance was formed. *b* Expected curve if no interaction between copolymers occurs. *c* Actual curve

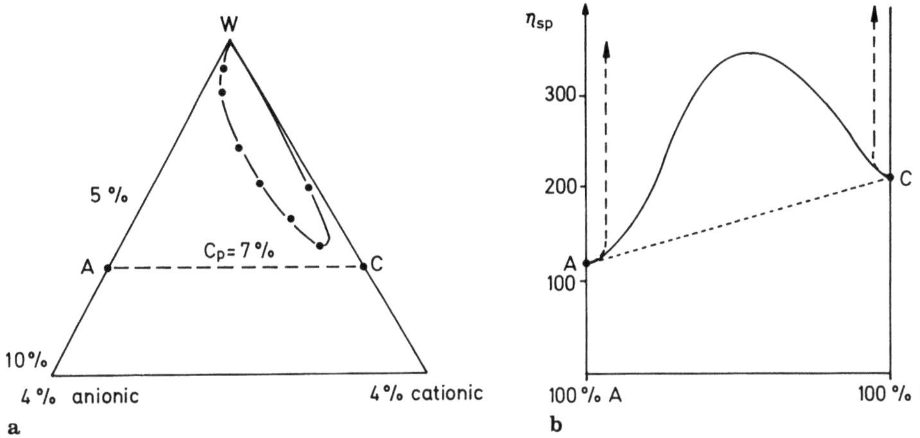

Fig. 9.9 a, b. Water solution of mixtures of P(AM-AMPS) 4% and P(AM-MSA) 4%. The phase diagram is given in (**a**). Along the line AC the system is homogeneous and the total polymer concentration is 7%. The specific viscosity of the corresponding mixtures is given in (**b**)

The behavior in more concentrated regimes was tested by viscometry. For a system P(AM-AMPS) 4%/P(AM-AMS) 4% with a total polymer weight concentration of 7% (that is in the compatible area, see Fig. 9.9a) we plot the viscosity according to the polymer composition (Fig. 9.9b). If there is no interaction between the two polymeric chains a linear evolution of the specific viscosity, according to the line A–C is expected. On the contrary, if the interaction between anionic and cationic units is considered as inducing a specific link, even reversible, a tridimensional framework should be, at least temporarily, but promptly, formed by adding one type of copolymer in the solution of the oppositely charged macromolecules. In these conditions a huge increase of viscosity (dotted lines in Fig. 9.9b) would be observed as soon as none of the polymers are very diluted. In fact only a relatively small increase in viscosity if effectively observed: the work to move a given chain in the polyanionic/polycationic environment is increased if compared with non interacting chains but no gelation effect could be observed. From this point of view the picture of polymer "complex" is not valid for these systems.

9.4 Polymer-Polymer Affinity and Phase Diagram

The thermodynamic data of polymer solutions were first derived from the approach of regular solutions of small molecules by Flory [29] and by Huggins [30]. In the original quasi-lattice model of Flory, the interaction parameter, χ, was representative of the enthalpic difference of polymer/polymer, solvent/solvent and polymer/solvent contacts. Progressively χ lost this simple signification and represented the shift with an ideal behavior (including entropic effects) but nevertheless the model was established for

- flexible chains
- short range (first neighbour) interactions
- globally repulsive interactions (endothermic mixing)
- interactions smaller than thermal agitation (that is a limiting range of χ values).

This approach was extended to polymer 1/polymer 2 systems [3–4]. They found that for the conditions mentioned above – the corresponding phase diagrams are schematized in Fig. 9.1 – the mixture is largely incompatible and when a separation occurs each polymer is mainly present in one of the two phases.

But, if an attractive effect exists between the units of the two kinds of polymers then an extra enthalpic term must be added to the free energy of mixing, calculated according to Flory–Huggins, ΔF_{FH}, to give the total free energy of mixing ΔF. More generally,

$$\Delta F = \Delta F_{FH} + \Delta F_{extra} \,.$$

If the affinity between the two types of polymers is large enough the enthalpy gain obtained by gathering macromolecules is more than balanced by the corresponding loss of entropy since the entropy of long chain molecules is never very large. So one can understand that a polymeric rich phase can be formed, in equilibrium with a diluted phase.

A similar situation can be expected for a self-associative polymer in solution: an aggregated phase in equilibrium with a dilute one is then expected, depending on the temperature [31].

A very close system is obtained by mixing an anionic polyelectrolyte to cationic micelles [32–33]. The corresponding phase diagram, given in Fig. 9.10 is very similar to the one depicted in Fig. 9.3b. More generally the neutralization of a polyelectrolyte by large multivalent counter-ions, without specific interaction, would be related by such a phase diagram.

Quantitatively the problem is difficult to solve because the effective experimental conditions are not compatible with the assumptions made for the calculations and because the analytical expression of extra free energy term due to attractive effects is generally unknown. For instance, diluted polyelectrolyte chains are generally considered as rodlike [36–37] and not as flexible chains, electrostatic interactions are long range interactions and can be larger than kT. More generally the interaction between polar species in water may drastically change the organization of water molecules around the reactive sites. So a term taking into account the change of water entropy should be introduced to express this hydrophobic effect.

The semidiluted phase is not necessarily the most complicated to be depicted. For instance for our copolymers with a content in ionic units of a few percent and a weight concentration of 5–10% it can be easily calculated, that the distance between ionic sites on a given chain is always smaller than the Bjerrum distance. Similarly in these conditions the electrostatic contribution to the persistence length [38–39] remains very small.

But the concentrated phase is in equilibrium with a diluted phase which is not consistently described by a Flory–Huggins model: it is difficult to imagine that a diluted solution of the two types of polyelectrolytes with an opposite global charge and a rather stiff conformation can be treated by the lattice formalism.

Practically some (not exhaustive) attempts to present the thermodynamic treatment of polyelectrolyte mixtures are given below but the difficulties mentioned above are not overcome.

One of the first treatments was proposed by Vorn and Overbeek [40–41]. The free enthalpy of mixing of the system is split into two terms.

$$\Delta F = \Delta F_{FH} + \Delta F_{DH}$$

the classical Flory–Huggins contribution and additionally the electrostatic contribution calculated according to the Debye and Hückel theory. To treat our aqueous mixtures of copolyacrylamides we used a similar approach [42], with a more complete expression of ΔF_{FH}. The term ΔF_{FH}, expressed by cell unit (volume a^3) of the lattice is:

$$\left.\frac{\Delta F^{FH}}{kT}\right|_{site} = \frac{\varphi_A}{N_A}\,\mathrm{Log}\,\varphi_A + \frac{\varphi_C}{N_C}\,\mathrm{Log}\,\varphi_C + \varphi_W\,\mathrm{Log}\,\varphi_W + \chi_{AC}\varphi_A\varphi_C + \chi_{AW}\varphi_A\varphi_W + \chi_{CW}\varphi_C\varphi_W$$

where φ_A, φ_C and φ_W are respectively the volume fraction of the copolymers (A, C) and of the solvent (W), χ_{ij} is the Flory interaction parameter between i and j species.

If φ_A and $\varphi_C \ll 1$ the above expression can be developed:

$$\left.\frac{\Delta F^{FH}}{kT}\right|_{site} = \frac{\varphi_A}{N_A}\,\mathrm{Log}\,\varphi_A + \frac{\varphi_C}{N_C}\,\mathrm{Log}\,\varphi_C + \frac{V_A}{a^3}\varphi_A^2 + \frac{V_C}{a^3}\varphi_C^2 + \frac{V_{AC}}{a^3}\varphi_A\varphi_C + \frac{1}{6}\frac{W^2}{a^6}(\varphi_A + \varphi_C)^3 + \cdots$$

with $V_A = (1/2 - \chi_{AW})a^3$
$V_C = (1/2 - \chi_{CW})a^3$
$V_{AC} = (1 - \chi_{AW} - \chi_{CW} + \chi_{AC})a^3$

In this expression of ΔF^{FH}, the complementary terms vanish when φ_A or φ_C tends to zero or are linear terms in φ_A and φ_C. The electrical contribution was calculated according to the same procedure rather than that of Voorn et al. [40].

$$\left.\frac{\Delta G^{el}}{kT}\right|_{site} = -\frac{1}{12\pi}\left(\frac{e^2}{\varepsilon kTa}\right)^{3/2} \varphi_{\pm}^{3/2}$$

where e, k are the classical physical constants. ε is the relative permittivity of the solvent, taken as $78.5\varepsilon_0$ (ε_0 = void permittivity) in all our experiments. The temperature T was 298 K. The term $\varphi_{\pm}$ represents the total volume fraction of all the ionic species

$$\varphi = \tau_A\varphi_A + \tau_C\varphi_C + \varphi_S$$

where τ_A and τ_C are the ionic contents (unit ratio per macromolecule) of the copolymers and φ_S the volume fraction of the other ionic species added in the solution. For all the volume fraction calculations, each species in the solution (monomeric unit, ion, water molecule) is considered as occupying a similar volume which is the cell volume a^3. This is again an approximation. We choose as cell volume the size of the water molecule, i.e. $a = 0.31$ nm. Modifications of this parameter only induce a shift of the curves in the phase diagram; there is no change in the shape of the curves.

Experimentally we determined a cloud point curve. We can consider that it is not very far from the spinodale, corresponding to the equation

$$\frac{\partial^2 \Delta F}{\partial \varphi_A^2} \cdot \frac{\partial^2 \Delta F}{\partial \varphi_C^2} - \left(\frac{\partial^2 \Delta F}{\partial \varphi_A \partial \varphi_C}\right)^2 = 0$$

that is:

$$\left[\frac{1}{N_A\varphi_A} + \frac{2V_A}{a^3} + \frac{W^2}{a^6}(\varphi_A + \varphi_C) - \frac{3}{4}\alpha\tau_A^2\varphi_{\pm}^{-1/2}\right]$$
$$\times\left[\frac{1}{N_C\varphi_C} + \frac{2V_C}{a^3} + \frac{W^2}{a^6}(\varphi_A + \varphi_C) - \frac{3}{4}\alpha\tau_C^2\varphi_{\pm}^{-1/2}\right]$$
$$-\left[\frac{V_{AC}}{a^3} + \frac{W^2}{a^6}(\varphi_A + \varphi_C) - \frac{3}{4}\alpha\tau_A\tau_C\varphi_{\pm}^{-1/2}\right]^2 = 0$$

where α is a numerical constant: $\alpha = \dfrac{1}{12\Pi}\left(\dfrac{e^2}{\varepsilon kTa}\right)^{3/2}$.

The predictions are qualitatively in very good agreement with the experimental results: formation of a diphasic zone (closed loop in the triangular diagram) if the ionic content of the copolymer is large enough and then a concentrated solution of the two polymers is in equilibrium with a very diluted phase. This area of incompatibility practically does not depend on the molecular weight (if it is large enough); it drastically increases with the ionic content of the copolymers and disappears by increasing the ionic strength of the medium (see Figs. 9.10 and 9.11). The asymmetry of the diagram is very sensitive to the symmetry of the pairs χ_{AW}, χ_{CW} and τ_A, τ_C (see Fig. 9.12).

However, a very careful quantitative description cannot be expected since the Debye–Hückel approach does not take in consideration the correlation of the ions born by the chains and the ΔF_{DH} is not correct. Similar attempts to give a separate expression of the electrostatic term were also done by Veis [43] and by Nakajima [8].

A different way was chosen to predict the behavior of the mixtures anionic polyelectrolyte/cationic surfactant [32, 33]. The formal Flory presentation is maintained but the interaction parameter, χ, is assumed to give a complete representation of all types of interactions (including electrical). In these conditions, χ, is a more or less adjustable parameter with values which can be out of the scope of the original Flory's definition. An important point, well developed

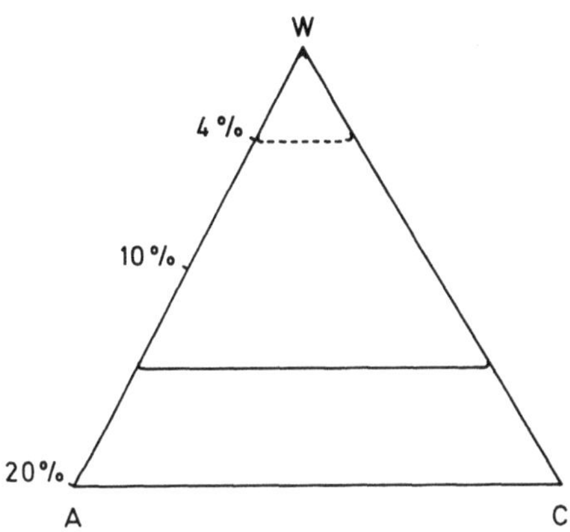

Fig. 9.10. Calculated phase diagrams (spinodale) for ionic contents $\tau_A = \tau_C = 5\%$ (*dotted line*) and 10% (*full line*). When parameters are $\chi_{AW} = \chi_{CW} = 0.45$; $\chi_{AC} = 0$, then the polymerization degree of each copolymer is 1000

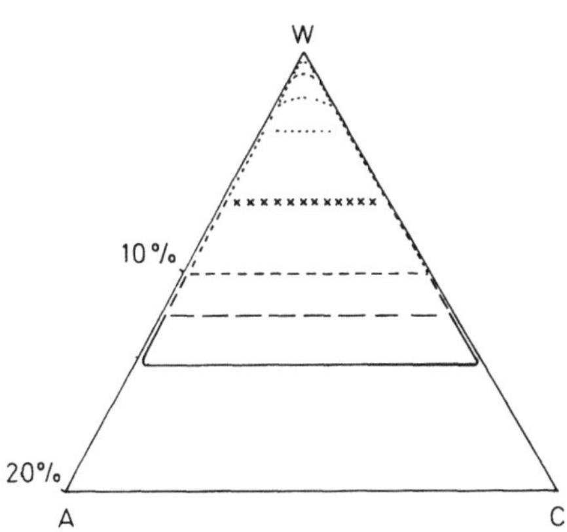

Fig. 9.11. Calculated phase diagrams: effect of the ionic strength. $\varphi_S = 0$ ——; $\varphi_S = 0.005$ ----; $\varphi_S = 0.01$ . . . .; $\varphi_S = 0.02$ ****; $\varphi_S = 0.03$ ○○○○ and $\tau_A = \tau_C = 10\%$. For other parameters see Fig. 9.10

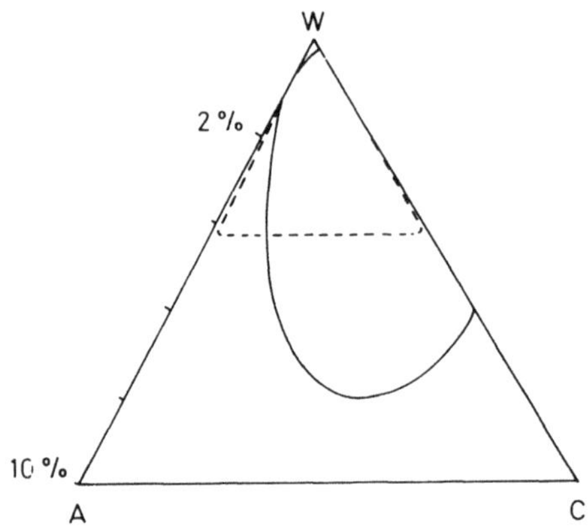

Fig. 9.12. Calculated phase diagram: effect of disymmetry on copolymer ionicity: $\tau_A = \tau_C = 5\%$ ---- and $\tau_A = 4\%$ $\tau_C = 6\%$ ——. For other parameters see Fig. 9.10

by Thalberg et al. [32, 34], is that the phase diagram does not correspond to a true ternary system. In other words if you consider the counter-ions of the two polyelectrolytes species, they constitute a small salt, the concentration of which is not necessarily identical in the two phases when demixing occurs. The thermodynamic equilibrium then must not be depicted in a triangle but in a pyramidal representation.

So, one of the main limitations for a valuable description of this kind of mixtures is the persistent lack of knowledge in the thermodynamic description of polyelectrolytes in solution. However, recently, new expressions of the free energy of a polyelectrolyte chain were proposed by Leibler, Pezron et al. [44]. More generally, a microphase separation for polyelectrolytes for which the macromolecular backbone is not soluble in water were proposed independently by Borue et al. [45] and Leibler et al. [46]. An approach, taking into account the role of polyelectrolyte counterions was also proposed [47].

9.5 Conclusion

The mixture of two oppositely charged low ionic density polyelectrolytes leads to the formation of a coacervate but in the polymer rich phase the chains are not paired, even locally as expected in a complex formation. The existence of this phase separation is simply explained by classical thermodynamic approaches but we need convenient theories for polyelectrolytes in water solution.

Acknowledgement. We are very grateful to Dr I. Iliopoulos for stimulating discussions and for his interest in this study.

9.6 References

1. Beijerinck NW (1910) Kolloid Z 7:16
2. Dobry A, Boyer-Kawenoki F (1947) J Polym Sci 2(1):90
3. Scott RL (1949) J Chem Phys 17(3):279

4. Tompa H (1949) Trans Farad Soc 45:1142
5. Bekturov EA, Bimendina LA (1981) Adv Polym Sci 41:99
6. Tsuchida E, Abe K (1982) Adv Polym Sci 45:2
7. Michaels AS, Miekka RG (1961) J Phys Chem 65:1765
8. Nakajima A, Sato H (1972) Biopolymers 10:1345
9. Nagata I, Morawetz H (1981) Macromolecules 14:87
10. Nozuyama K, Ueno Y, Kato T, Sekine Y (1986) Makromol Chem 187:1159
11. Senuma M, Kuwabara S, Kaeriyama K, Hase F, Shimura Y (1986) J Appl Polym Sci 31:1687
12. Dautzenberg H, Linow KJ, Philipp B (1981) Acta Polym 32(4):225
13. Dautzenberg H, Linow KJ, Philipp B (1982) Ibid 33(11):619
14. Kotz J, Linow KJ, Philipp B, Dautzenberg H (1986) Ibid 37(2):108
15. Weight H, Dautzenberg H, Koetz J, Linow KJ, Phillip B, Rother G (1991) ACS Atlanta meeting "Soluble polymer complexes" symposium
16. Djadoun S, Goldberg R, Marawetz H (1977) Macromolecules 10:1015
17. Gupta MK, Bansil R (1981) Polym Prepr Am Chem Soc Div Polym Chem 22(2):375
18. Feng XD, Guo XQ, Qiu KY (1988) Makromol Chem 189:77
19. Mabire F, Audebert R, Quivoron C (1984) Polymer 25:1136
20. Lecacheux D, Lesec J, Quivoron C (1982) J Appl Polym Sci 27:4867
21. Lesec J, Volet G (1990) J Appl Polym Sci Appl Polym Symp 45:177
22. Tsuchida E, Osada Y, Sanada K (1972) J Polym Sci Polym Chem Ed 10:3397
23. Tsuchida E, Osada Y, Abe K (1974) Makromol Chem 175:583
24. Abe K, Ohno H, Tsuchida E (1977) Makromol Chem 178:2285
25. Philipp, B, Linow KJ, Dautzenberg H (1984) Acta Chim Hungarica 117(1):67
26. Dautzenberg H, Dautzenberg H (1985) Acta Polym 36(2):102
27. Kotz J, Kunze J, Linow KJ, Philipp B (1986) Polym Bull 15:247
28. Desbrieres J, Rinaudo M (1981) Eur Polym J 17:1265
29. Flory PJ (1942) J Chem Phys 10:51
30. Huggins M (1942) Ann NY Acad Sci 43:1
31. de Gennes PG (1991) 7th Int Conf Surface and Colloid Science Compiegne (France) 7–13 July
32. Thalberg K (1990) Polyelectrolyte-surfactant interactions. Thesis Lund University (Sweden)
33. Thalberg K, Lindman B, Karlstrom G (1990) J Phys Chem 94:4289
34. Thalberg K, Lindman B, Karlstrom G (1991) J Phys Chem, 95:6004
35. Thalberg K, Lindman B, Bergfeldt K (1991) Langmuir 7:2893
36. Manning GS (1967) J Chem Phys 47(6):2010
37. Oosawa F (1971) "Polyelectrolytes" M. Dekker Inc New York
38. Odijk T (1977) J Polym Sci Polym Phys Ed 15:477
39. Skolnick J, Fixman M (1977) Macromolecules 10:944
40. Voorn MJ (1956) Rec Trav Chim 75:317:405:427:925
41. Michaeli I, Overbeek JTG, Voorn MJ (1957) J Polym Sci 23:443
42. Frugier D (1988) "Anionic and cationic copolymers with low charge density:synthesis and study of their complex coacervation". Thesis Univ P et M Curie (Paris)
43. Veis A (1983) In: Wilson, Prosser (eds) Developments in ionic polymers Vol 1 (London) Chap 7
44. Pezron E, Leibler L, Lafuma F (1989) Macromolecules 22:2656
45. Borue V, Erukhimovich I (1990) Macromolecules 23:3625
46. Joanny JF, Leibler L (1990) J Phys (France) 51:545
47. Khokhlov A, Nyrkova IA (1992) Macromolecules 25:1493

10 Basic Properties of Soluble Interpolyelectrolyte Complexes Applied to Bioengineering and Cell Transformations

V.A. Kabanov

Interpolyelectrolyte complexes (IPEC) are formed as result of cooperative coupling reactions between two oppositely charged polyions. Nonstoichiometric IPEC formed from polyions of different length, in dilute solution can be considered as peculiar amphophilic block-copolymers. The recent data on kinetics and mechanisms of polyelectrolyte coupling and IPEC interchange reactions as well as the factors which determine interchange direction and position of equilibria in IPEC containing systems are discussed. In particular, it is shown that in some cases such interactions are remarkably selective and may be effectively controlled by minor changes in composition of the reaction media and polyelectrolyte constitution. These basic properties can be applied to design enzyme/IPEC systems where catalytic activity is controlled by interpolyelectrolyte interactions as well as highly effective IPEC based pseudohomogeneous immune diagnostic systems. It is also shown, that coupling of DNA plasmid with carbochain polycations results in the formation of membrane-active IPEC species which may be much more active in transformation of cells, than the pure plasmid.

10.1 Introduction

Interpolyelectrolyte complexes (IPEC) form a special class of polymer bicomponent compounds [1–4]. IPEC can be obtained by direct mixing of solutions of polycationic and polyanionic components

$$(\rangle\!\!-A^-b^+)_n + (\rangle\!\!-B^+a^-)_m \rightleftarrows$$
$$[(\rangle\!\!-A^-\,{}^+B\!\!-\!\langle)_x \cdot (\rangle\!\!-A^-b^+)_{n-x} \cdot (\rangle\!\!-B^+a^-)_{m-x}] + xa^- + xb^+ \qquad \text{I}$$

The degree of conversion, θ, in reaction I is determined as the ratio of a current or equilibrium number of interpolyion salt bonds to the ultimate one ($\theta = x/m$ when $n \geq m$, and $\theta = x/n$ when $n < m$). If one of the interacting polyelectrolytes is a weak polyacid or a weak polybase θ can be easily controlled by variation of the pH.

$$(\rangle\!\!-COOH)_n + (\rangle\!\!-\overset{|}{\underset{|}{N}}H^+)_m \underset{H^+}{\overset{HO^-}{\rightleftarrows}}$$
$$[(\rangle\!\!-COO^-\,{}^+H\overset{|}{\underset{|}{N}}\!\!-\!\langle)_x \cdot (\rangle\!\!-COOH)_{n-x} \cdot (\rangle\!\!-\overset{|}{\underset{|}{N}}H^+)_{m-x}] \underset{H^+}{\overset{HO^-}{\rightleftarrows}}$$
$$(\rangle\!\!-COO^-)_n + (\rangle\!\!-\overset{|}{\underset{|}{N}})_m \qquad \text{II}$$

Macromolecular Complexes
in Chemistry and Biology
Eds. Dubin/Bock/Davis/Schulz/Thies

Therefore, in such cases IPEC can be prepared in equilibrium conditions by gradual change of pH, and a current θ value can be determined in a good approximation from potentiometric titration data using the following equations [2]:

$$\theta = (m_b/V + [H^+] - \sqrt{K_a C_0})/C_0 \quad \text{(in case of a weak polyacid)} \qquad (10.1)$$

$$\theta = (m_a/V + [HO^-] - \sqrt{K_b C_0})/C_0 \quad \text{(in case of a weak polybase)} \qquad (10.2)$$

where m_a or m_b is the molar amount of an added acid or base, V is the current volume of the reaction mixture, K_a and K_b are intrinsic dissociation constants of polyacid and polybase, C_0 is a concentration of one of polyelectrolytes.

Figure 1 represents typical experimental dependencies of θ on pH, calculated from titration curves of weak polyelectrolytes in the presence of oppositely charged polyion counterparts. θ-pH curves are compared with pH dependencies of degree of ionization, α, of individual polyelectrolyte components. A value of $\Delta pH(\theta, \alpha)$ is the direct measure of the free energy of cooperative stabilization of IPEC (ΔG_{cs}) at given θ:

$$\Delta pH(\theta, \alpha) = (\Delta G'(\alpha) - \Delta G''(\theta)]/2.3RT = \Delta G_{cs}/2.3RT \qquad (10.3)$$

where $\Delta G'(\alpha)$ and $\Delta G''(\theta)$ are ionization free energies of a corresponding polyelectrolyte in the absence and in the presence of a polyion counterpart at a given value of α or $\theta = \alpha$, respectively. Equation (10.3) is written assuming that all ionized groups of an IPEC polyelectrolyte component being charged currently form ionic bonds with its polyionic counterpart, that is close to the real situation, when the total ratio of ionizable groups $z = m/n = 1$.

10.2 Kinetic and Equilibrium Properties of Interpolyelectrolyte Complexes

Whether IPEC are soluble [2, 4–6] or of limited solubility in aqueous media is determined by a ratio of ionizable groups, z, and a relative length of oppositely charged polyions as well as by a water-phase composition (nature and concentration of a simple salt, pH, presence and concentration of organic additives etc.). IPEC formed at $z = 1$ are always insoluble at $\theta \to 1$, but have limited

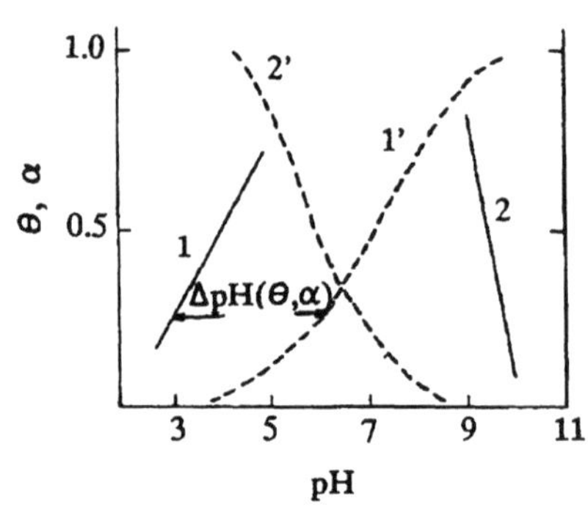

Fig. 10.1. The dependence of degree of conversion, θ, in polyion addition reactions on pH for the following systems: *1* polyacrylic acid/poly(dimethylaminoethyl methacrylate hydrochloride); *2* poly(dimethylaminoethyl methacrylate)/poly(sodium acrylate). *1'* and *2'* are the potentiometric titration curves for polyacrylic acid and poly(dimethylaminoethyl methacrylate); 20 °C

solubility in aqueous media. Water-soluble IPEC can be obtained from any oppositely charged polyelectrolytes, when certain conditions are met [2,4,5]. For example, IPEC are water-soluble at $\theta \to 1$, if the degree of polymerization of one polyion component is higher (host polyelectrolyte, HPE), than that of another one (guest polyelectrolyte, GPE), and if a molar ratio of GPE and HPE repeating units (r.u.), $\varphi = [\mathrm{GPE}]/[\mathrm{HPE}] < \varphi_c < 1$. (At uniform distribution of GPE by HPE $\varphi = z$). Such IPEC species may be considered as a peculiar block-copolymer containing hydrophilic single-stranded and relatively hydrophobic double-stranded segments [2, 4]. A critical value φ_c determines the ultimate extent of populating HPE by GPE chains, above which the hydrophilicity of single-stranded segments is not enough any more to hold IPEC species in solution. The value φ_c for most of the investigated systems varies over the range 0.5–0.2 and depends basically on chemical structure of the polyelectrolyte constituents.

A cooperative character of multisite interpolyion binding makes IPEC extremely stable with respect to dissociation of the original polyelectrolyte components within the pH region enclosed by the potentiometric curves. The dissociation constant of IPEC at θ close to 1 sharply decreases while the length of GPE increases and reaches practically zero even for relatively short oligomers. Thus in case of quaternized oligo(ethyleneimine)s

$$\mathrm{H(CH_2{-}CH_2{-}\underset{\displaystyle CH_3}{\overset{\displaystyle R}{N^+}}{-})_nH} \qquad (\mathrm{R} = -\mathrm{CH_3} \text{ or } -\mathrm{CH_2{-}C_6H_5})$$

interacting with poly(methacylate) polyanions (PMA) dissociation becomes negligible starting from about octamer and higher [3, 7]. Similar results were obtained for a series of oligo(phosphate)s interacting with poly(dimethylaminoethyl methacrylate) [8] and oligo(2,5-ionen)s interacting with poly(acrylate) polyanions (PA) [9]. Therefore, one can assume that after a certain "critical" length of polyions is exceeded their IPEC practically does not dissociate to the original polyion components.

The luminescence quenching measurements are very useful for quantitative study of formation and transformations of soluble IPEC. Most of the measurements [10–16] were made using fluorescence tagged poly(methacrylate) (PMA*) [17] poly-anions containing one anthryl or pyrenyl tag per 350–1500

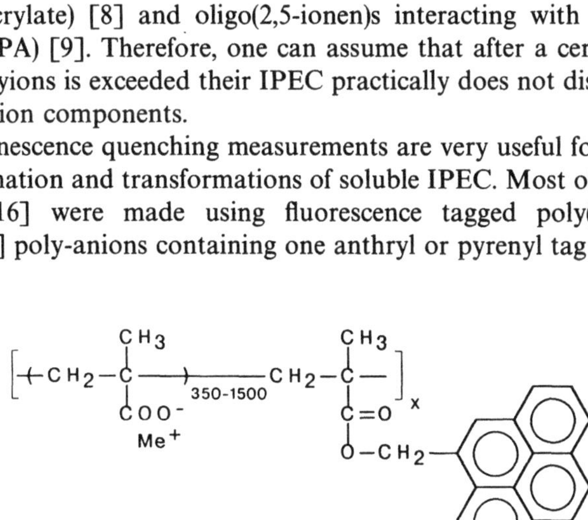

monomer units and 1-ethyl-4-vinylpyridinium (EVP)/vinylpyridine copolymer polycations (Q–P4VP). EVP unit is a quencher of fluorescence of anthryl and

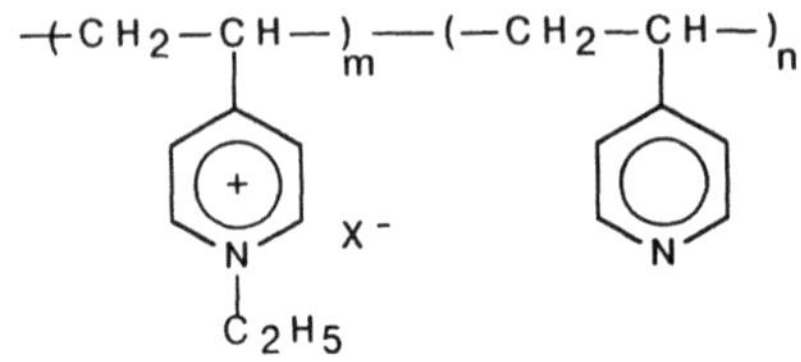

pyrenyl groups. Q–P4VP was prepared by quaternization of poly(vinylpyridine) fractions with ethyl bromide [18]. EVP r.u. content was controlled by the degree of quaternization, $\beta(\%) = [m/(m+n)] \cdot 100$.

Chemical transformations corresponding to schemes I and II, in terms of traditional classification represent ion exchange and neutralization reactions, respectively. However, with respect to polyion interaction these processes can be considered as *polyion addition reactions.*

The ability to participate in polyion interchange reactions is one of the basic properties of IPEC [10–13]:

$$\mathrm{IPEC(HPE_1/}n\mathrm{GPE) + HPE_2} \rightleftarrows$$

$$\mathrm{IPEC(HPE_1/(}n-x\mathrm{)GPE) + IPEC(HPE_2/}x\mathrm{GPE)} \qquad \text{III}$$

where n and x are numbers of GPE chains. If HPE_1 is identical with HPE_2, reaction III with respect to IPEC-polyion interaction represents the *polyion exchange reaction.* If HPE_1 and HPE_2 are chemically different, reaction III can be considered as a *polyion substitution reaction.* GPE exchange can also proceed as intra-IPEC rearrangement involving various segments of HPE within the same IPEC species, that imparts to such species a large extent of structural variability.

A representative polyion addition reaction was studied [14] using the fraction of pyrenyl tagged PMA*Na and the fraction of Q–P4VP. Formation of IPEC (PMA*/Q–P4VP) results in a decrease of relative fluorescence intensity I/I_0 of the tags due to coupling of PMA* tagged units with Q–P4VP quencher (I_0 is fluorescence intensity of an original PMA* solution at a given ionic strength). The reaction mixture was prepared by quick injection of 0.02 ml of Q–P4VP aqueous solution into a cell containing 1.5 ml of PMA*Na aqueous solution under continuous stirring. The homogenization time didn't exceed 2–4 s. I/I_0 was measured using JI-3CS (France) spectrofluorimeter at $\lambda_{ex} = 342$ nm, $\lambda_{reg} = 395$ nm. Figure 10.2 shows that mixing of the reagent solutions in absence of a simple salt is followed by an immediate drop of I/I_0 to a constant value corresponding to a certain number of rapidly established Q–P4VP/PMA* contacts (curve 1). However, this decrease is considerably smaller than that corresponding to the equilibrium system containing IPEC (PMA*/Q–P4VP) of $\varphi = 0.2$ (dotted line 1′). Mixing of the same reagent solutions but at [NaCl] = 0.03 M resulted in a much larger fall of I/I_0, value

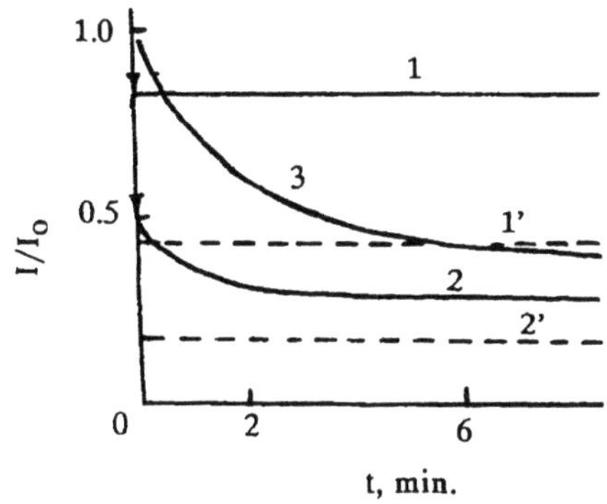

Fig. 10.2. Typical kinetic curves of polyion addition and exchange reactions. *1.* PMA*Na ($P_w = 4.4 \times 10^3$, 1 tag per 600 repeating units) + Q–P4VP ($P_w = 2.0 \times 10^2$, $\beta = 95\%$) in the absence of low molecular salt, [PMA*Na] = 86 nM, z = [Q–P4VP]/[PMA*Na] = 0.2; *2.* similar reaction but in presence of [NaCl] = 0.03 M; *3.* IPEC(PMA/Q–P4VP) + PMA*Na, [PMA*Na] = [IPEC] = 86 nM, P_w(PMA) = 4.4×10^3, $\varphi = 0.20 \cdot 20°$C, pH = 10, $\theta \to 1$. *1'* equilibrium level for curve *1*; *2'* equilibrium level for curve *2*

and its gradual decrease tending to the equilibrium level (compare curve 2 and dotted line 2′). Stop flow measurements have shown, that the fast stage of the polyion addition reaction corresponding to the falls on the kinetic curves is actually accomplished in less than 5 μs ("dead" time of the instrument), i.e. the lower limit of the corresponding bimolecular rate constant K_{2D} is of 10^9 M^{-1} s^{-1}; that is close to the estimated rate constant of diffusion collisions of the polyion coils [14]. Such collisions result in formation of random ionic bonds between the oppositely charged polyions. Polyanion/polycation associates formed remain "frozen" in a nonequilibrium state, while concentration of a simple salt is close to 0 (Fig. 10.2, curve 1). However, in presence of a simple salt intra- and interassociate rearrangements occur, which result in increase of the amount of interpolyion salt bonds and formation of equilibrium IPEC (PMA*/Q–P4VP) species (Fig. 10.2, curve 2) [14]. These rearrangements very likely proceed via intra- and inter-IPEC exchange involving oppositely charged polyion segments. The pure case of such exchange reactions is represented by kinetic curve 3 (Fig. 10.2). Pyrenyl tags additionally stabilize PMA*/Q–P4VP structure comparatively to that of PMA/Q–P4VP. Therefore, PMA* completely substitute PMA in IPEC(PMA/Q–P4VP); this can be considered as an example of molecular recognition. In fact, curve 2 can be considered qualitatively as superposition of curves 1 and 3.

In the case of soluble IPEC it is shown [15] that polyion exchange reaction consisting in transfer of a single GPE chain

$$\text{IPEC(PMA/Q–P4VP)} + \text{PMA*Na} \xrightarrow{K_2}$$

$$\text{IPEC(PMA*/Q–P4VP)} + \text{PMANa} \qquad \text{III}$$

proceeds as a second order reaction with respect to concentrations of IPEC and

Table 10.1 Rate constants of polyion exchange reaction between various IPEC(PMA·PVP) and PMA*Na. pH = 10, $q \to 1$, 20 °C. ($[PMA^*]_0 = [PMA]_0$ values relate to moles of polyanions; $P_w(PMA) = P_w(PMA^*)$, PMA* contains 1 perenyl tag per 350 repeating units)

No	$[PMA^*]_0$ / nM	$\bar{P}_w/10^3$ PVP	$\bar{P}_w/10^3$ PMA	β/%	[NaCl] / M	K_2 / $M^{-1}\cdot s^{-1}\cdot 10^5$
1.	3	0.6	3.9	95	0.08	5.1
2.	7	0.6	3.9	95	0.08	5.0
3.	15	0.6	3.9	95	0.08	4.7
4.	40	0.6	3.9	95	0.06	1.2
5.	40	0.6	3.9	95	0.07	2.6
6.	40	0.6	3.9	95	0.08	5.1
7.	40	0.6	3.9	95	0.09	8.7
8.	33	0.1	4.4	95	0.04	2.0
9.	33	0.1	1.0	95	0.04	2.0
10.	28	0.6	4.4	95	0.07	1.5
11.	28	0.32	4.4	95	0.07	6.0
12.	28	0.2	4.4	95	0.07	11.0
13.	28	0.1	4.4	95	0.07	23.0
14.	23	0.2	2.8	89	0.035	0.2
15.	23	0.2	2.8	80	0.035	0.7
16.	23	0.2	2.8	69	0.035	2.6
17.	23	0.2	2.8	62	0.035	16.2
18.	23	0.2	2.8	55	0.035	36.1

PMA* species. The kinetic curves are perfectly linearized in coordinates of Eq. (10.4)

$$q/(1-q) = K_2[PMA^*]_0 \cdot t \qquad (10.4)$$

where q is degree of conversion, $[PMA^*]_0 = [PMA]_0 = [IPEC]_0$ are initial molar concentrations of the reacting macromolecular species. That gives a possibility of quantitative determination of the values of absolute rate constants of GPE transfer from one HPE to another. The experimental data are represented in Table 10.1.

One can see that, in agreement with Eq. (10.4), K_2 values are actually not dependent on the concentration of the macromolecular reagents (ex. No 1–3, 6). It is also important that K_2 is not dependent on the length of HPE (ex. No 8, 9) and strongly dependent on the length of GPE (ex. No 10–13). In the studied experimental conditions K_2 values are of the order 10^4–10^6 $M^{-1}s^{-1}$, i.e. 3–5 orders less that the estimated value of K_{2D}. The scope of these data suggests that interchange proceeds via a ternary (PMA. . .GPE. . .PMA*) short-lived intermediate ("transition state"), formed by interpenetrating IPEC (PMA/Q–PVP) and PMA* coils at their random collisions in the solution. In the above mentioned experimental conditions far not an every collision becomes a productive one: 10^3–10^5 of them are required in average for one act of polyion transfer to occur. The strong dependence of K_2 on the ionic strength (ex. No 4–7), and on the linear charge density of Q–P4VP chains (ex. No 14–18) are in consistence with the interchange mechanism proposed.

Alongside with these data it should be pointed out that interchange reaction can be fully restricted even in presence of a simple salt, when electrostatic interaction between the oppositely charged polyions is strongly modulated by some other attraction forces such as hydrophobic interaction. The striking example is given by comparison of the behavior of two IPECs, one formed by PA (or PMA) and linear poly(ethyleneimine) (PEI), the other by PA (or PMA) and poly(*tret*-butylaziridin) (PTBA) in aqueous solution [19]. Both polycationic components are characterized by equal linear charge density

$$(\text{-}\overset{+}{N}H_2\text{-}CH_2\text{-}CH_2\text{-})_n \qquad (\text{-}\overset{+}{N}H\text{-}CH_2\text{-}CH_2\text{-})_n$$

$$\text{PEI} \qquad\qquad CH_3\text{-}C(CH_3)\text{-}CH_3 \text{ (on N)} \quad \text{PTBA}$$

Nevertheless, while PEI easily participates in polyion interchange reactions, PTBA chains at elevated temperatures (40–60 °C) are fixed in IPEC species even at a relatively high ionic strength, apparently, due to intra-IPEC interaction of bulky hydrophobic groups.

It is also shown that in case of polyion substitution reactions not only kinetics but also a position of equilibria is extremely sensitive to a structure and a length of polyion counterparts as well as to a nature and concentration of small counterions [8, 9, 13]. In particular, sulfonate- and sulfate-containing polyanions usually replace carboxylate-containing polyanions in corresponding IPEC with various polycations [9]. The rate of polyion substitution reaction between IPEC (PMA*/Q–P4VP) and poly(potassium vinyl sulfate) (PVSK) drastically depends on the concentration of a simple salt [9]:

$[NaCl]/(M \times 10^3)$	0	2	4	6	8	10
$t_{1/2}/s$	no reaction	600	60	13	5	< 1.5

($P_w(PMA^*Na) = 3 \times 10^3$; $P_w(PVSK) = 1.3 \times 10^3$; $P_w(Q\text{–}P4VP) = 2 \times 10^2$; $[IPEC(PMA^*/Q\text{–}P4VP)] = 0.004$ M(r.u.), $\varphi = 0.2$, $[PVSK] = [Q\text{–}P4VP]$; pH = 10, at 20 °C).

A dramatic effect of chain length of a competitive polyanion and a nature of counterions on direction of polyion substitution reaction was observed in [16]. It is found that the equilibrium

$$\text{IPEC(PMA}^*/n\text{Q–P4VP)} + x\text{PP} \rightleftarrows$$
$$\text{IPEC(}x\text{PP}/n\text{Q–P4VP)} + \text{PMA}^* \qquad \text{IV}$$

where PP are polyphosphates of various chain length: 20, 50, 70 and 230 r.u., PMA* is anthryl-tagged PMA ($P_w = 4100$, 1 tag per 500 r.u.), P_w(Q–P4VP) = 170 in 0.3 M NaCl solution is completely shifted to the right in case of the longest (230 r.u.) and to the left in case of the shortest (20 r.u.) PP chains (at fixed base-molar concentrations of all polymeric components). It has intermediate positions in case of intermediate PP chain lengths. At fixed PP chain length (70 r.u.) the equilibrium IV is completely shifted to the right in 0.3 M LiCl, to the left in 0.3 M KCl and has an intermediate position in 0.3 NaCl solution. The latter phenomena have no analogy in nonpolymeric systems and may be related to important regulatory processes involving biopolymers in living cells. The thermodynamic explanation is given in [16]. Its essence is the following. Let us represent the total free energy change (ΔG_t) in reaction V as the sum of two terms:

$$\Delta G_t = \Delta G_p + \Delta G_{ci} \tag{10.5}$$

Here, ΔG_p corresponds to the change of Q–P4VP contact interaction with PMA* as opposed to that with PP, while ΔG_{ci} is the difference of free energies of interaction of counterions with the segments of PMA* and PP participating in the substitution reaction.

Let us consider at first the case of reaction IV, when the length of both polyanions (4100 and 230 r.u.) is higher than the length of Q–P4VP (170 r.u.) as in Scheme IV′

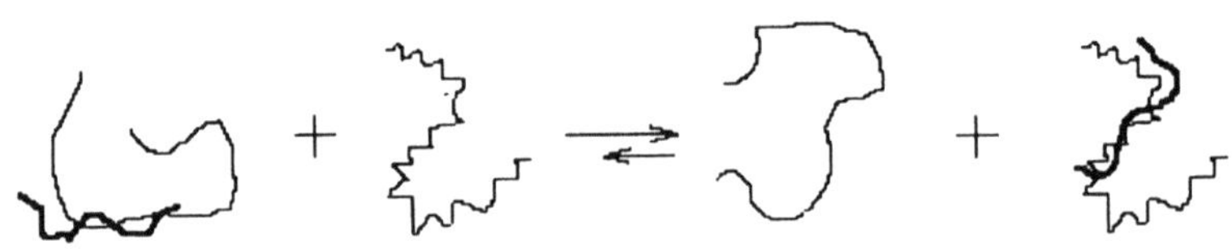

IV′

IPEC(PMA*/Q–P4VP) PP PMA* IPEC(PP/Q–P4VP)

Then, the number of polymeric species in the system is not dependent on the position of the equilibrium and remains constant. As it follows from the above mentioned experimental data the equilibrium of Scheme IV′ is shifted to the right in NaCl solution. This means that the total change of free energy in IV′ version of the reaction IV is negative ($\Delta G_t' = \Delta G_p' + \Delta G_{ci}' < 0$). However, the decrease of the chain length of PP results in the shift of the equilibrium to the opposite side as in Scheme IV″

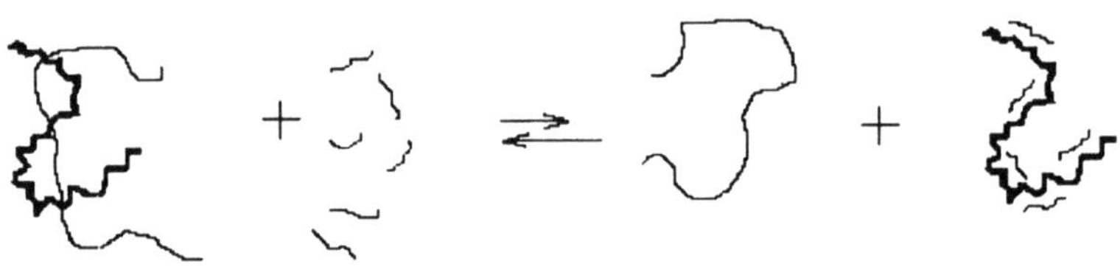

IV″

IPEC(PMA*/Q–P4VP) PP PMA* IPEC(PP/Q–P4VP)

This means that ΔG_t changes sign. In other words the total change of free energy in reaction IV″ becomes positive ($\Delta G_t'' = \Delta G_p'' + \Delta G_{ci}'' > 0$). As far as the total number of electrostatic contacts between the oppositely charged polyelectrolytes remains unchanged ($\theta \to 1$), ΔG_{ci} does not depend on the chain length, i.e. $\Delta G_{ci}' = \Delta G_{ci}''$. Therefore, the observed effect can be attributed to an increase of ΔG_p, i.e. $\Delta G_p'' > \Delta G_p'$. Indeed, from the qualitative consideration it is clear that the transition from left to right in case IV″ is less favorable with regard to the entropy than in case IV′ because of the decrease of the total number of polymeric species.

Now, let us consider the case when the alkaline metal counterions are varied at fixed length of the polyions. Then, ΔG_p remains unchanged and the fact of a reverse of the sign of ΔG_t, when Li^+ is changed for K^+ is caused by the change of ΔG_c. This can be explained by a difference in free energies of interactions between PMA and PP polyanions and various alkali metal cations. Of course, such difference is due to the contribution of a specific (of donor-acceptor type) interaction superimposed on the Coulomb interaction. Indeed, the critical salt concentration resulting in dissociation of IPEC(PMA/Q–P4VP) increases in the series: LiCl < NaCl < KCl. The sequence is reversed for IPEC(PP/Q–P4VP).

Thus, the equilibrium in a reaction of type V can be controlled by changing counterions and the degree of polymerization of the participating chains.

The role of GPE in IPEC can be played by charged globular species such as protein molecules [21, 22]. It is significant to note that linear polyions characterized by a relatively high charge density usually replace GPE proteins, "squeezing" them from the original IPECs [22, 23]

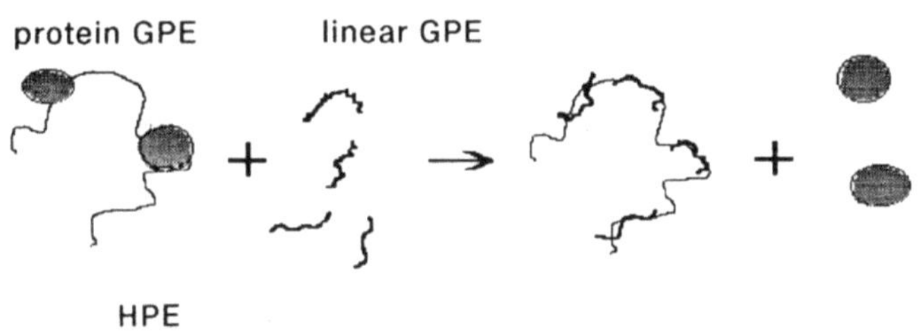

V

Soluble IPEC species in which HPE component is a weak polyacid or a weak polybase can be compacted and precipitated by a pH change; i.e. suppressing ionization of single-stranded blocks [2, 4, 24]. For example, IPEC formed by PMANa (HPE) and Q–P4VP (GPE) is soluble at pH > 6 but sharply segregates below this value (Fig. 3). Phase separation in IPEC(PMANa · Q–P4VP)/H_2O system can be also induced by addition of a low molecular salt [2, 4, 25–27] (Fig. 10.4). The transition is also very sharp. This phenomenon which has been studied in detail can be described by the following Scheme:

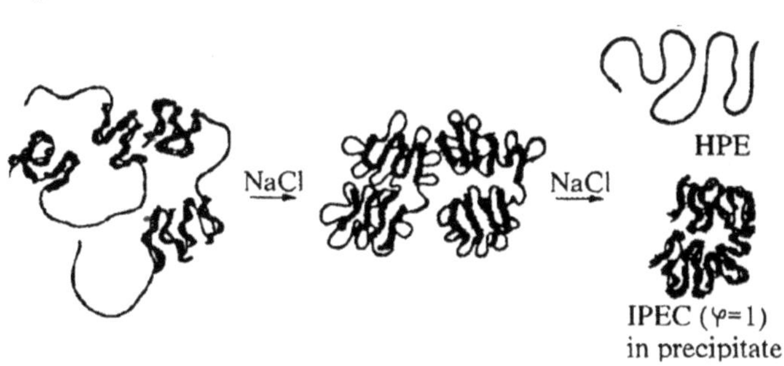

VI

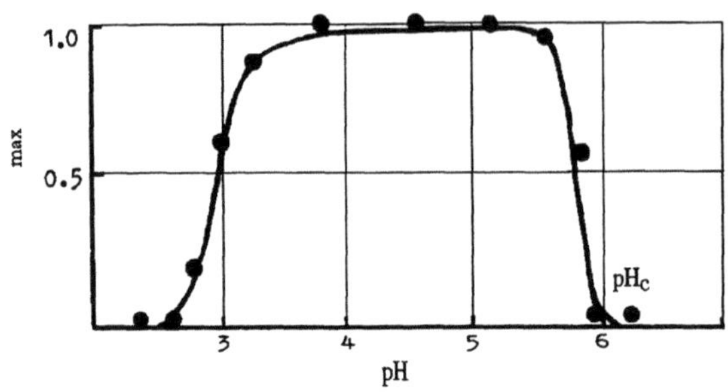

Fig. 10.3. The pH dependence of relative turbidity of IPEC(PMANa·Q–P4VP) solution; P_w(PMANa) = 4100, P_w(Q–P4VP) = 200, [Q–P4VP] = $2 \cdot 10^{-3}$ M, $\varphi = 0.33$, [NaCl] = 0.02 M, 20 °C. (Clarification of the solution at pH < 2.5 is caused by complete dissociation of IPEC to poly(methacrylic acid) and Q–P4VP)

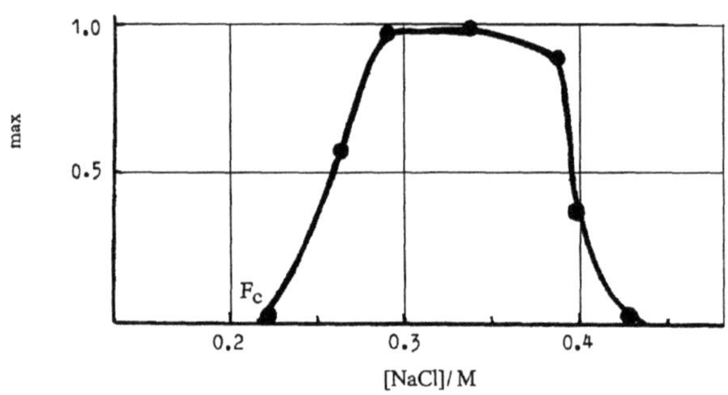

Fig. 10.4. The dependence of relative turbidity of PIC(PMANa·Q–P4VP) solution on the concentration of the low molecular salt. (IPEC composition, concentration and characteristics of the components are equal to those in Fig. 10.3); pH = 7.5, 20 °C. (Clarification of the solution at [NaCl] > 0.4 M is caused by IPEC dissociation to the original polyelectrolyte components)

At the first stage single-stranded HPE blocks shrink because of the shielding of their charges by low molecular counterions. At the second stage when a certain critical value of ionic strength (I_c) is exceeded, GPE disproportionation occurs resulting in formation of compact insoluble stoichiometric ($\varphi = 1$) IPEC species, which quantitatively precipitate, while practically pure HPE remains in solution. Disproportionation proceeds via a foregoing polyion exchange mechanism.

Critical pH (pH_c, Fig. 10.3) and I_c (Fig. 10.4) values depend on the composition of original soluble IPEC: the higher is φ the lower are pH_c (if HPE is a poly(carboxylic acid) and I_c. However, qualitatively the dependences represented in Figs. 10.3 and 10.4 are similar at all $\varphi < \varphi_c$. It is essential that the phase separation observed under a change of pH or salt concentration is quite reversible.

10.3 Interpolyelectrolyte Complexes as Protein Carriers

The basic properties of IPECs can be applied to design enzyme/IPEC systems, whose catalytic activity is controlled by the foregoing interpolyion interactions. Some promising results of this approach were recently reviewed in [28].

The proposed strategy consists in the covalent bonding of an enzyme to a GPE or HPE chain and incorporating the conjugate into IPEC by complexing it with a proper HPE or GPE correspondingly. For example, a typical procedure using P4VP based GPE includes quaternization of a certain fraction of P4VP units by 2-bromoethanol and the rest by ethylbromide [29] to obtain the copolymer (Q′–P4VP) of the following structure:

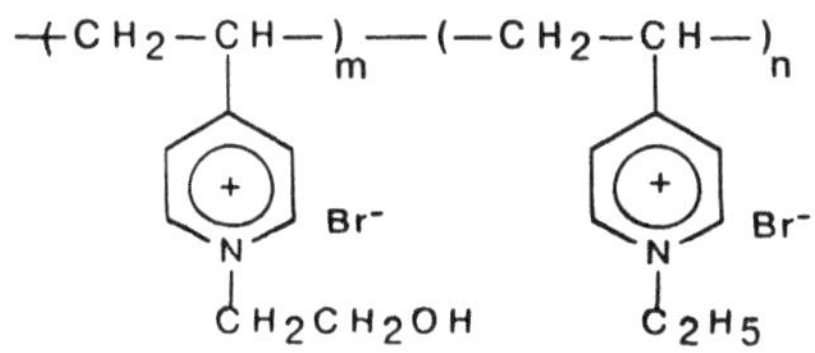

Then the enzyme is chemically bound to hydroxyl groups of the copolymer using *s*-trichlorotriazine [29]. Another procedure involves carboxyl containing polymers or copolymers using water-soluble carbodiimides as coupling reagents [30]. The above mentioned procedures have been used to prepare penicillin amidase (PAase), α-chymotrypsin (CT), alcoholdehydrogenase (ADHase) and urease conjugates with P4VP based polycations [30, 31] and PAase and CT conjugates with PMANa [31]. These conjugates were then complexed with complementary polyelectrolyte partners to obtain a series of IPEC entrapped enzymes: (PAase·Q′–P4VP/PMANa) [32], (ADHase·Q′–P4VP/PMANa) [32], (ADHase·Q′–P4VP/PMANa) [29], (urease·Q′–P4VP/PMANa) [31], (PAase·PMANa/Q′–P4VP) [30], (CT·PMANa/Q′–P4VP) [33], (urease·PMANa/Q′–P4VP) [31].

Some characteristics of enzymes entrapped in soluble IPEC such as pH profiles differ in comparison with those of native enzymes, mainly because of the influence of charged single-stranded HPE segments. For example, pH optima of PAase, CT and ADHase bound to Q′–P4VP and complexed with PMANa are shifted by 0.2–0.4 pH units to the alkaline region. The shift is higher the lower is a φ value, i.e. the smaller is a fraction of HPE repeating units blocked by GPE chains. At the same time the intrinsic characteristics of the enzymes such as the catalytic rate constant, k_{cat}, and Michaelis constant, K_M, are not changed much (Table 10.2 [34]). The reaction catalyzed by the enzymes incorporated in soluble IPEC are not diffusion controlled [34].

The phase separation diagrams of enzyme-containing IPEC in most cases practically coincide with those of the corresponding free IPEC (Fig. 10.5). Only in the case of IPEC carrying bulky urease globule consisting of 6 protein subunits one can observe a shift of the upper branch at phase separation in the

Table 10.2. Kinetic parameters of hydrolysis of benzyl penicillin by various preparations of penicillin-amidase. Conditions pH 7.5, 25 °C

PAase preparation	$k_{cat} \cdot sec$	$K_M \cdot \mu M^{-1}$
Native enzyme	50	9
PAase · Q'–P4VP	33	9
IPEC(PMANa/PAase · Q'–P4VP)	20	8
PAase immobilized in cellulose triacetate fibers	—	430
PAase immobilized in poly(acrylamide) gel	—	530

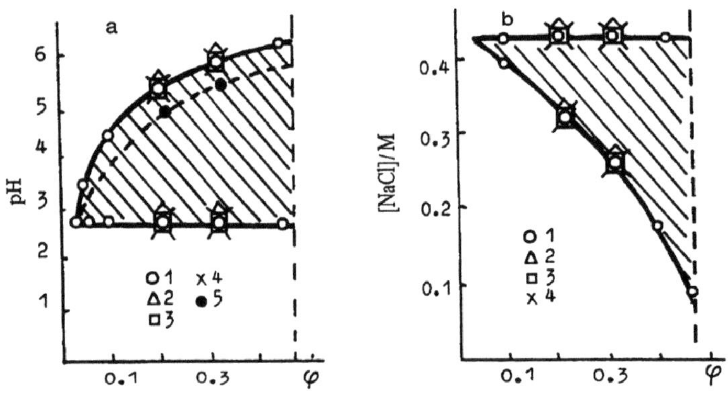

Fig. 10.5 a,b. Phase separation φ-pH (**a**) and φ-[NaCl] (**b**) diagrams of IPEC(PMANa/Q'–P4VP) and (PMANa/enzyme·Q'·P4VP) systems. P_w(PMANa) = $4 \cdot 10^3$, P_w(Q'–P4VP) = $2 \cdot 10^2$, [PMANa] = 10^{-2} M, 25 °C; (**a**) [NaCl] = 0.05 M, (**b**) pH = 7.5. (1) IPEC(PMANa/Q'–P4VP), (2) IPEC(PMANa/PAase · Q'–P4VP), (3) IPEC(PMANa/CT · Q'–P4VP), (4) IPEC(PMANa/ADHase · Q'–P4VP), (5) IPEC(PMANa/urease · Q'–P4VP). (1 bound enzyme molecule per 3 Q'–P4VP macromolecules in averages)

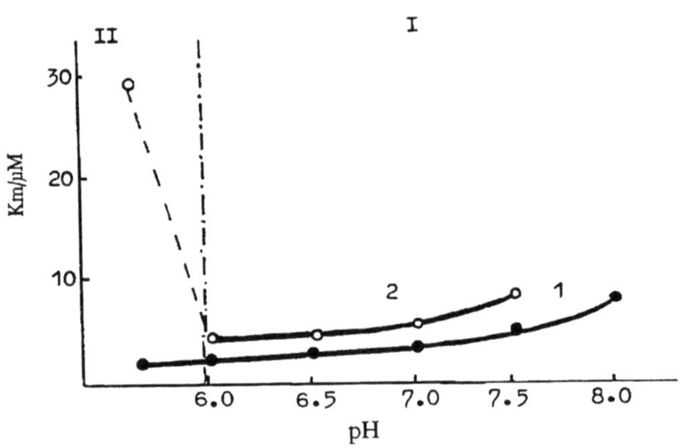

Fig. 10.6. The pH dependence of K_M in benzyl penicillin hydrolysis (*1*) by native PAase and (*2*) by IPEC(PMANa/PAase · Q–P4VP), φ = 0.33. 25 °C

φ–pH diagram by 0.5 pH unit to the acidic region [31]. Figure 10.6 shows that the observed K_M of PAase in IPEC (PMANa/PAase · Q′–P4VP) remains practically constant in the whole pH-range, where the complex is soluble and only slightly differs from K_M of native PAase. However, it sharply increases in the pH region (below pH 6), where the complex is compacted and precipitates (compare with Figs. 10.3 and 10.5). Of course, it does not mean that phase separation suppresses the binding capacity of the enzyme; it simply reflects the drastic decrease of a local substrate concentration in the vicinity of the active sites. Correspondingly, the rate of enzymatic reaction also sharply decreases.

As follows from Fig. 10.5 enzymes entrapped in soluble IPEC can easily be separated from the reaction system, i.e. from the reaction products by appropriate change of pH or ionic strength. Compacting and transition of enzyme/IPEC species into an insoluble state immediately terminates or strongly inhibits a catalytic reaction. Scheme VII illustrates the molecular mechanism of these transitions,

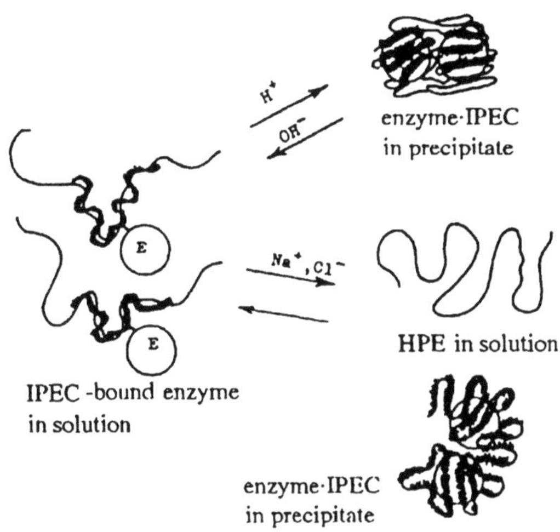

VII

which are fully reversible and can be repeated many times without any damage or distortion to the enzyme. Reiterated dissolving of enzyme containing IPEC each time is followed by quantitative recovery of its catalytic activity; that is an advantage in comparison with other reversibly soluble immobilized enzyme systems [35–38] described.

An enzyme catalyzed reaction, such as ester hydrolysis, resulting in formation of additional ionic species, i.e. an increase of ionic strength, I, can be reversibly self-terminated, if I reaches its critical value I_c. For a given enzyme–IPEC system the following equation is valid

$$I_c = I_0 + \Delta I_t = \text{const} \qquad (10.6)$$

where I_0 is ionic strength in the initial reaction system, ΔI_t is the increase of ionic strength, caused by formation of the ionic reaction product, which is required to self-terminate. In other words the desirable degree of conversion of a substrate

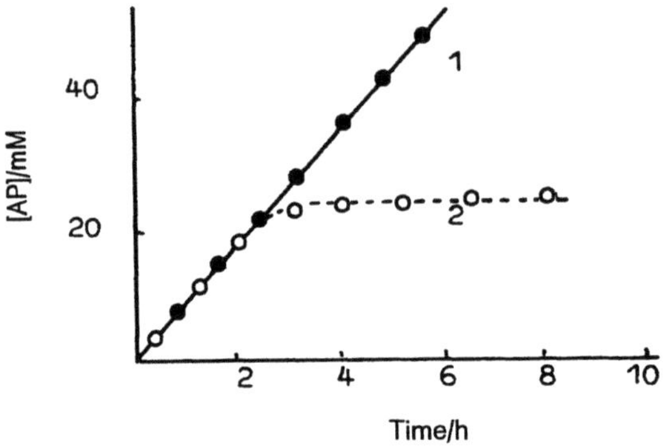

Fig. 10.7. Kinetic curves of hydrolysis of benzyl penicillin catalyzed by IPEC(PMANa/PAase · Q′–P4VP), $\varphi = 0.33$, 0.1 M phosphate buffer, pH 7.5, 25 °C; (*1*) $I_0 = 0.20$ M, (*2*) $I_0 = 0.25$ M

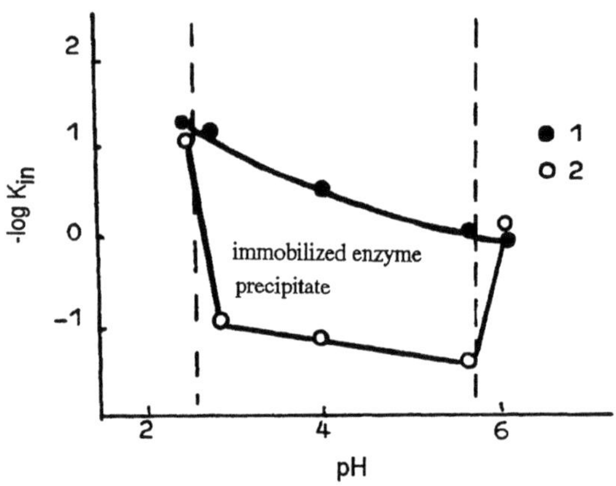

Fig. 10.8. The pH dependence of the inactivation first order rate constant (K_{in}) of native PAase (*1*) PAase entrapped in IPEC(PMMNa/PAase · Q′–P4VP) (*2*), 60 °C

can be programmed simply by variation of I_0 or other parameters (φ, pH) affecting I_c. A good example is hydrolysis of sodium benzyl penicillin catalyzed by soluble IPEC(PMANa/PAase · Q′–P4VP) system resulting in formation of sodium 6-aminopenicillinate (AP) and sodium phenyl acetate (Fig. 10.7) [34]. These results may be considered as a mimic of feedback and self-regulating processes in living cells.

Reversible collapse of enzyme·IPEC species can also be used to control enzyme stability with respect to denaturation processes. Thus, thermal stability of PAase in soluble IPEC (PMANa/PAase · Q′–P4VP) is at pH > 6.0 practically equal to that of native PAase. However, compacting of the complex species at pH < 6.0 (see Figs. 10.3 and 10.5a) results in more than a 10-fold decrease in the inactivation rate constant (K_{in}), i.e. in considerable enhancement of thermal stability, as it is seen in Fig. 10.8. At the same time a sharp decrease of the apparent catalytic activity (V/K_M, where V is the reaction rate) is also observed (Fig. 10.6, curve 2). Dissolving of enzyme · IPEC species at pH < 2.9 is followed by immediate loss of the additional stabilization effect, so that K_{in} becomes equal again to that of native enzyme [39] (Fig. 10.8, compare curves 1 and 2). Conformation transitions of enzyme · IPEC species with variation of pH are represented schematically in Fig. 10.9. Stabilization of the enzyme globule

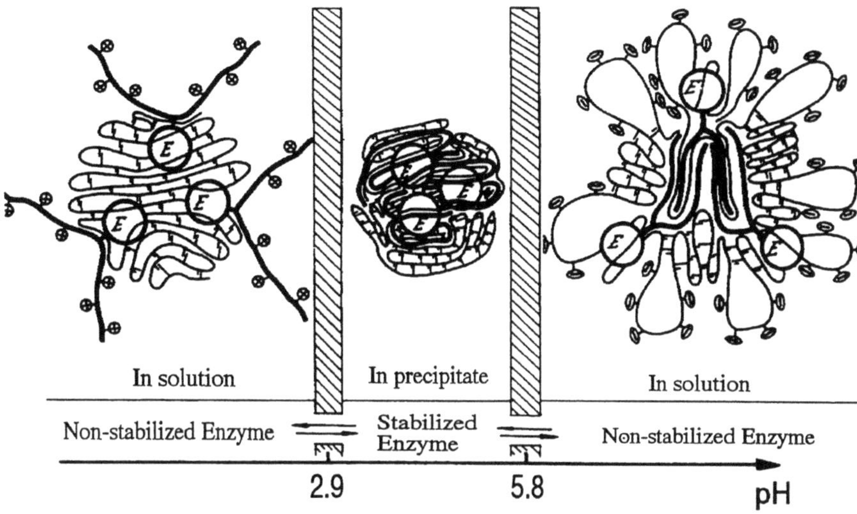

Fig. 10.9. Schematic representation of pH-induced conformation transitions in IPEC(HPE/enzyme·GPE) system [28]

immobilized in the compact IPEC species is most probably caused by casehardening and shielding its tertiary structure in the dense insoluble IPEC core crosslinked by multipoint systems of hydrogen and ionic bonds formed between all polymeric partners of the complex. Reversible stabilization against denaturation correlated with a loss of catalytic activity of IPEC-incorporated enzymes, induced by the foregoing changes in the reaction systems, may mimic spore formation [40–42], protecting some living cells against unfavorable environmental deviations.

It is significant to note that all the above-mentioned phenomena are typical only for IPEC formed by complexing HPE and enzyme·GPE polycation conjugates. In the reverse case of enzyme·HPE conjugates complexed with GPEs the behavior of the resulting enzyme·IPEC systems is somewhat different. Thus only a 2-fold decrease of K_{in} (instead of 10-fold) is observed for IPEC(PMANa·PAase/Q′–P4VP) system, probably because in such a case the enzyme is not localized within the dense core but in the swollen "shell" of the insoluble IPEC species [31].

Enzyme incorporation in IPEC may be effectively used to control availability of catalytic sites for inhibitors, especially those of high molecular mass. In the latter case the active sites of the IPEC-incorporated enzyme can be protected from inhibition not only in the precipitate but also in solution. This is illustrated by the data obtained from investigating the interaction of IPEC-incorporated CT with basic pancreatic inhibitor of trypsin (BPIT) (M.m. 6500) [43]. Figure 10.10a shows that CT attached to Q′–P4VP GPE and complexed with PMANa HPE is at least partly protected from BPIT even at pH 7.5, i.e. pretty far from the pH value (< 6) corresponding to phase separation (Fig. 10.10a, curve 1). At pH 6, IPEC-entrapped CT being still in solution is considerably protected

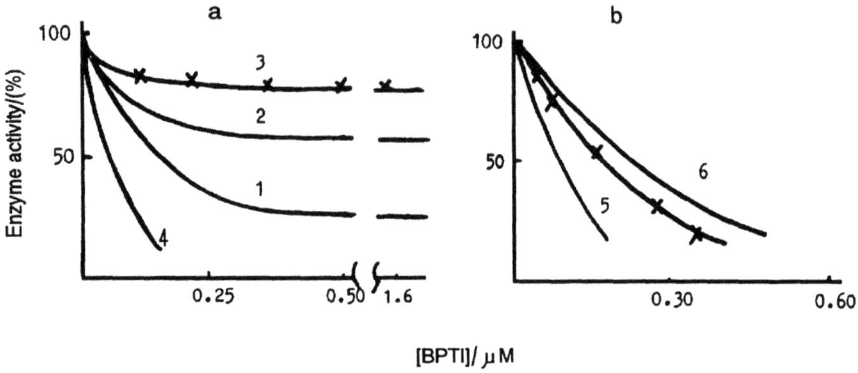

Fig. 10.10a, b. The dependences of the residual activity of various α-chymotrypsin preparations on the concentration of basic pancreatic inhibito of trypsin (BPTI); (**a**) IPEC (PMA/CT · Q–P4VP′): (1) in solution at pH 7.5; (2) at pH 6.0; (3) in precipitate at pH 5.7; (4) native enzyme at pH 6.0 (**b**) IPEC (CT · PMA/Q–P4VP′): (5) in solution at pH 6.0; (6) in precipitate at pH 5.7. The data obtained at pH 5.7 but in the presence of Ca^{+2}-salt are indicated by crosses

(Fig. 10.10a, curve 2), and it is fully protected in precipitate at pH 5.7 (Fig. 10.10a, curve 3). Protection is much less pronounced both in solution and in precipitate in the case of CT · PMANa (HPE) conjugate complexed with Q′–P4VP GPE (Fig. 10.10b). It, evidently, shows that HPE bound enzyme molecules are located on average closer to a periphery of the IPEC species than those bound to GPE.

Another promising approach may consist of using various enzyme-polyion conjugates self-assembling multienzyme IPEC systems. At least one example of such system containing urease and glutamate-dehydrogenase (GDHGase) is described [44]. Urease and GDHGase are covalently bound to PMANa(HPE) and Q′–P4VP(GPE), correspondingly. Then IPEC(urease · PMANa/GDHase · Q′–P4VP) has been obtained by electrostatic coupling the conjugates in aqueous solution. It is found that the lag-period of the reaction cycle of transformation of 2-oxoglutaric acid into glutamic acid catalyzed by bienzyme · IPEC species is drastically decreased in comparison with that observed in the reaction system containing the same amounts of the native enzymes. Unfortunately, that research was not continued.

Finally, it is worth emphasizing that the above-mentioned ability of linear polyions to quantitatively replace protein molecules in IPECs (Protein/HPE) (Scheme V) may be used for significant improvement of certain analytical procedures and in particular, for designing new effective immunochemical reagents. A specific antibody covalently bound to a polyion may serve as a selective binding site for a corresponding protein antigen. Other antigen molecules can be non-specifically adsorbed by polyion segments to form IPEC. However, interaction of such IPEC with equimolar amounts of an oppositely charged linear polyion results in replacement of all non-specifically adsorbed antigens and precipitation of a new IPEC containing only the antigen specifically bound to the complementary antibody.

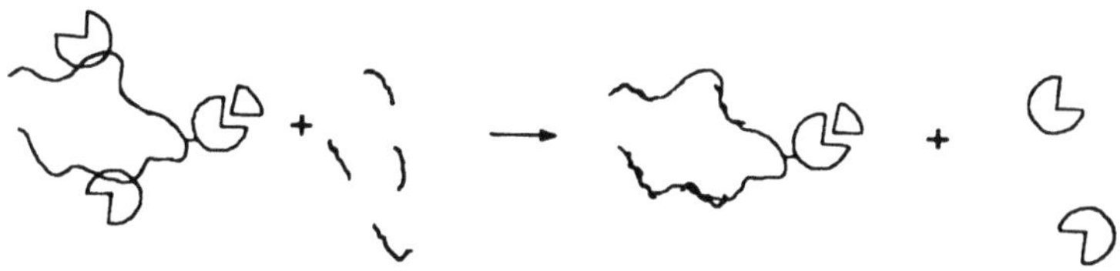

VIII

This principle is applied for developing highly effective IPEC based pseudohomogeneous immune diagnostic systems [45–48].

Recently [49] the advanced procedure for the synthesis of protein-polyelectrolyte conjugates using reversed micelles as microreactors has been proposed, which permits easy control of their composition and results in high yields. Hence, it may stimulate some additional efforts in the study and application of IPEC incorporated proteins.

10.4 Complexes of DNA with Synthetic Polycations for Cell Transformation

Recently it was suggested that the modification of biopolymers with hydrophobic residues enhance their binding to cell membranes and may promote them to penetrate into cells [50]. This approach was successfully applied for protein [51] and oligonucleotide [52] introduction into intact mammalian cells.

It is expected that binding to a membrane is also an essential step for penetration of nucleic acids into the cell [53, 54]. Such binding of nucleic acids may be achieved by chemical modification with hydrophobic residues as well. Thus, the general strategy described in this Section arises from the results obtained in [50–52]. However, in the case of nucleic acids, which represent polyanions with relatively high charge density, another route for introduction of a hydrophobic moiety can be proposed.

It is known [55] that nucleic acids as any other polyanionic compounds interact with linear polycations, in aqueous solutions forming soluble or insoluble IPECs. If the polycation chain has a hydrophobic backbone, its "sticking" to the nucleic acid polyanion accompanied by compensation of the phosphate group charge, results in formation of a hydrophobic site. The length and the number of such hydrophobic sites is determined by the length (degree of polymerization) of a polycation and by the IPEC composition, φ, i.e. the molar ratio of polycation repeating units and nucleic acid phosphate groups.

The applicability of this approach for efficient gene transfer was recently demonstrated by *B. subtilis* transformation with plasmid DNA incorporated into soluble IPEC with carbochain polycations [56–58] including those additionally modified by fatty residues. Behr et al. [59] independently used a similar approach for the efficient transformation of various mammalian cells by DNA complexes with lipopolyamines which, actually, represent a variety of IPEC.

Our experiments [56–58] have been performed using Q–P4VP and Q′–P4VP containing 3 mol.% of *N*-cetyl groups (Q″–P4VP) and poly-L-lysine (PL).

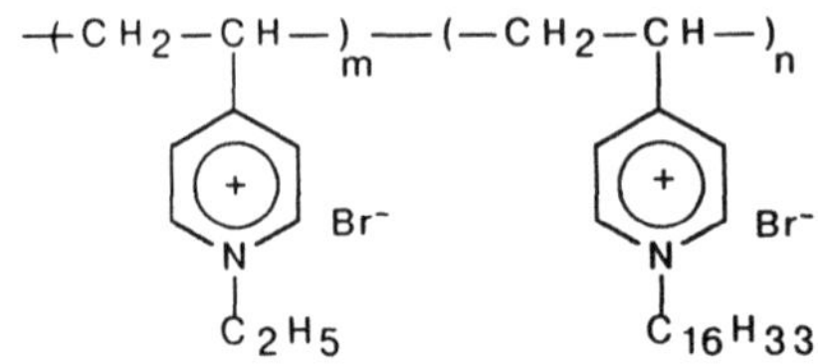

P4VP quaternary salts were prepared by quaternization of fractionated P4VP (degrees of polymerization $P_n = 18$, $P_w = 200$ and 400) with corresponding alkylhalides, purified and characterized as described in [60]. PL, $M_w = 1.5–3.0 \cdot 10^4$, was purchased from Sigma Company. The polycomplexes were formed by mixing the aqueous solutions of DNA and the polycations. DNA, Q–P4VP and Q″–P4VP concentrations in the systems obtained were determined by UV-absorbance measurements at 260 nm. The contour lengths of polycations ($P_w = 400$ corresponds to $M_w \sim 8 \cdot 10^4$) were considerably lower than the contour length of the DNA ($M_w \sim 3 \cdot 10^7$).

The characteristics of IPECs formed by Lambda DNA and Q–P4VP or Q″–P4VP were studied by ultracentrifugation techniques. Ultracentrifugation of DNA and its polycomplexes was carried out at 20 °C in a Beckman E. analytical ultracentrifuge at 20000 rpm. The scanning was measured at 260 nm.

Only one step is observed from the sedimentation curves of the samples obtained by mixing of the DNA and Q–P4VP solutions in the range $\varphi = 0–0.5$. The sedimentation coefficient corresponding to this step continuously increases from $18 \cdot 10^{-13}$ s for the free DNA to $23 \cdot 10^{-13}$ s (Fig. 11). These data give

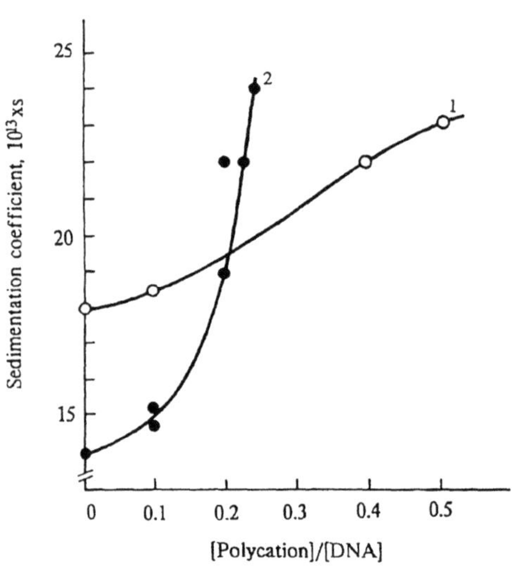

Fig. 10.11. The dependencies of sedimentation coefficients of soluble IPEC formed by Lambda DNA with Q–P4VP (1) and Q″–P4VP (2), on the [polycation]/[DNA] base mole ratio [58].

evidence of formation of the soluble nonstoichiometric IPEC, the composition of which, φ, is equal to the initial ratio of the polymer components ($0 < \varphi < 0.5$). In other words, Q–P4VP chains are completely complexed and uniformly distributed among the DNA molecules. Further addition of Q–P4VP causes a typical disproportionation phenomenon [25]. Parallel with the soluble IPEC ($\varphi = 0.5$, $S = 23 \cdot 10^{-13}$ s) the insoluble complex with a higher polycation content is formed, which precipitates in the ultracentrifuge at low rotation rates. The regularities of Q″–P4VP interaction with DNA are analogous to those observed for Q–P4VP. However, in this case the region of formation of the soluble IPEC corresponds to a narrower interval of $\varphi = 0$–0.25 (Fig. 10.11). Moreover, the uniform distribution of the Q″–P4VP chains among the DNA molecules is observed only at relatively low ionic strength. At higher ionic strength (for example in 10 mM phosphate buffer, pH 7.0, containing 0.15 M NaCl) a series of IPECs with various composition as well as free DNA are revealed in the sedimentation curve. (The difference in the sedimentation coefficients of the free DNA observed in Fig. 10.11, probably, arises from the difference in the ionic strengths.)

One can follow the formation of the IPEC by gel electrophoresis technique. The DNA and its IPEC were analyzed using gel electrophoresis measurement in 0.8% agarose gel in standard TRIS-borate buffer, pH 8.6 [61]. The densitograms were obtained using the laser densitometer "Ultrascan" (LKB). As determined from the densitograms represented in Fig. 10.12, incorporation of the plasmid DNA in the soluble complexes with polycations causes a noticeable shift of the bands, corresponding to the linear forms but not to the cyclic form of

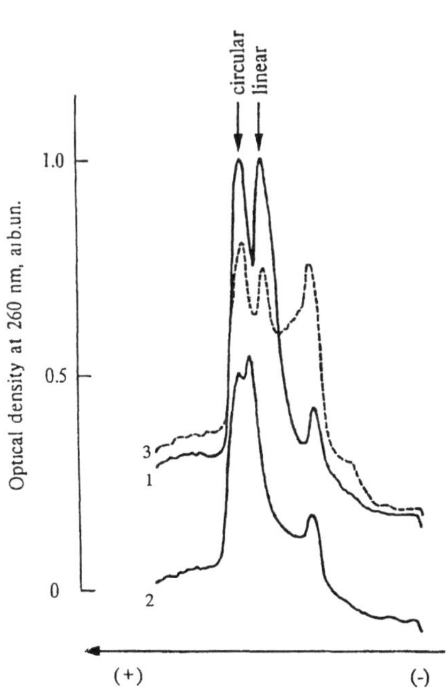

Fig. 10.12. The electrophoresis profiles of the native plasmid pBC16 (1) and its IPECs ($\varphi = 0.2$) with Q–P4VP (2) and Q″–P4VP (3). The arrows show the positions of the cyclic and linear forms of the plasmid [58].

the plasmid. The disproportionation phenomenon is revealed for the plasmic Q″–P4VP complex. In this case, near the bands of free DNA (its circular and linear forms) a broad band corresponding to insoluble IPEC is observed.

Addition of PMANa solution to a solution of DNA-polycation IPECs results in replacement of DNA by PMA polyanions, characterized by a higher linear charge density, and release of free DNA, which is detected by sedimentation as well as by electrophoresis data (not shown in Figs.). This substitution reaction is also typical for other IPEC's (see Sect. 2); a particular case is the separation of DNA or its fragments formed in the course of various chemical or biochemical treatments of IPEC(DNA/polycation).

Biological action of DNA necessarily requires specific recognition by competent enzymes. In this regard, restriction of DNA, incorporated into IPEC, by specific restrictases was studied.

2 microlitres of the solution, containing 2 units of the nuclease (Alu I or Pvu II), were added to 1.7–2.0 μg of plasmid DNA (pTZ19 plasmid) in the free form or incorporated into IPEC (plasmid DNA/polycation) dissolved in 30 μl of buffer, (the nucleases and the corresponding buffers for the DNA digestion were obtained from Boehringer). The system obtained were incubated at 37 °C. The reaction was stopped by addition of 1 μl of 60 mM EDTA, pH 8.0. Then the DNA-containing IPECs were decomposed by replacement of DNA with the threefold molar excess of PMANa as related to the polycation. The digestion products were analyzed by gel electrophoresis technique as described above. It was found, that the products of restriction of free and IPEC incorporated pTZ19 plasmid with Alu I, an enzyme with broad specificity recognizing a multitude of restriction sites in the DNA, are exactly the same. This means that the polycation in the soluble IPEC structure does not prevent the recognition of DNA by the enzyme. This fact is in agreement with above-mentioned (Sect. 2) dynamic behavior of polyions incorporated into soluble IPEC, which was first observed for soluble IPEC(DNA/PL) in an early study by Miller and Bach [55]. The polycation chains are not fixed permanently on any certain fragments of the DNA polyanions. They can transfer from one site to another via IPEC polyion interchange mechanism, opening the DNA sequences required for restriction.

Figure 10.13 presents the data on the kinetics of restriction of the pTZ19 plasmid with Pvu II. This enzyme recognizes only one site in the plasmid molecule. The treatment with Pvu II results in opening of the circular form of the plasmid, i.e. in formation of its linear form. As it is seen from Fig. 10.13 the relatively rapid cleavage of the native DNA is observed under the experimental conditions. The picture observed for the IPEC incorporated DNA is somewhat different. In this case the kinetics of restriction is characterized by two different phases. The first phase, probably, corresponds to the rapid cleavage of the DNA sites which are not covered by the polycation chains and are therefore easily available for the enzymes. During the second phase relatively slow restriction of the residual DNA proceeds. It is reasonable to assume that the rate determining step during the second phase is the opening of restriction sites for the enzyme resulting from polycation transfer from one DNA fragment to another. As is seen in Fig. 10.13, DNA plasmid restriction does not occur in its IPEC at

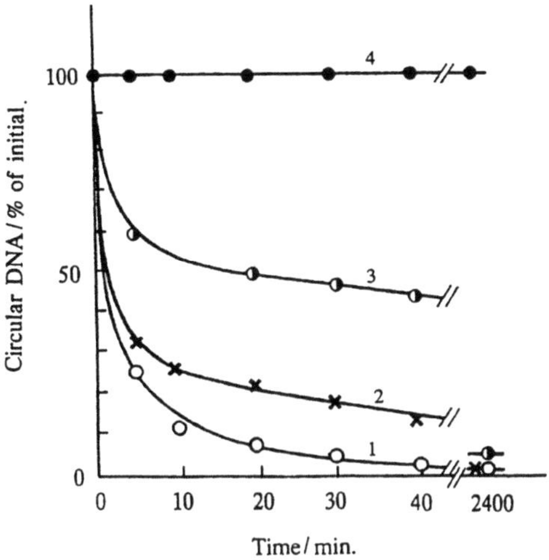

Fig. 10.13. Kinetics of Pvu II catalyzed restriction of cyclic pTZ19 plasmid (1) and its IPECs with Q–P4VP of various compositions: (2) $\varphi = 0.2$; (3) $\varphi = 0.3$; and $\varphi = 0.4$ (4).

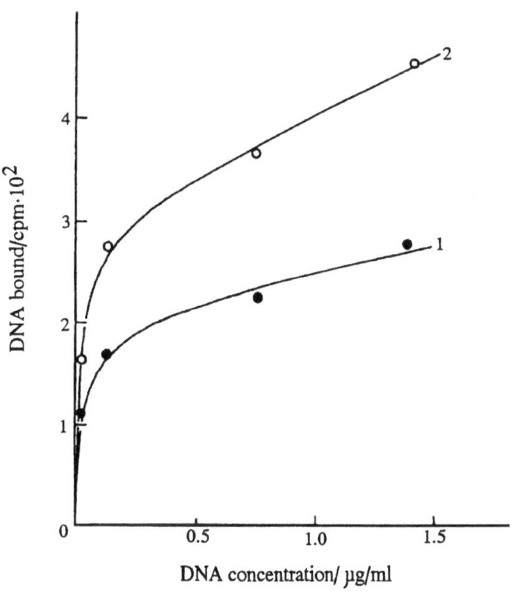

Fig. 10.14. The binding isotherms of C^{14}-labeled pTZ19 plasmid incorporated in IPEC with Q–P4VP (1), and with Q″–P4VP (2) by isolated *B. subtilis* membranes. The IPEC composition $\varphi = 0.2$.

$\varphi = 1.0$. This means that the DNA incorporated into the insoluble complex is not accessible to nuclease attack.

The study of binding of free and IPEC incorporated C^{14}-labeled pTZ19 plasmid with isolated *B. subtilis* membranes was carried out as described in [56–58]. It is seen in Fig. 10.14 that binding of DNA complexed with Q–P4VP and Q″–P4VP is 1.5–1.8 fold higher than that of free DNA.

It is known that the transmembrane transport of plasmid DNA is a rate determining step of *B. subtilis* genetic transformation [53]. To estimate the efficiency of the DNA transfer through a cell membrane the experiments were carried out [56–58] with transformation of the competent *B. subtilis* by pBC 16 plasmid containing a determinant of stability against tetracycline. It is found

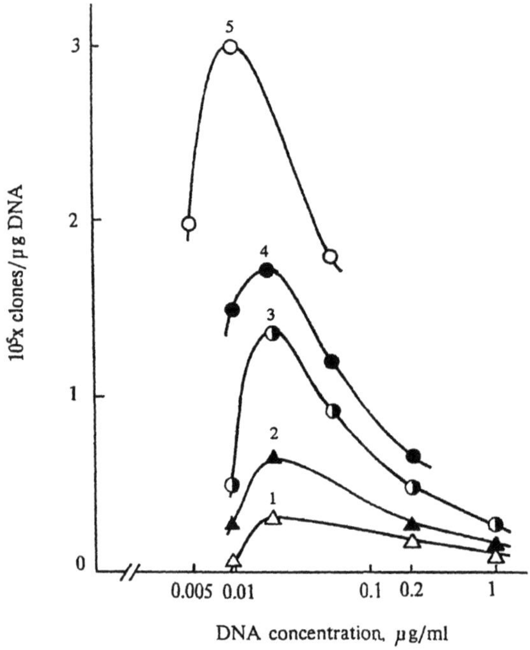

Fig. 10.15. Transformation of competent *B. subtilis* cells by free pBC16 plasmid (1) and by its IPECs with Q″–P4VP (2) or with Q–P4VP (3–5). Degrees of polymerization of the polycations are: 400 (2, 3); 200 (4); and 18 (5). The IPEC composition $\varphi = 0.2$ [58].

that incorporation of the plasmid into IPECs results in a considerable increase in the transformation efficacy (Fig. 10.15). The observed effect depends on the degree of polymerization of the polycation: at constant IPEC composition φ the transformation efficacy increases with decrease of the polycation length (Fig. 10.15, curves 3–5). It also increases with increasing φ (not shown in Fig. 10.15).

The approach [56–59] described in this Section can undoubtedly be used for efficient transformation of procaryotic and eucaryotic cells in vitro. However, its significance would be much higher, if it can be employed for genetic correction in vivo. The following arguments support this theory.

(1) The polycation complexed with DNA represent a "building block", which can easily be conjugated with any target-recognizing molecule using a standard chemical technique, e.g. with an antibody against a certain receptor of a tagged cell. This might be a way to address DNA packed in IPEC species to the target cell. Using this strategy Wu et al. [64] have realized in vivo transformation of liver cells by a plasmid incorporated into IPEC with PL covalently coupled with a liver-specific hormone.
(2) Incorporation of DNA molecules into IPEC with φ close to 1 can block DNA cleavage by nuclease presented in living organisms, e.g. in the blood flow. This might be a method to protect DNA on its route to the target cells.

The important advantage of DNA containing IPEC arises from the fact that they are thermodynamically equilibrated self-assembling species, which can be obtained by simple mixing of DNA and polycation solutions. The IPEC species can be regarded as a mimic of a virus with the "core" formed by DNA (or RNA) and the "cover" containing receptor-recognizing molecules, which can interact with the cell membrane and "infect" the cell.

10.5 Conclusion

This paper presents just a few interesting examples illustrating the possibilities of applying the properties of soluble IPEC in order to affect biological macromolecules and biological systems. At present, we are probably at the beginning of the whole new approach, which seems to be rather attractive, based even on a very general consideration. IPEC formed by non-natural polyelectrolytes, being highly co-operative supramolecular compounds in this sense, are somewhat similar to self-assembling supramolecular complexes of biopolymers. So they can interfere and interact with each other in aqueous media using similar principles and are controlled by similar specific factors. Therefore, this is the case when biomimetic species may be transferred into real biofunctioning systems operating not only *in vitro* but also *in vivo*.

10.6 References

1. Kabanov VA (1973) Pure and Appl Chem Macromol Chem 8:121
2. Kabanov VA, Zezin AB (1982) Soviet Sci Rev, Sec B, Chem Rev 4:207
3. Abe K, Tsuchida (1982) Adv Polym Sci 45:1
4. Kabanov VA, Zezin AB (1984) Makromol Chem, Suppl 6:259
5. Tsuchida E, Osada Y, Sanada K (1972) J Polym Sci, Polym Chem Ed 10:3397
6. Gulyaeva ZhG, Poletaeva OA, Kalachev AA, Kasaikin VA, Zezin AB (1976) Visokomol Soedin A18:2800
7. Tsuchida E, Osada Y (1974) Macromol Chem 175:603
8. Kharenko AV, Starikova EA, Lucenko VV, Zezin AB (1976) Visokomol Soed A18:1604
9. Gulyaeva ZhG, Zansokhova MF, Razvodovskii YeF, Yefimov VS, Zezin AB, Kabanov VA (1983) Visokomol Soed A25:1238
10. Izumrudov VA, Savitskii AP, Bakeev KN, Zezin AB, Kabanov VA (1984) Makromol Chem, Rapid Commun 5:709
11. Izumrudov VA, Bronich TK, Zezin AB, Kabanov VA (1985) J Pol Sci, Pol Let Ed 23:439
12. Kabanov VA, Zezin AB, Izumrudov VA, Bronich TK, Bakeev KN (1985) Makromol Chem Suppl 13:137
13. Zezin AB, Izumrudov VA, Kabanov VA (1989) Frontiers of Macromolecular Science. Blackwell Sci Publ, pp 219–225
14. Bakeev KN,. Izumrudov VA, Zezin AB, Kabanov VA (1988) Dokladi Akad Nauk SSSR 299:1405
15. Bakeev KN, Izumrudov VA, Kuchanov SI, Zezin AB, Kabanov VA (1988) Dokladi Akad Nauk SSSR 300:132
16. Izumrudov VA, Bronich TK, Saburova OS, Zezin AB, Kabanov VA (1988) Makromol Chem Rapid Commun 9:7
17. Krakovyak MG, Anufrieva EV, Skorokhodov SS (1972) Visokomol Soedin A14:1127
18. Pavlova NR, Kirsh YuE, Kabanov VA (1979) Visokomol Soedin A21:2062
19. Kabanov VA, Zezin AB, Rogacheva VB, Grishina NV, Goethals EJ, Van de Velde M (1986) Makromol Chem 187:1151
20. Kabanov VA, Giriyakova MV, Kargov SI, Zezin AB, Izumrudov VA (1993), Dokladi Akad Nauk 329:66
21. Kabanov VA, Evdakov VP, Mustafaev MI, Antipina AP (1977) Molek Biol 11:582
22. Kabanov VA, Zezin AB, Mustafaev MI, Kasaikin VA (1980) In: Goethals EJ (ed) Polymeric amines and ammonium salts. Pergamon Press, Oxford/New York, p 173

23. Mustafaev MI, Kabanov VA (1980) Farmakol i Toksikol 43:395
24. Izumrudov V, Kasaikin VA, Ermakova LN, Zezin AB (1978) Visokomol Soed A20:400
25. Kabanov VA, Zezin AB (1984) Pure and Appl Chem 56:343
26. Izumrudov VA, Kharenko OA, Kharenko AV, Gulaeva ZG, Kasaikin VA, Zezin AB, Kabanov VA (1980) Viskomol Soed A22:692
27. Kabanov VA, Zezin AB, Rogacheva VB, Izumrudov VA, Ryzhykov SV (1982) Dokladi Akad Nauk SSSR 268:1419
28. Zezin AB, Izumrudov VA, Kabanov VA (1989) Makromol Chem, Macromol Symp 26:249
29. Margolin AL, Izumrudov VA, Shviadas VK, Zezin AB, Kabanov VA, Berezin IV (1981) Biochem Biophys Acta 660:359
30. Margolin AL, Izumrudov VA, Sherstiuk SF, Zezin AB, Shviadas VK (1983) Molek Biol 17:1001
31. Margolin AL, Sherstiuk SF, Izumrudov VA, Zezin AB, Kabanov VA (1985) Eur J Biochem 146:625
32. Margolin AL, Izumrudov VA, Shviadas VK, Zezin AB (1982) Biotechn Bioeng 24:237
33. Margolin AL, Sherstiuk SF, Izumrudov VA, Shviadas VK, Zezin AB, Kabanov VA (1985) Dokladi Akad Nauk SSSR 284:997
34. Margolin AL, Sherstiuk SF, Izumrudov VA, Shviadas VK, Zezin AB, Kabanov VA, Berezin IV (1980) Dokladi Akad Nauk SSSR 253:1508
35. Charles M, Roughlin RW, Hasselberg FY (1974) Biotechn Bioeng 16:153
36. Coughlin RW, Aizawa M, Charles M (1976) Biotechn Bioeng 18:199
37. Leemputten EV, Horisberger M (1976) Biotechn Bioeng 18:587
38. Patent USA No 4088538 (1978)
39. Margolin AL, Sherstiuk SF, Izumrudov VA, Shviadas VK, Zezin AB, Kabanov VA (1983) Dokladi Akad Nauk SSSR 272:230
40. Swerdlow BM, Setlow B, Setlow P (1981) J Bacteriol 148:20
41. Warth AD (1980) J Bacteriol 143:27
42. Bradbury JH, Foster JR, Hammer L (1981) Biochim Biophys Acta 678
43. Izumrudov VA, Margolin AL, Sherstiuk SF, Shviadas VK, Zezin AB, Kabanov VA (1983) Dokladi Acad Nauk SSSR 269:631
44. Izumrudov VA, Margolin AL, Shviadas VK, Zezin AB (1985) Abstracts of the 5th All Union Simposium on Enzymology Engineering, Kabuleti (Russia) 1:13
45. The method of express-analysis of biologically active substances, Russia Author Sartificate No 1200681, 1985
46. The method of immunochemical detection of viruses, USSR Author Sartificate No 1394708, 1988
47. Dzantiev BB, Blintsov AN, Civileva LS, Berezin IV, Egorov AM, Izumrudov VA, Zezin AB, Kabanov VA (1988) Dokladi Akad Nauk SSSR 302:222
48. Dzantiev BB, Blintsov AN, Bobkova AF, Izumrudov VA, Zezin AB, Kabanov VA, Atabekov IG (1990) Dokladi Akad Nauk SSSR 311:1482
49. Kabanov AV, Levashov AV, Khrutskaya MM, Kabanov VA (1990) Makromol Chem 191:2801
50. Kabanov AV, Nametkin SN, Levashov AV, Martinek K (1985) Biol Membrany 2:985
51. Kabanov AV, Levashov AV, Alakhov VYu, Kravtrova TN, Martinek K (1989) Coll Czech Chem Commun 54:835
52. Kabanov AV, Vinogradov SV, Ovcharenko AV, Krivonos AV, Melik-Nubarov NS, Kiselev VI, Severin ES (1990) FEBS Lett 259:327
53. Grinus LP (1986) Transport of macromoleculs in bacteria. Nauka Publishers, Moscow
54. Smith HO, Danner DB (1981) Annu Rev Biochem 50:41
55. Miller IR, Bach D (1968) Biopolymers 6:169
56. Kabanov AV, Chikindas ML, Astafieva IN, Glukhov AI, Gordeeva SA, Izumrudov VA, Zezin AB, Levashov AV, Severin ES, Kabanov VA (1989) Dokl Acad Nauk SSSR 306:226
57. Kabanov VA, Kabanov AV, Astafieva IN (1991) ACS Polymer Preprints 32:592
58. Kabanov AV, Astafieva IV, Chikindas ML, Rosenblat GF, Kiselev VI, Severin ES, Kabanov VA (1991) Biopolymers, 31:1437
59. Behr J-P, Demeneix B, Loeffler J-P, Perez-Mutul J (1989) Proc Natl Acad Sci USA 86:6982
60. Kirsh YuE, Rakhnyanskaya AA, Lukovkin GM, Kabanov VA (1974) Eur Polymer J 10:393
61. Maniatis T, Fritsch EF, Sambrook J (1984) Molecular Cloning: A Laboratory Manual, Mir Publishers, Moscow
62. Spizizen J (1958) Proc Natl Acad Sci USA 44:1072
63. Prozorov AA (1980) Genetic Transformation and Transfection, Nauka Publishers, Moscow
64. Wu GY, Wu CH (1988) J Biol Chem 263:14621

11 Conformation Presumption for Polysaccharide–Polylysine Complexation

Slavomír Bystrický and Anna Malovíková

A short presentation of the electrostatic interactions of acidic polysaccharides with D- and L-enantiomers of polylysine is given. As polysaccharide components pectate and alginates rich in L-guluronate and D-mannuronate, respectively were used. The results of the study by means of circular dichroism showed that the complex formation is governed by the stoichiometric ratio of the charges of both interacting components. The conformation of polysaccharide used for complexation was proved to be the decisive factor influencing the interaction efficiency.

11.1 Introduction

The circular dichroism (CD) measurement has proved to be an extremely sensitive method for investigating the complex-forming interactions of polysaccharides and polypeptides [1–3]. Formation of the complex is generally associated with the change of secondary structure of polypeptides in the sense of a transition from less ordered to regular structures. CD-monitoring enables quantitative evaluation of complex-forming efficiency of dilute solutions. We have studied the interaction of D- and L-enantiomers of polylysine with natural anionic polysaccharides, pectate and alginate [4–6]. The main chain of the pectate molecule consists of α (1–4) D-galacturonan ($\sim$ 85%) randomly interrupted with L-rhamnose units. The neutral sacccharides ($\sim$ 15%) are situated mainly in the form of short side chains. Alginates are linear chains composed of blocks of (1–4)-linked β-D-mannuronate and α-L-guluronate and alternating sequences of these residues. We used two alginates with high contents of mannuronate and guluronate. The interaction which is mainly electrostatic is substantially influenced by charge density of the polyions, ionic strength of the solution and type of the ionic groups. However, the problem of stereochemistry, namely the spatial arrangement of charges which also conditions the interaction, has not been studied so far. The proposed model of the complex consists of an α-helical polypeptide core surrounded by polysaccharide superhelix [4].

11.2 Complex Formation

The interaction of polylysines with polysaccharides was examined in one or two measurement series: Increasing amounts of the polypeptide were added to excess

Macromolecular Complexes
in Chemistry and Biology
Eds.: Dubin/Bock/Davis/Schulz/Thies

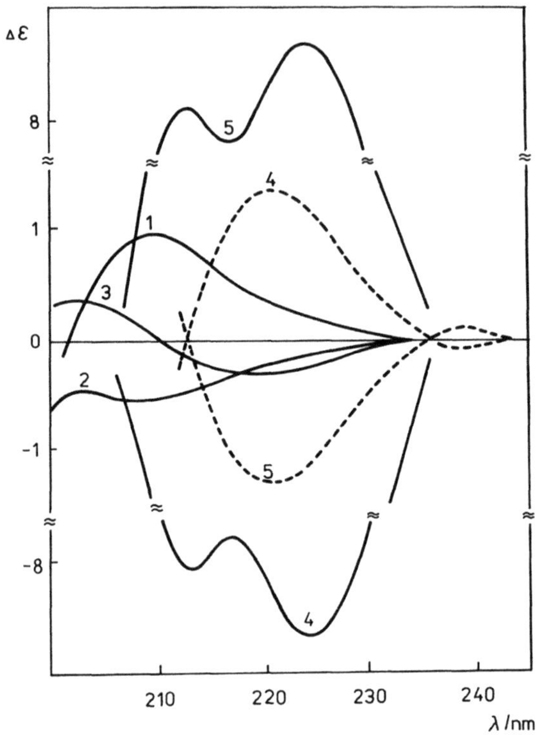

Fig. 11.1. The CD spectra of the individual acidic polysaccharides and polypeptides: (*1*) potassium pectate; (*2*) alginate rich in L-guluronate; (*3*) alginate rich in D-mannuronate; (*4*) poly(L-lysine); (*5*) poly(D-lysine); (- - -) at neutral pH; (——) at pH 11.3

polysaccharide in solution (0.3 mmol$\cdot$l^{-1}) and, vice versa, increasing amounts of the polysaccharide were added to excess polypeptide (0.3 mmol$\cdot$l^{-1}). The CD spectra of components used for complexation are presented in Fig. 11.1. Polylysine in neutral solution has a spectrum corresponding to a "charged coil" structure, while at pH = 11.3 the spectrum is characteristic of a complete helical arrangement. The obtained CD spectra of the mixture were corrected by subtracting the CD of all the polysaccharide present in the solution and additionally in the second series, the CD of the excess polypeptide in the charged-coil arrangement was substracted. The spectra corrected in this way (Figs. 11.2–11.6) represent the CD of that part of the polypeptide in solution having an equivalent amount of carboxyl counterions on polysaccharide required for interaction.

11.3 Pectate–Polylysine Interaction

The CD spectra of D- and L-polylysines in the presence of potassium pectate are given in Figs. 11.2 and 11.3. The spectral shape of complexes with poly(L-lysine) clearly indicates the presence of induced α-helical structure.

Measurements in both series clearly show (Fig. 11.2) a regular increase of values corresponding to the increasing amount of the constituent added. The identity

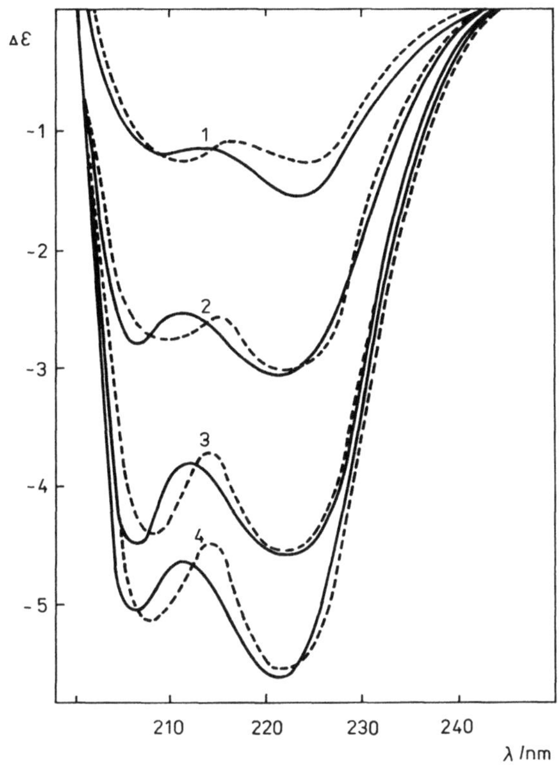

Fig. 11.2. The CD spectrum of poly(L-lysine) in complex with potassium pectate: (——) pectate with addition of 20% (*1*), 40% (*2*), 60% (*3*) and 70% (*4*) poly(L-lysine); (- - -) poly(L-lysine) with addition of 20% (*1*), 40% (*2*), 60% (*3*) and 70% (*4*) pectate

of the corresponding spectra of both series further demonstrates that the interaction of poly(L-lysine) with pectate is a quantitative one; the 1:1 ($COO^-:NH_3^+$) complex was formed regardless the excess of either component. The complex-formation proceeds, as a co-operative process of two polymeric structures.

Figure 11.3 shows the CD spectra of poly(D-lysine) added stepwise to the excess of potassium pectate. While the unordered structure of poly(D-lysine) is characterized by the negative sign of CD (Fig. 11.1), the complex-formation, inducing regular arrangement, is reflected in the positive sign of CD. As seen from the spectra the shift to the positive region is surprisingly small. In contrast to poly(L-lysine)–pectate interaction where a very intensive conformational change of polypeptide connected with the maximum complex formation has been observed, such small change of poly(D-lysine) conformation caused by very low complexation efficiency has not been noticed so far.

The CD spectra of poly(D-lysine) reflect a markedly low degree of complexation. The prevalent part of poly(D-lysine) remains in unchanged random coil arrangement. The extent of complexation may be judged from comparison with the CD spectrum of poly(D-lysine) at pH = 11.3 (Fig. 11.1). Complexation efficiency of poly(L-lysine) with pectate is almost 100% while with poly(D-lysine)

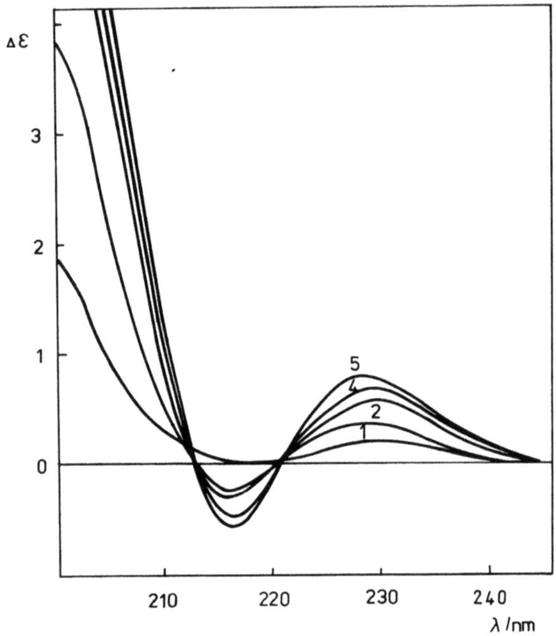

Fig. 11.3. The CD spectrum of poly(D-lysine), in the presence of potassium pectate: potassium pectate with addition of 20% (*1*), 40% (*2*), 60% (*3*), 80% (*4*) and 100% (*5*) poly(D-lysine)

it is only about 10%. This difference in the interaction of pectate with enantiomers of polylysine points at a new, until now unobserved factor, i.e. the chiral discrimination of adaptable conformation of polysaccharide in solution. So far, the conformation of D-galacturonan in solution has not been determined precisely. Conformational polymorphisms allows two-fold and three-fold helical symmetry [7, 8].

11.4 Polyguluronate Rich Alginate–Polylysine Interaction

The complexation was further studied with an alginate consisting of ~ 75% L-guluronate. As may be seen from Fig. 11.4, the CD spectra are practically identical with those of poly(L-lysine) in a disordered arrangement (Fig. 11.1). In this case, the interaction connected with inductive formation of the α-helical structure has not taken place. Consequently, the L-guluronate sequence of alginates does not enter the complexation with poly(L-lysine).

Figure 11.5 shows the CD of poly(D-lysine) in the presence of alginate rich in L-guluronate. The shift of spectra to the positive region, more distinct with increased concentration of polypeptide, indicates that the part of poly(D-lysine) interacts with alginate. However, the spectral shape and low values of CD spectra show that major part of polypeptide has not entered the complexation.

The reason why the L-guluronan does not interact with poly(L-lysine), whereas the D-galacturonan interacts very intensively, lies in the spatial struc-

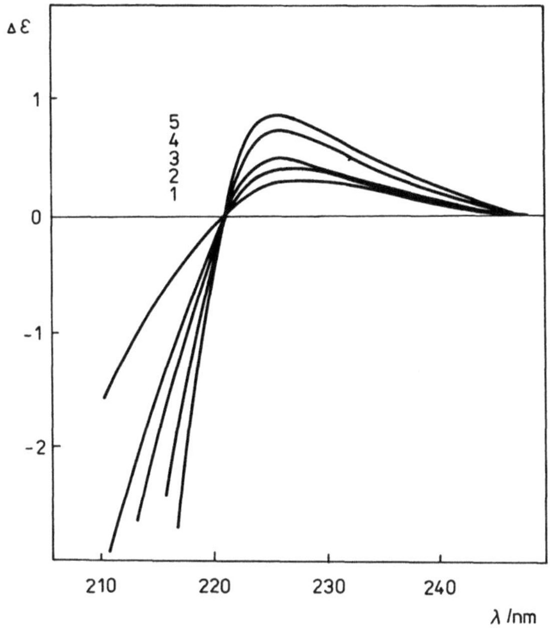

Fig. 11.4. The CD spectrum of poly(L-lysine) in the presence of alginate rich in L-guluronate: alginate with addition of 20% (*1*), 40% (*2*), 60% (*3*), 80% (*4*) and 100% (*5*) poly(L-lysine)

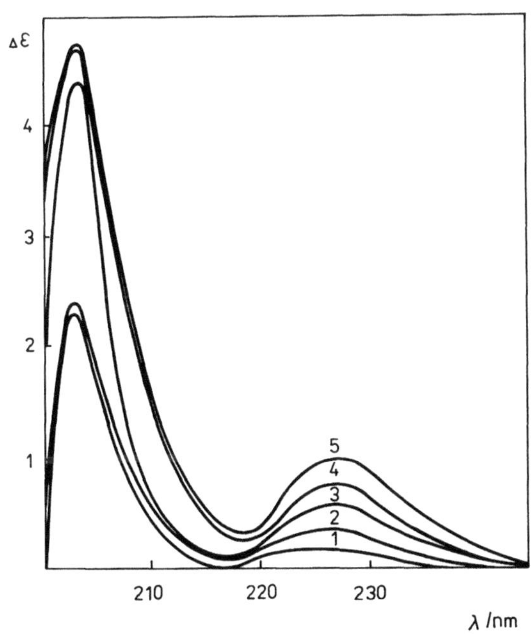

Fig. 11.5. The CD spectrum of poly(D-lysine) in the presence of alginate rich in L-guluronate: alginate with addition of 20% (*1*), 40% (*2*), 60% (*3*), 80% (*4*) and 100% (*5*) polylysine

ture of both polysaccharides. The conformations of L-guluronan and D-galacturonan in solution have not so far been determined. The extrapolation of the conformation found in the solid state to solutions is not reliable. The reason is evident from the polymorphic-phase transition of calcium D-galacturonan [8]. On the basis of the results of complexation achieved in this study, it may be suggested that the L-guluronan chain is more rigid, i.e. less adaptable to the changes in the surrounding medium than the D-galacturonan chain.

11.5 Polymannuronate Rich Alginate–Polylysine Interaction

The CD spectra of both enantiomers of D- and L-polylysine in the presence of alginate rich in D-mannuronate are shown on Figs. 11.6 and 11.7. In the case of interaction with poly(D-lysine), marked bands characteristic of an α-helical structure are observed in the positive region of the spectrum. Here the complex-forming interaction is effective. On the other hand, with poly(L-lysine) the CD in the negative region is of low intensity. The shape and the low values of CD testify that only a small part of poly(L-lysine) is involved in the interaction inducing regular helical arrangement.

The interaction with poly(D-lysine) appears to be more effective (approximately 50%). The complexation efficiency with enantiomeric poly(L-lysine) is about 20% only. This difference, smaller than with pectate, reflects the fact that the D-mannuronate arrangement in alginate is rather flexible due to the $\beta(1\text{–}4)$

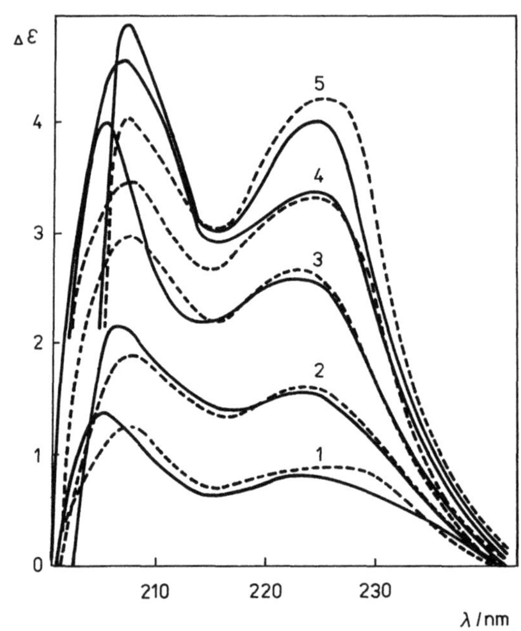

Fig. 11.6. The CD spectrum of poly(D-lysine) in the presence of alginate rich in D-mannuronate: (- - -) alginate with addition of 20% (*1*), 40% (*2*), 60% (*3*), 80% (*4*) and 100% (*5*) poly(D-lysine); (——) poly(D-lysine) with addition of 20% (*1*), 40% (*2*), 60% (*3*), 80% (*4*) and 100% (*5*) alginate

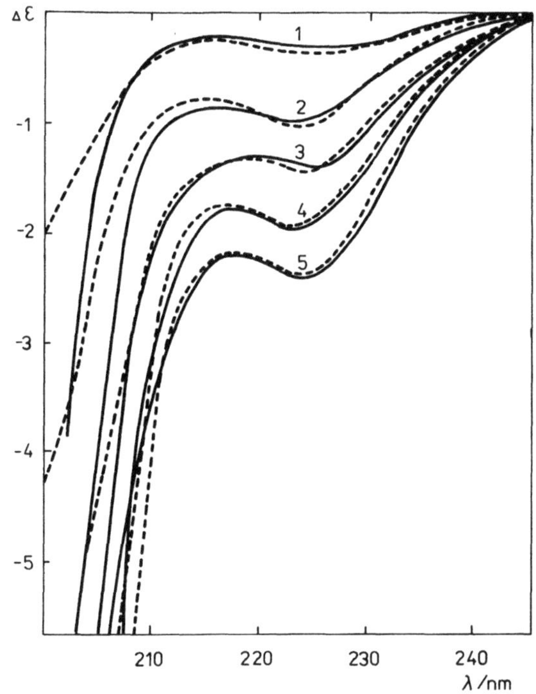

Fig. 11.7. The CD spectrum of poly(L-lysine) in the presence of alginate rich in D-mannuronate: (- - -) alginate with addition of 20% (*1*), 40% (*2*), 60% (*3*), 80% (*4*) and 100% (*5*) poly (L-lysine); (——) poly(L-lysine) with addition of 20% (*1*), 40% (*2*), 60% (*3*), 80% (*4*) and 100% (*5*) alginate

diequatorial bond. It was found by X-ray method for the salts of D-mannuronan three-fold helical structure [8] that was quoted as lefthanded [9]. Here the enantiomeric effect of the interaction studied was shown to be reverse when compared with pectate. While pectate effectively interacted with the L-form of polylysine, alginate rich in D-mannuronate interacted more effectively with the D-form. On the basis of the results obtained we have arrived at the conclusion that D-mannuronan in water solution adopts a conformation of screw-sense reverse to that of D-galacturonan.

11.6 Conclusion

It is known that L-guluronan in solid state adopts a strictly two-fold helical conformation [8, 10]. The polysaccharide with such conformation is not able to interact with polylysine effectively due to incompatibility of charge densities. The inability of alginate rich in L-guluronate to interact with either poly(D-lysine) or poly(L-lysine) confirms that the structure of L-guluronan itself is also strongly rigid in water solution. This conclusion is supported by the last observed interaction with the cationic molecule rhutenium red [11].

The results of the interaction probe with pectate (D-galacturonan) and D-mannuronan pointed out that for the effective interaction some unrestricted conformational freedom of polysaccharide backbone is necessary. The direction of this freedom correlates with the assumed screw-sense of helical conformation of polysaccharide in solution.

From comparison of complexation efficiencies of acidic polysaccharides with polylysine enantiomers the sense of conformational freedom of polysaccharide may be deduced. Our experiments have clearly shown that D-mannuronan in solution tends to adopt a left-handed helical structure, while D-galacturonan may be characterized by conformational freedom in a right-handed sense.

Generally, for acid polysaccharides having each pyranose unit charged, the following chiral discrimination is valid; Polysaccharides naturally adopting a right-handed helical structure in solution readily enter the complex-forming interaction with poly(L-lysine), while with poly(D-lysine) they interact very weakly or not at all. Polysaccharides with left-handed helical arrangement in solution interact strongly with the D-form of polylysine and weakly with the L-form.

11.7 References

1. Hopfinger AJ (1977) Intermolecular interactions and biomolecular organization. J. Wiley, New York, p 240
2. Stone AJ, Epstein P (1977) Biochim Biophys Acta 497:298
3. Shinoda K, Hayashi T, Nakajima A. (1976) Polymer J 8:216
4. Bystrický S, Kohn R, Sticzay T, Bláha K (1985) Collection Czechoslov Chem Commun 50:1097
5. Bystrický S, Malovíková A, Sticzay T (1990) Carbohydr Polym 13:283
6. Bystrický S, Malovíková A, Sticzay T (1991) Carbohydr Polym 15:299
7. Walkinshaw MD, Arnott S (1981) J Mol Biol 153:1075
8. Mackie W (1971) Biochem J 125:89
9. Atkins EDT (1977) Pure Appl Chem 49:1135
10. Mackie W, Perez S, Rizzo R, Taravel F, Wignon M (1983) Int J Biol Macromol 5:329
11. Murano E, Paoletti S, Cesaro A, Rizzo R (1990) Anal Biochem 187:120

12 Interpolymer Complexes and their Ion-Conduction

E. Tsuchida and S. Takeoka

An interpolymer complex can be prepared as a white precipitate immediately after the mixing of aqueous solutions of poly(oxyethylene) (POE) and poly(acrylic acid) (PAA), or poly(methacrylic acid) (PMAA). Hydrogen bonds between ether oxygens and carboxylic groups in the chain are formed in the complex by cooperative interaction. The ratio of the binding groups to the total of potentially interacting groups (θ), and the stability constant (K) of the polymer–polymer complexes were strongly dependent on chain length, temperature and medium, as well as on enthalpy and entropy changes of the complex formation; hydrophobic interaction was a significant factor for the stabilization of the complexes.

The complexes were applied to solid polymer electrolytes because POE has the ability to transport ions even in a completely dry state. The complex provided the amorphous matrix for oligomer POE; high mechanical stability being maintained by hydrogen bonding. Ions migrate through the oligomer POE region and show 4.3×10^{-6} S/cm at 25 °C. The complex from partially neutralized PAA and POE showed single ionic conduction of 4.9×10^{-7} S/cm for Na^+ at 25 °C. These materials were applied to electrochromic devices.

12.1 Introduction

In view of the importance of molecular associations in biology, such as the formation of two- or three-stranded polynucleotides, supermolecular assemblies in virus shells and muscle contraction, many polymer–polymer complexation systems involving synthetic macromolecules have been investigated in the last decade [1]. Interpolymer complexation can be made due to secondary binding forces, solvation, steric factors, and interpenetration. As reflected by the reaction mechanisms in vivo, specific interactions among inter- and intrapolymers are necessary for the display of a functionality regardless of the polymer chain structure. However, in spite of the importance of these interactions, relatively few studies on assemblies formed through interactions between synthetic polymer chains have been reported. Since synthetic polymers have simpler structures than biopolymers, the functional phenomena occurring in complicated reactions in living cells may be more easily understood. Therefore, suggestions about the design of the functions on the polymer chains are given to study the mechanism of complexation between synthetic macromolecules.

The important points to study in the formation process of interpolymer complexes are as follows; (1) to establish the basic treatment and theory for the formation mechanism of interpolymer complexes by accumulating experimental

Macromolecular Complexes
in Chemistry and Biology
Eds.: Dublin/Bock/Davis/Schulz/Thies

information; (2) to clarify the "polymer effect" which is the distinctive feature of the polymer reaction systems; (3) to find uses for polymer complexes as new kinds of polymer materials exhibiting specific physical and chemical properties and functionalities which are entirely different from those of the component polymers. In this paper, the interpolymer complexes through hydrogen bonding are applied to solid polymer electrolyte which conducts ions in a completely dry state.

12.2 Classification of Interpolymer Complexes

Macromolecular chains, as stated in the previous chapter, may undergo interactions in solution except in the ideal state. In this chapter, the association phenomena of more than two different macromolecular chains in solution caused by secondary binding forces such as electrostatic interactions are discussed. The obtained associates are generally called "interpolymer complexes". These complexes are divided into four classes on the basis of their main interaction forces, i.e. (1) polyelectrolyte complexes, (2) hydrogen-bonding complexes, (3) stereocomplexes and (4) charge-transfer complexes.

12.2.1 Polyelectrolyte Complexes

Polyelectrolyte complexes are formed by mixing oppositely charged polyelectrolytes, i.e. polyanions and polycations, due to Coulomb forces. Simultaneously, microions are released almost quantitatively. Polyelectrolyte complexes are divided into four subclasses based on a combination of strong and weak polyelectrolytes. These polyelectrolyte complexes are denoted and analyzed by their electrochemical behavior, owing to the released microions, the hydrodynamic properties, the changes of molecular weight, the conformation and radius of gyration, and the physical and chemical properties of the products [2–4]. In general, the composition of polyelectrolyte complexes depends on the degree of dissociation of the polyelectrolyte components, Therefore, in the system of strong polybase-strong polyacid, the composition of the obtained complexes is unity. On the other hand, the composition markedly depends on the degree of dissociation when weak polyelectrolytes are used.

12.2.2 Hydrogen-Bonding Complexes

Complexes containing hydrogen bonds (hydrogen-bonding complexes) are formed by combination of polymers bearing proton-accepting units and proton-donating units. The formation of complexes containing hydrogen bonds is detected by the same methods as in the case of polyelectrolyte complexes. Moreover, spectroscopic methods such as Infrared, Raman and Nuclear Magnetic Resonance are used as efficient analytical methods. The studies of hydrogen-bonding complexes are listed in Table 12.1. In the following chapters, the formation and properties of this kind of complex are described.

Table 12.1. Combinations of the components in interpolymer complexes formed through hydrogen bond

Polymer A	Polymer B	Ref.
poly(carboxylic acid)s	poly(oxyethylene)	5–18
poly(carboxylic acid)s	poly(*N*-vinyl-2-pyrrolidone)	19–29
poly(carboxylic acid)s	poly(vinyl alcohol)	30–37
poly(carboxylic acid)s	poly(acrylamide)	38, 39
copolymers containing carboxylic acid unit	copolymers containing 4-vinylpyridyl unit	40, 41
poly(carboxylic acid)s	poly(1,2-dimethoxyethylene)	42, 43
poly(carboxylic acid)s	poly(dimethyltetramethylene-phosphorictriamide)	44–46
poly(carboxylic acid)s	poly(vinylmethylether)	47
poly(carboxylic acid)s	poly(vinylbenzo-18-crown-6)	48
poly(vinyl alcohol)	poly(*N*-vinyl-2-pyrrolidone)	49
poly(vinyl alcohol)	poly(acrylamide)	50
polyarylate	poly(oxyethylene)	51

12.2.3 Stereocomplexes

Stereocomplexes are generated by combination of isotactic with syndiotactic poly(methyl methacrylate) (PMMA), mainly through van der Waals forces. Stereocomplexes exhibit a specific higher-order structure, e.g. syndiotactic PMMA with a beta-pleated sheet structure interwound in the grooves of the helical structure of isotactic PMMA [52, 53].

The analytical methods for the determination of stereocomplexes are the same as those mentioned previously, especially the utilization of the hydrodynamic properties of the complex solution and the physicochemical properties of the products. In polar organic solvents, the complex is obtained as a precipitate and its composition is 1/2 according to the molar ratio of the units [iso-PMMA]/[synd-PMMA]. In contrast, in nonpolar aromatic solvents, the complexes are obtained as a gel but its composition is also 1/2. The complexes are soluble in $CHCl_3$ and CH_2Cl_2.

12.2.4 Charge-Transfer Complexes

Charge transfer complexes are formed in systems of electron-accepting polymers and electron-donating polymers through charge-transfer interaction [54–57]. Charge-transfer complexes are mainly characterized by their ultraviolet and visible spectra. The ratio of their repeating units is almost unity.

Only few theoretical studies on the formation of such intermolecular complexes have been performed up to now. This may be due to the following facts:

1) The contribution of each secondary binding force cannot be clearly separated
2) Synthetic polymers essentially exhibit random molecular weight, tacticities, sequences, and secondary structures
3) Biopolymers contain too many kinds of functional groups so that characterization if individual bonds is not possible.

12.3 Formation of Interpolymer Complexes from PAA with POE [10]

12.3.1 Formation of Complex from PAA with POE

Figure 12.1 shows the relationship between the turbidity changes of mixed aqueous solutions and the molar ratio of poly(acryric acid) (PAA) to POE ($\bar{M}_w = 1.4 \times 10^6$) in mole units (in this paper, concentrations of the polymers are represented in mole units) at various pH levels. At pH 3, where dissociation of the carboxylic groups of PAA is extremely restrained, the transmittance of the mixed solution decreases with an increase in the concentration of added PAA up to [PAA]/([PAA] + [POE]) = 0.5, and then it increases again. On the contrary, at pH = 5, where a fairly substantial percentage of carboxylic groups is dissociated, the transmittance of the mixed solutions is always more than 90%, which is the same as the original solution. These facts suggest that the complex is not formed in the high pH region, but at low pH; the composition of the formed complex is 1/1 in PAA/POE.

12.3.2 Critical pH for the Interpolymer Complexation [10]

12.3.2.1 Measurements from the Yield of Precipitate

The dependence of precipitate yield of the interpolymer complex on pH is shown in Fig. 12.2. It is found that the ability to form interpolymer complexes from poly(carboxylic acid) with POE differs with the type of carboxylic acid. In each system the yield of the precipitate increased drastically at a certain pH value, and at less than this pH value the complex is obtained almost stoichiometrically. These pH values may be called critical pH values for precipitation of the complex. They are 3.0, 2.3, and 1.9 for the PMAA, styrene-maleic acid copolymer (PSMA), and PAA systems, respectively. This state can be explained from the dissociation of each poly(carboxylic acid). Dissociation constants (pK_a)

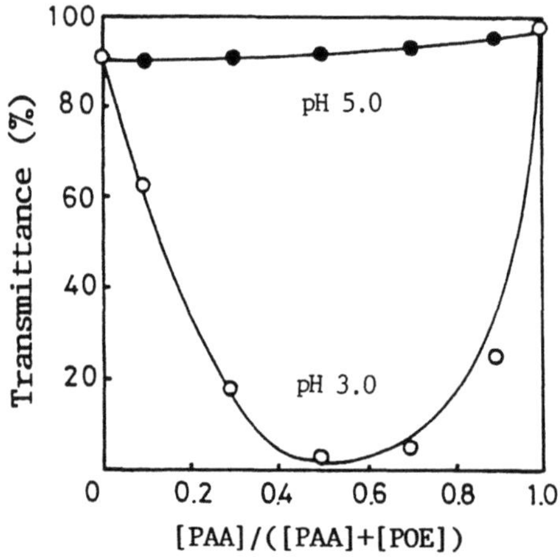

Fig. 12.1. Formation of PAA–POE complex as measured from transmittance of the solution: [POE] = [PAA] = 5×10^{-2} (unit mol/l); $\bar{M}_n(PAA) = 1 \times 10^5$, $\bar{M}_n(POE) = 1.4 \times 10^6$

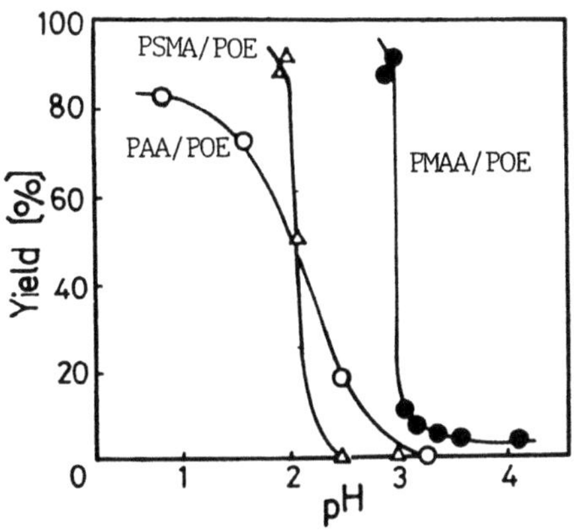

Fig. 12.2. Critical pH for precipitate yield in poly(carboxylic acid)–POE systems: poly(carboxylic acid)/[POE] = 1/1; concentration, 2×10^{-2} (unit mol/l); $\bar{M}_n(PAA) = 1 \times 10^5$, $\bar{M}_n(PMAA) = 1.5 \times 10^5$; $\bar{M}_n(PSMA) = 5 \times 10^3$; $\bar{M}_w(POE) = 1.4 \times 10^6$

obtained from the Henderson-Hasselbach equation [58] are $pK_a(PMAA) = 7.3$, $pK_a(PSMA) = 6.5$, and $pK_a(PAA) = 5.6$. The order of the constants coincides well with that of the critical pH values. The magnitude of the constant indicates the difficulty of the dissociation of carboxylic groups on polymer chain. Therefore, it is considered that the critical pH value increases with an increase in the pK_a of poly(carboxylic acid). In the PMAA and PSMA systems yields of the complexes increase drastically at their critical pH values and their maxima amount reach about 100%, while in the PAA system the increase of the curve is slow and does not reach 100%. This can be explained by the difference in the hydrophobic characters of the component polymers. PMAA and PSMA show comparatively high hydrophobicity because they have methyl groups and phenyl groups in their side chains. For this reason, interaction between their hydrophobic groups and the main chain of POE should participate in complexation as well as in hydrogen bonding. Therefore, the cooperativity on complexation becomes larger. This effect will be described thermodynamically later.

12.3.2.2 Measurements from the Solution Properties [10]

As described above, critical pH was measured from the yield of complex precipitate. However, when the solution properties such as the viscosity and pH were measured in detail, critical pH values for complexation existed in the higher pH region. Figure 12.3 shows the relationship between the mixed unit molar ratio, [POE]/[PMAA], and the viscosity and pH change of the mixed solution. When the initial pH of the PMAA solution is lower than 5.5, the viscosity decreases slowly until the unit molar ratio is equal to unity, and after this point it increases again with an increase in the POE ratio. Corresponding to this change, the pH increases linearly up to the point where the ratio is equal to unity. It is well known that when poly(carboxylic acid) forms a complex with POE through hydrogen bonding, protons in solution are introduced into the complex, resulting in an increase in pH [59]. This explains the pH change in this

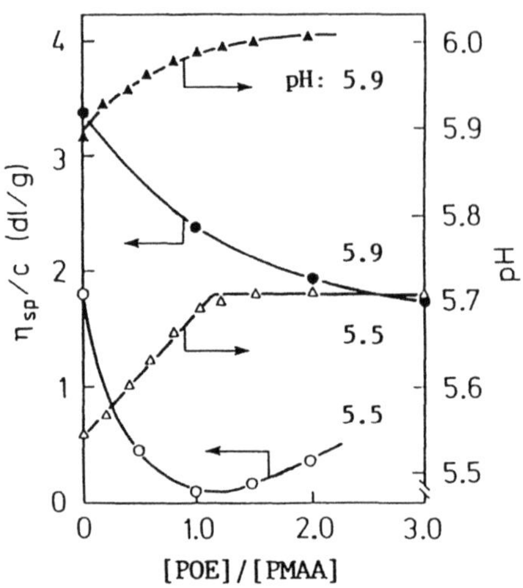

Fig. 12.3. Complexation of PMAA–POE as a function of the initial pH. [PMAA] = 5 $\times 10^{-3}$ (unit mol/l); T = 25 °C

Table 12.2. Dissociation constants (pK_a)

Monomer	pK_a	Polymer	pK_a
AA	4.3	PAA	6.8
MAA	4.6	PMAA	7.3
AcOH	4.8	PMAA–POE	7.9
		PMAA–2X	4.3

2X :

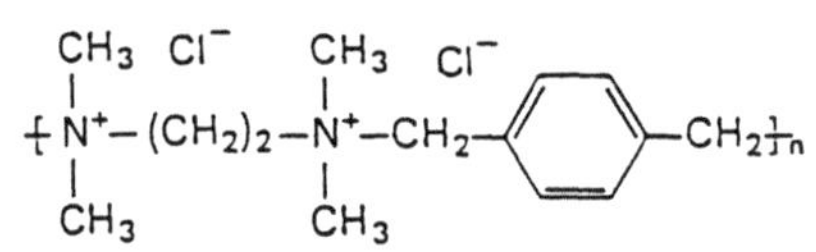

system up to [POE]/[PMAA] = 1. The decrease in the viscosity is explained in terms of the shrinking of the polymer chains, because electrostatic repulsion is weakened by complexation. The following increase in viscosity after [POE]/[PMAA] = 1 is due to an excess of POE. It is evident that the increase in viscosity in this region is not due to the destruction or formation of the complexes, since no pH change is observed in this region. That is, in the region of pH < 5.5, PMAA and POE form an equimolar complex.

On the other hand, an entirely different set of phenomena is observed in the pH region higher than 5.6. That is, the viscosity and pH of the mixed solutions gradually change with an increase in the amount of POE. This suggests that a stable complex with a definite composition may not be formed. In the high dissociation of carboxylic acid, the complexation is considered to be a reversible equilibrium with regard to dissociation into two component polymers. In this case, since complexes already formed are unstable and dissociation cannot be ignored, a decrease in the viscosity or an increase of pH should be observed at [POE]/[PMAA] > 1.

The above complexation mechanism is schematically shown in Eqs. (12.1) and (12.2).

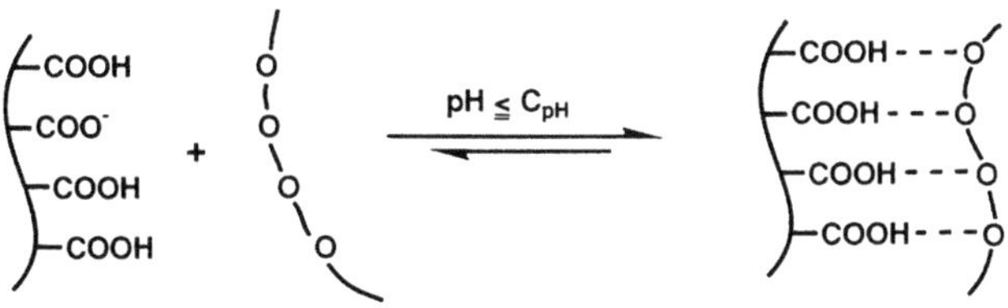

(12.1)

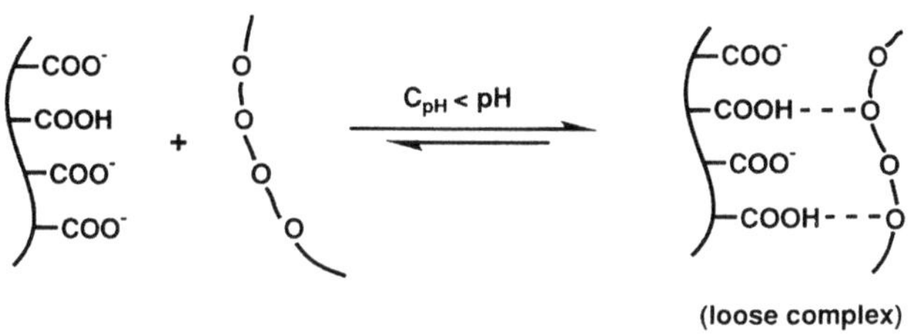

(12.2)

The existence of a certain amount of undissociated carboxylic groups is necessary for PMAA to form a stable complex with POE through hydrogen bonding: if this condition is satisfied at a certain pH (so called "critical" pH), a stable complex is formed irreversibly. At complex formation, dissociated carboxylic groups are also influenced by the neighboring undissociated groups and change to undissociated ones by taking protons from the solution. Since most of the carboxylic groups on PMAA are able to take part in the complexation, the composition of the complex [PMAA]/[POE] is equal to unity.

Table 12.2 shows the apparent dissociation constants (pK_a) of PMAA and the PMAA-POE complex obtained from the Henderson–Hasserbach equation. The addition of POE to the aqueous PMAA solution makes the pK_a higher than that of only PMAA, while the addition of polycations (2X) makes it lower. Moreover, the effect of complexation with POE on the change of pK_a is more pronounced when the chain length of POE is larger. These results suggest that the dissociation of carboxylic groups on PMAA is restricted by complexation with POE and that the ability of complexation of POE with PMAA is stronger for POE with longer chain length.

12.3.3 Critical Chain Length for the Complexation

12.3.3.1 Interpolymer Complex with Hydrogen Bonding

The critical chain lengths of POE for the complexation with poly(carboxylic acids) (PMAA and PAA) were studied and the influence of the hydrophobic interaction on the formation of the interpolymer complex through hydrogen bonding is discussed in this section.

Figure 12.4 shows the dependence of the reduced viscosity of the mixed solution at [POE]/[PMAA] = 1 on the chain length of POE in a water-methanol mixed solvent. In every complexation the reduced viscosity decreases drastically at a certain chain length; i.e. at a critical chain length (L_c) of POE (L_c

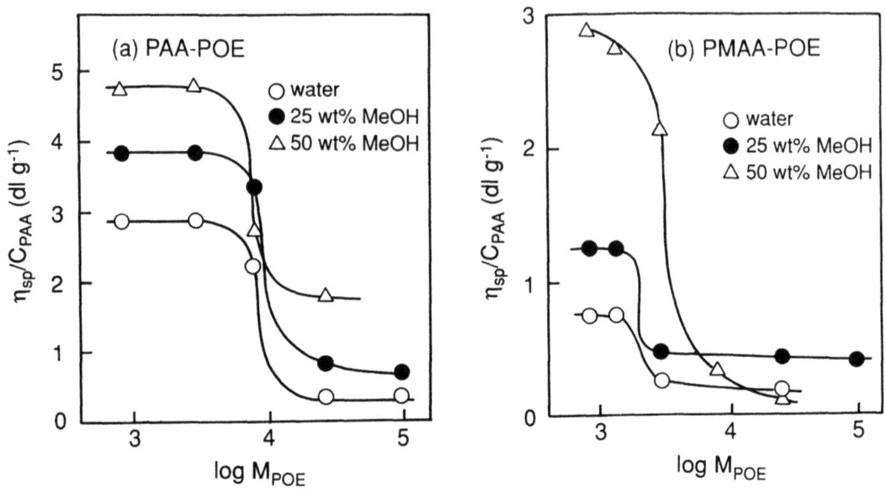

Fig. 12.4. Critical chain length of POE for complexation in a water–methanol mixed solvent. [poly(carboxylic acid) = [POE] = 3×10^{-3} (unit mol/l); at 25 °C; $\bar{M}_n$(PAA) = 1.0×10^5; $\bar{M}_n$(PMAA) = 1.5×10^6

is obtained as the chain length at the inflection point of each viscosity curve). The L_c in the PAA–POE systems are higher than that in the PMAA–POE system. This indicates that complexation of PMAA with POE is more stable than with PAA. Since the interaction force between PMAA and POE contains hydrophobic interaction between the methyl groups of PMAA and the main chain of POE (besides hydrogen bonding), L_c in the PMAA system increases with increasing methanol content of the mixed solvent, while that in the PAA system it is independent on the methanol content. The existence of hydrophobic interaction in the complexation of PMAA with POE can be confirmed from this result and the hydrophobic interaction between PMAA (especially α-methyl group) and POE is weakened or destroyed by addition of methanol.

12.3.3.2 Polyelectrolyte Complex [60]

The stabilities of polyelectrolyte complexes were studied for a series of polycations with chain lengths from one to six units and the hydrophobic effect was studied by comparing two different polycations, having methyl (I) and benzyl (II) side groups.

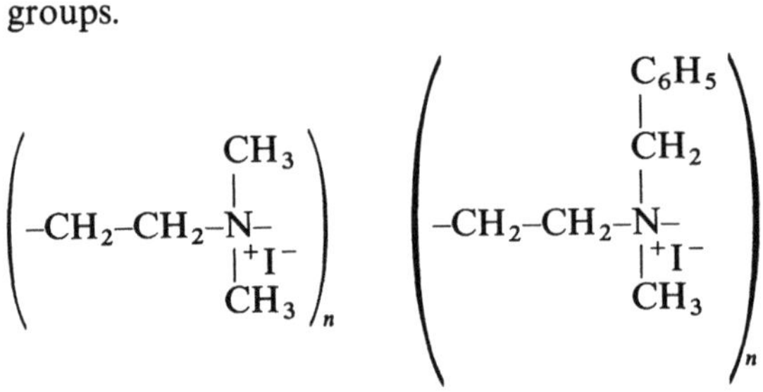

$1 \leq n \leq 6$

I II

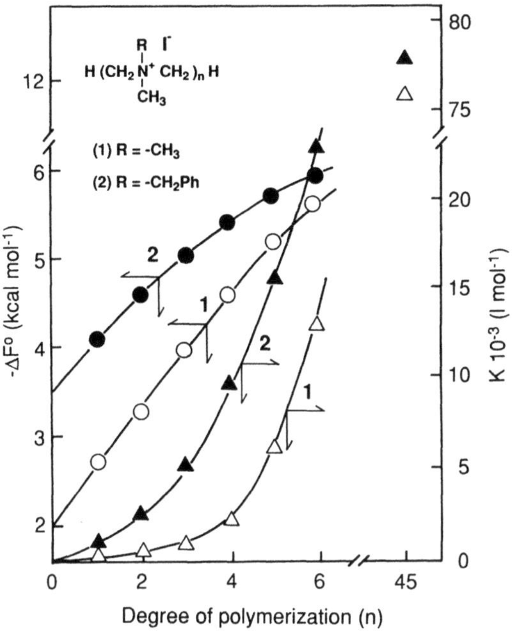

Fig. 12.5. Change of the stability constant (K) and the free energy ($-\Delta F^0$) in the formation of polyelectrolyte complexes with various degree of polymerization of oligocation. $\bar{M}_w$(PMAA) = 5.3×10^4

It should be noted that the amount of released microions corresponds to the degree of complexation of polymer chains [61]. Thus, in the case of complexation of equimolar amounts of components, the stability constant (K) of the complexes defined by Eq. (12.3) is calculated as follows [62].

$$[\text{PMAA}] + [\text{Cation}] \overset{K}{\rightleftarrows} [\text{Complex}] + [\text{HI}] \tag{12.3}$$

As shown in Fig. 12.5, the stability constant (K) of the complex increases exponentially when the degree of the polymerization of oligocations is above 4. The free energy change in the complex formation decreases almost linearly up to the degree of polymerization below 4. These relationships are described by Eqs. (12.4) and (12.5):

$$K = A * e^{Bn} \tag{12.4}$$

$$-\Delta G^0 = \alpha * n + \beta \tag{12.5}$$

where α is comparable with a cooperative coefficient and β represents the basic bonding constant. In the range of chain length of these oligocations, the standard free energy change (ΔG^0) of complex formation decreases by about 0.6 kcal/mol upon the addition of one cation residue. The difference in β (about 1.6 kcal/mol) between two different types of cations may be due to the variation of the hydrophobicities of benzyl and methyl residues. Thus, when the total free energy change exceeds the kinetic energies of the polymer chain components, a stable intermolecular complex is formed. When using low molecular weight compounds whose degree of polymerization is below four, the stable complex is not formed.

In the case of interpolymer complex formation hydrogen bonding, critical chain lengths for POE in the complexation with PMAA and PAA are 40 and 200, respectively. This shows that cooperativity in the complexation through Coulomb bonding is significantly larger than that in the complexation through hydrogen bonding.

12.4 Thermodynamics of Interpolymer Complexes from PAA (or PMAA) with POE

12.4.1 Degree of Linkage [60]

A simple method for evaluating the fraction of binding groups of the complex (degree of linkage θ) and the enthalpy and entropy changes for complexation is briefly outlined here for the case of complexes formed by cooperative hydrogen bonding. The procedure for the calculation of the degree of linkage (θ) is given by Eq. (12.6) for complexes with a stoichiometric composition:

$$\theta = 1 - ([H^+]/[H^+]_0)^2 \tag{12.6}$$

where $[H^+]$ and $[H^+]_0$ are the proton concentration in the presence and the absence of the complementary macromolecule, respectively. The value of θ for various PMAA–POE and PAA–POE complexes as a function of temperature are summarized in Fig. 12.6a and b, respectively. As predicted from the viscosity change for the complex PMAA with POE, POE with molecular weights of 1000 or lower do not form the complex, while those with the molecular weights of 2000 and 3000 favor the formation of the complex above a certain temperature. In these complexes, 30–50% of the interacting groups remain unbound even at high temperatures. Furthermore, the complex with POE of a molecular weight of 2000 seems to undergo partial redissociation above 55 °C. This shows that at this molecular weight the hydrophobic interaction helps the complex formation. Complexes prepared from POE with molecular weights of 7500 and 20000 have values of nearly 1.0 and are stable over the temperature range investigated.

The complexation of POE–PAA was quite different from that of POE with PMAA. POE with molecular weights of 7500 or lower do not form any complexes, while POE with molecular weight of 20000 forms a complex but with a low value of θ. The difference between the complexation ability of PMAA and PAA could not be explained from the difference in dissociation, since the dissociation constants of PMAA and PAA are approximately equal. An endothermic complexation favored by a rise in temperature in the PMAA–POE system could be interpreted in terms of hydrophobic interaction between the α-methyl groups of PMAA and the ethylene backbone of POE which might provide additional stability in the complex.

The temperature dependence on the stability constant (K) in the complexation of PMAA–POE (with molecular weight of 2000) at various concentra-

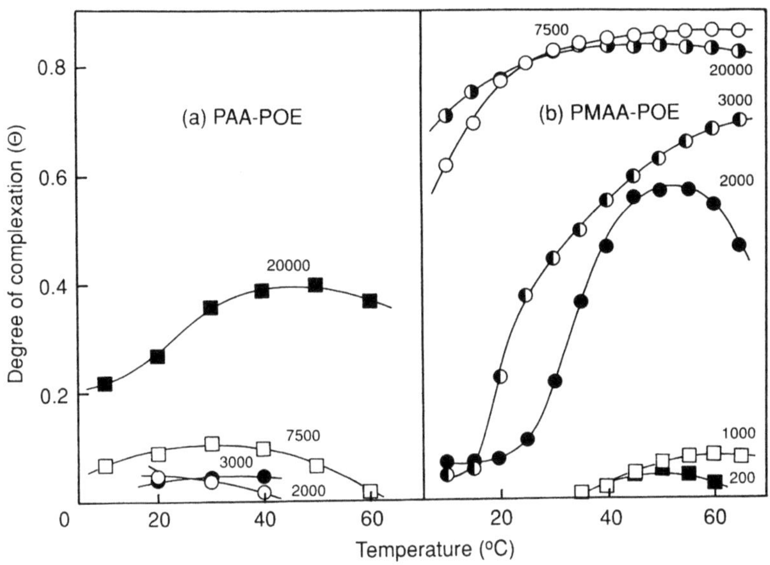

Fig. 12.6. Dependence of degree of complexation (θ) on temperature for the complexation of (**a**) PMAA and (**b**) PAA with POE various molecular weights; [PMAA] = [PAA] = [POE] = 5.8 $\times 10^{-3}$ (unit mol/l); $\bar{M}_w$(PMAA) = 6.6×10^4; $\bar{M}_w$(PAA) = 4.0×10^4

tions of ethanol in water are listed in Table 12.3. The constant is given by Eq. (12.7) for complexes with stoichiometric composition:

$$K = \theta / C_0(1 - \theta)^2 \tag{12.7}$$

The data indicate that the addition of ethanol reverses the temperature dependence of the complexation at ethanol concentrations greater than 17 vol%. The complexation of PMAA–POE in the presence of an appropriate amount of ethanol is realized only in the low temperature range. Table 12.3 also shows that complexation is completely inhibited at concentrations of ethanol greater than 37 vol%.

12.4.2 Enthalpy and Entropy Changes [60]

The stability constant (K) can be used to calculate the enthalpy and the entropy changes for the complexation from Eqs. (12.8) and (12.9):

$$\mathrm{d}(\ln K)/\mathrm{d}(1/T) = -\Delta H^0/R \tag{12.8}$$

$$\Delta S^0 = -(\Delta G^0 - \Delta H^0)/T \tag{12.9}$$

where ΔG^0 ($= -RT \ln K$) is the free energy change for the complexation. The results of various complexation systems are shown in Figs. 12.7 and 12.8. Positive values of both ΔH^0 and ΔS^0 for these systems are characteristic of hydrophobic interaction. If the hydrophobic interaction is significant in the

Table 12.3. Stability constants (K) of the complexes at various temperatures

Complex system	Molecular weight of PEO or PVPo $\times 10^{-3}$	Stability constant K (l/mole) 10 °C	20 °C	30 °C	40 °C	50 °C	60 °C
PMAA–PEO	0.2	2	0	0	4	10	5
	1.0	3	0	0	5	14	16
	2.0	15	16	62	280	520	440
	3.0	10	68	250	470	790	1130
	7.5	760	2600	4700	6200	6800	7200
	20.0	1450	2800	4200	5000	5000	4580
PAA–PEO	2.0	1	10	8	3	0	0
	3.0	2	7	9	8	1	0
	7.5	13	18	24	20	13	3
	20.0	64	85	150	180	190	160
PMAA–PVPo	10.0	950	920	1260	1850	3920	7560
	40.0	1460	1700	1780	2200	5010	11 900
	160.0	14 500	12 200	14 300	18 000	22 200	27 400
	360.0	15 500	18 700	24 700	33 500	43 500	48 600
PMAA–PEO in EtOH–H_2O mixed solvent (EtOH %)							
10%		22	55	162	198	171	131
17%		328	268	207	118	61	49
23%	2.0	—	208	72	39	33	20
37%		0	0	0	0	0	0
50%		0	0	0	0	0	0

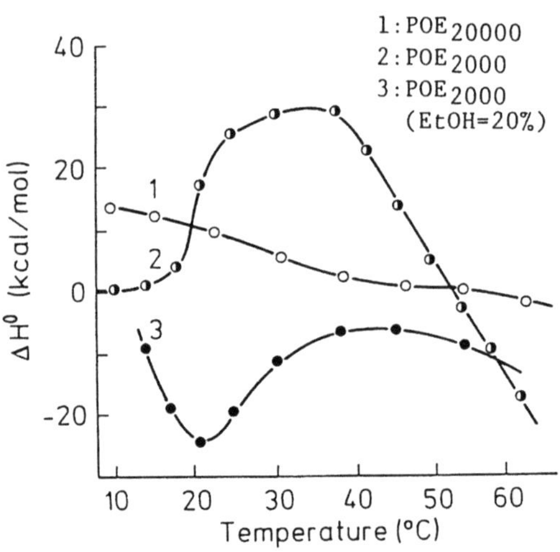

Fig. 12.7. Temperature dependence of the standard enthalpy changes (ΔH^0) for the complexation of PMAA–POE with various molecular weights; $\bar{M}_w$(PMAA) = 6.6×10^4

stabilization of the resulting complexes, ΔH^0 should be reversed by breaking or weakening the hydrophobic interaction by adding ethanol to the complex solution. As shown in Fig. 12.7, the sign of the ΔH^0 is minus when 25% ethanol is added. This indicates that the stability constant decreases with an increase in temperature, which is a typical characteristic of hydrogen bonding.

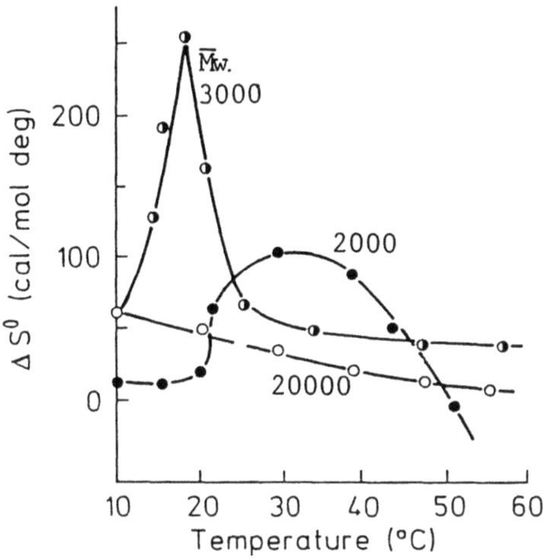

Fig. 12.8. Temperature dependence of the standard entropy changes (ΔS^0) for the complexation of PMAA – POE with various molecular weights. $\bar{M}_w$(PMAA) = 6.6×10^4

The maxima of ΔS^0 at 20 °C and 30 °C in the mixed solution of PMAA and POE with molecular weight of 2000 and 3000 are related to the entropy change at complex formation (Fig. 12.8). Positive values of ΔS^0 in aqueous systems suggest the release of water during complexation. The complexation system of PMAA and POE with a molecular weight of 20 000 does not exhibit any pronounced values of ΔH^0 and ΔS^0. This shows that a complex was already formed in the temperature range that was not investigated.

12.5 Selective and Substitution Interpolymer Complexation [63]

12.5.1 Selective Complexation of a Ternary Polymer Mixture

The selective interpolymer complexation is described as follows:

$$P1 + P2 + P3 \longrightarrow (P1\text{–}P2) \text{ complex} + P3 \qquad (12.10)$$

In this case, P2 is poly(methacrylic acid) (PMAA) and P1 and P3 are the polymers which can interact with PMAA through various kinds of secondary binding forces;

$$P1 + P2 \longrightarrow (P1\text{–}P2) \text{ complex} \qquad (12.11)$$

$$P3 + P2 \longrightarrow (P3\text{–}P2) \text{ complex} \qquad (12.12)$$

In the case of the OEI–POE–PMAA system [OEI = oligo(ethyleneimine)], from the elemental analysis as summarized in Table 12.4, OEI was discovered in the precipitate at the low pH region. Moreover, in the higher pH region, (pH = 10), the precipitate is not obtained. From these results, the selective formation of interpolymer complexes can be represented by the following equations.

(1) (pH < 3)

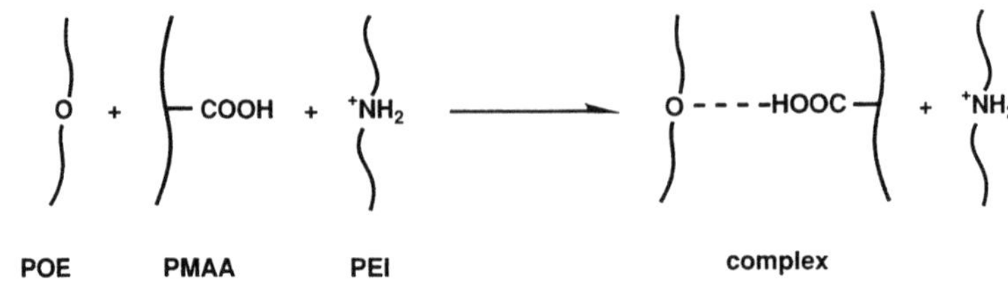

(12.13)

(2) (3 < pH <7)

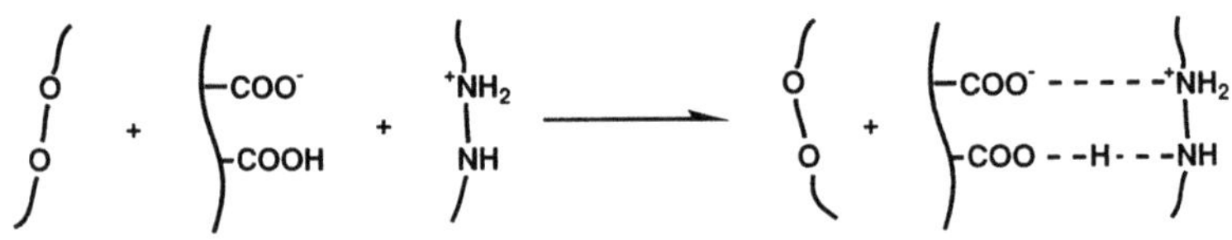

(12.14)

(3) (7 < pH)

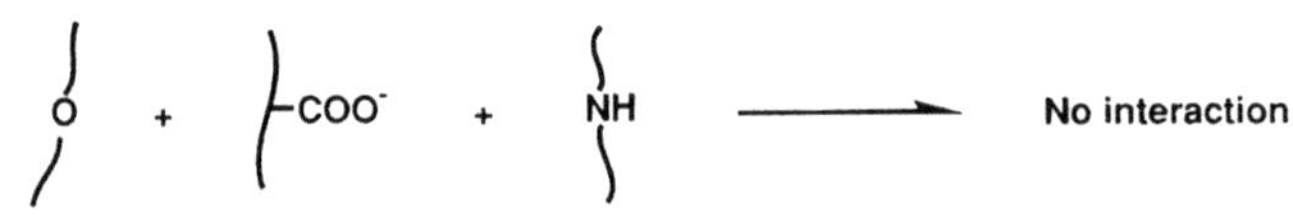

(12.15)

Table 12.4. Selective interpolymer complexation in the PMAA–POE–OEI[a] systems

Sample			Solvent	pH	Yield (mg)	Elemental analysis C	H	N	C/N
PMAA	POE		H_2O	2.0	—	55.37	7.75	0	
PMAA		OEI	H_2O	5.0	—	55.79	8.59	10.85	5.14
PMAA	POE	OEI	H_2O	2.0	29.5	55.1	7.8	0.6	88.2
PMAA	POE	OEI	H_2O	5.0	25.9	69.3	8.5	10.3	6.7
PMAA	POE	OEI	H_2O	9.0	0				

[a] OEI = $(NHCH_2CH_2)_n$

That is, (1) when the pH is less than 3, OEI is nearly completely protonated but PMAA is scarcely dissociated. PMAA preferentially forms a complex with POE through hydrogen bonding but not with OEI. (2) In the pH region 3 to 7, both PMAA and OEI are partially ionized. Therefore, PMAA forms a polyelectrolyte complex with OEI. (3) In the pH region above 7, PMAA is almost dissociated, while OEI are not protonated; thus, neither the polyelectrolyte complex nor the complex formed through hydrogen bonding are present.

12.5.2 *The Ability of Selective Interpolymer Complexation*

Using other polymers, e.g. poly(acrylamide) (PAAm), poly(*N*-vinyl-2-pyrrolidone) (PVPo), and poly(vinyl alcohol) (PVA), the ability of selective interpolymer complex formation was studied. The order is as follows:

$$POE_{1.4\times10^6} > PAAm > PVPo > POE_{2900} > PVA$$

On the basis of these results, the complexation may be described by the following scheme Eq. (12.16):

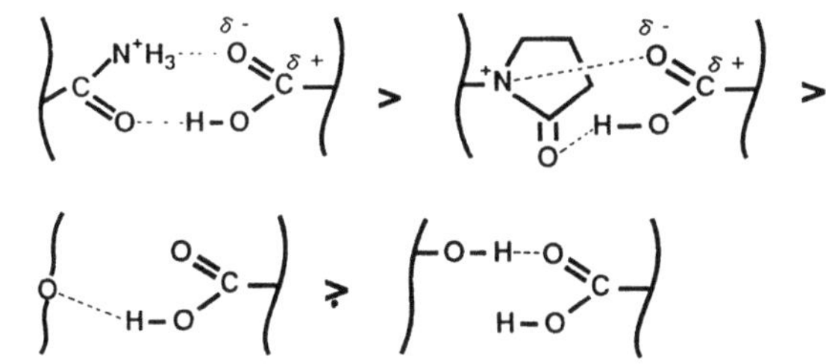

(12.16)

PAAm or PVPo binds PMAA more strongly than POE or PVA, since PAAm forms a complex with PMAA not only through hydrogen bonding but also through ion-dipole interaction between the partially protonated amide groups of PAAm and the C=O dipoles of PMAA. Moreover, hydrophobic interaction might be involved in selective complexation. The hydrophobicity of PVPo is the largest for all of these polymers. Thus, the interaction between PVPo and PMAA is larger than other combinations. The PMAA–POE complex is preferentially formed in comparison with PAA–POE complex because of the higher hydrophobic interactions between PMAA and POE. On the other hand, PVA has a strong tendency to aggregate with itself. Therefore, the complex formation of PMAA with PVA is difficult if PVA aggregates are not destroyed.

12.5.3 *Selective Complexation Between PAA and POE with Different Chain Lengths*

The selective interpolymer complexation based on the difference in the chain length of POE is shown in Table 12.5. The reduced viscosities of the original solutions of POE with different chain lengths ($\bar{M}_w = 1400$, 8000, and 1.4×10^6) at 30 °C are 0.072, 0.193, and 0.346 (dl/g), respectively. When PMAA was added to the mixed solution of POE_{1400} and POE_{8000}, the complex precipitate was obtained. The yield was 63.4 mg which was the total amount of PMAA and either component of POE. The relative viscosity of the final supernatant solution is 0.074, which is very similar to that of the solution of POE_{1400}(0.072). A similar phenomenon was observed in the system of PMAA, POE_{1400}, and $POE_{1.4\times10^6}$. On the other hand, when PMAA was added to a mixture of the three kinds of POE, the relative viscosity of the supernatant solution was 0.124 which is very similar to that of the equimolar mixture of POE_{8000} and

Table 12.5. Effect of the chain length on the selective interpolymer complexation in the PMAA–POE system

Init. Cont. (mg)				Yield (mg)	η_{sp}/c[a] (dl/g)
PMAA	POE				
	$Mw \sim 1400$	8000	1.4×10^6		
43.6	22.0	22.0	—	64.8	0.074
43.6	22.0	—	22.0	65.2	0.076
43.6	—	22.0	22.0	66.7	0.187
43.6	22.0	22.0	22.0	67.1	0.124

[a] η_{sp}/c (dl/g): 0.072 (POE_{1400}), 0.193 (POE_{8000}), 0.346 ($POE_{1.4 \times 10^6}$)

POE_{1400}(0.134). This can be described as follows;

$$POE_{1.4\times10^6} + POE_{8000} + POE_{1400} + PMAA \longrightarrow (POE_{1.4\times10^6} - PMAA)_{comp} + POE_{8000} + POE_{1400} \tag{12.17}$$

As mentioned previously, there is a critical chain length (a minimum chain length for a polymer to form a stable complex) for complexation. A critical chain length in the PMAA–POE system is about 40 with regard to the degree of polymerization of POE [10]. Since POE ($\bar{M}_w = 1400$, $P_n = 32$) is below the critical chain length, it can not form a complex with PMAA. However, although both $POE_{1.4\times10^6}$ and POE_{8000} have chain lengths over the critical chain length, only $POE_{1.4\times10^6}$ forms a complex with PMAA. From these results, it is clear that the complexation ability of POE with PMAA depends on the chain length of the component polymer even beyond the critical length because of the cooperativity interaction between each active site.

12.5.4 Substitution of Interpolymer Complexation

As mentioned previously, selective interpolymer complexation depends on the magnitude of the total bond energy between each pair of polymers. Therefore, if P1 can interact with P2 more strongly than P3, the interchain macromolecular substitution reaction of P3 and P1 should take place on the addition of P1 to the P2–P3 complex solution. In this system, a cooperative interpolymer substitution reaction is

$$(P2\text{–}P3)\ \text{complex} + P1 \longrightarrow (P1\text{–}P2)\ \text{complex} + P3 \tag{12.18}$$

Such substitution reactions may proceed via two types of mechanisms:

(1) Open system

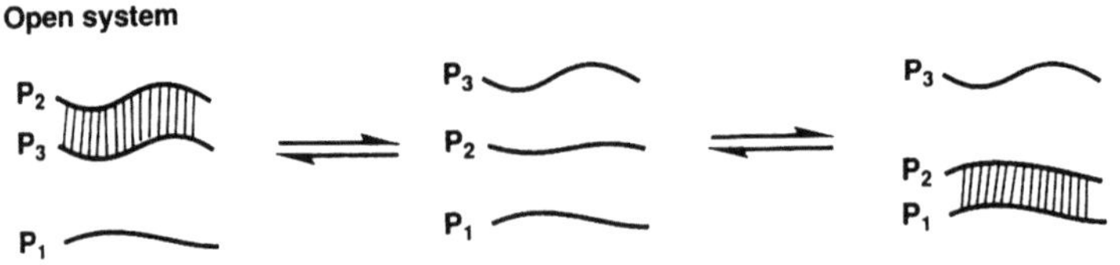

(12.19)

(2) Closed system

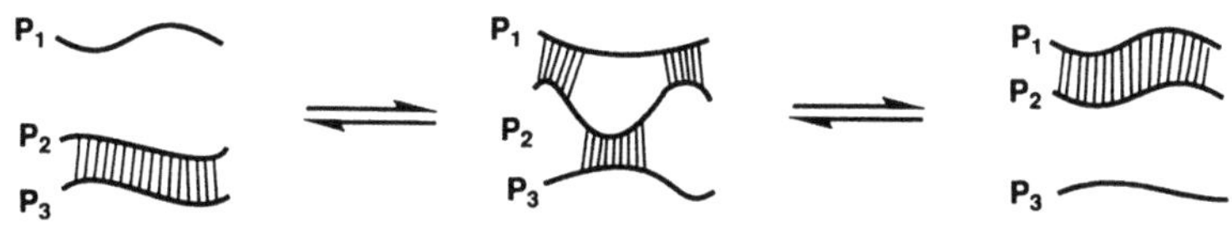

(12.20)

In the open system, the complex once formed is dissociated into two component polymers according to the changes in environmental factors, and then a different pair of component polymers forms another type of complex. On the contrary, in a closed system, before the complex is completely destroyed, the third component polymer interacts with the complex to form a ternary polymer complex, and the first component polymer is dissociated from the ternary complex. In the closed system, the substitution reaction, especially the releasing process of the first component polymer, seems to be accelerated when the interaction force between a parent polymer matrix chain and the third component polymer chain is stronger.

Kabanov et al. [9] studied the interchain substitution reaction

$$(\mathrm{PAA^*\text{–}POE})_{\mathrm{comp}} + \mathrm{PMAA} \longrightarrow (\mathrm{PMAA\text{–}POE})_{\mathrm{comp}} + \mathrm{PAA^*} \quad (12.21)$$

by means of polarized luminescence using PAA labeled with a luminescent probe (PAA*). When the reaction advances, $1/P^*$ increases from the value of PAA* in the complex to that of free PAA*. They also reported that the rate of the substitution by PMAA decreases with increasing chain length of POE; finally, with a much longer POE chain length the rate is independent of the chain length. These results may be explained from the assumption that, with short chain length, the substitution reaction takes place after almost all of the first complex (PAA–POE) is dissociated because of the easy dissociation of this complex (open system). In contrast, POE with a longer chain length causes the substitution reaction to proceed via a ternary complex (closed system) since dissociation of the first complex does not practically occur at all [64].

The results of another substitution reaction of the 2X–POE–PMAA system are shown in Fig. 12.9. The POE–PMAA complex and the 2X–PMAA complex are preferably formed in the low or high pH region, respectively, as shown from turbidity measurements in Fig. 12.9b; in the intermediate pH region the 2X–POE–PMAA ternary polymer complex is formed since the turbidity is very low but still higher than that of each polymer component (see Fig. 12.9a). Thus, in the 2X–POE–PMAA system a substitution reaction takes place through a closed system. The decrease in the yield of the precipitation of the ternary polymer complex is well explained by the following scheme:

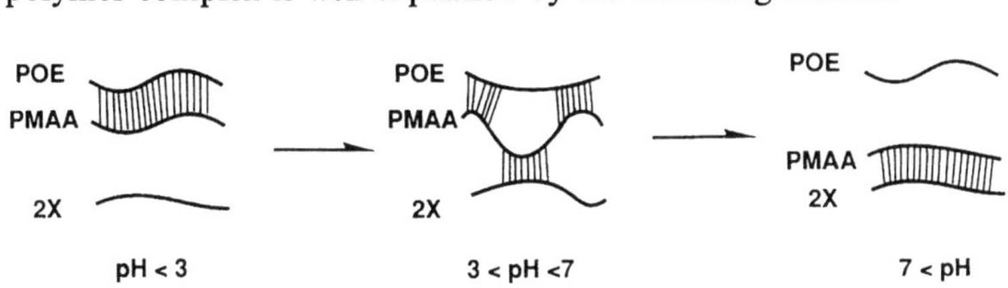

(12.22)

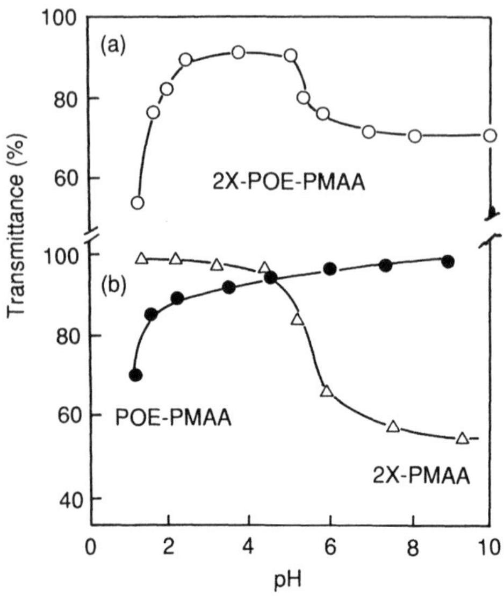

Fig. 12.9a, b. pH dependence of the cooperative interchain substitution in the system 2X–POE–PMAA measured from turbidity change: (**a**) ternary component polymers; (**b**) two component polymers of the three, [PMAA] = [POE] = [2X] = 2.5×10^{-3} (unit mol/l)

Hence, part of the non-bonding chains of the polymer components makes the complex itself hydrophilic so that the ternary complex becomes soluble in aqueous medium.

12.6 Solid Properties of a Hydrogen-Bonding Complex

Smith et al. [7] reported the unique properties of the interpolymer POE–PAA complex. Figure 12.10 shows the temperature dependence of the stiffness of the complex in the solid state. Two transitions of the stiffness are observed for a mixture with 75% POE. The first break around 15 °C is attributed to a compatible mixture of the two polymers. The resulting new phase is an amorphous complex and has a lower softening temperature than pure PAA. The second break means that POE crystalline phase is melting. The additional PAA has effectively eliminated the POE crystalline phase, apparently leaving only a single amorphous phase (50% in Fig. 12.10). On further addition of PAA, a series of similar stiffness–temperature curves approach the position of PAA. Surprisingly, with a POE content of 10%, the stiffness values are substantially above those of each polymer component. At this composition a maximum of the glass transition temperature was observed (Fig. 12.11). T_g decreases significantly with an increase in POE content up to 50% and shows a constant value, which is the T_g of an amorphous POE–PAA complex. On the other hand, T_g of pure amorphous POE might be observed more apparently when the POE content is larger.

Figure 12.11, also shows the dependence of the heat stability on the composition of the complex. The heat stabilities (retention rate in weight) are plotted against the composition of the polymer components using a film sample

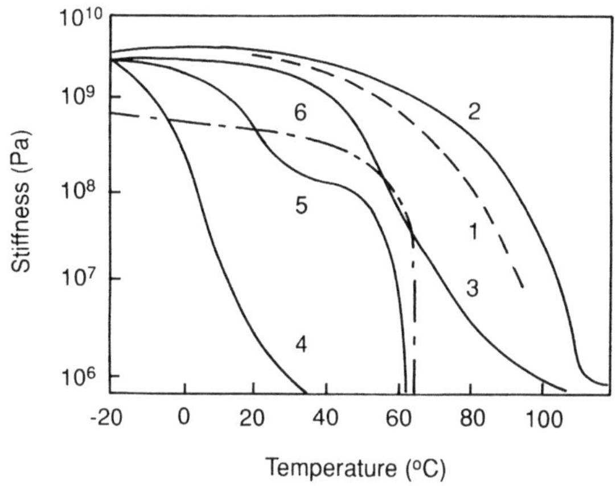

Fig. 12.10. Temperature dependence of stiffness of the PAA–POE complexes with various POE content (wt%): (1) 0, (2) 10, (3) 25, (4) 50, (5) 75, (6) 100

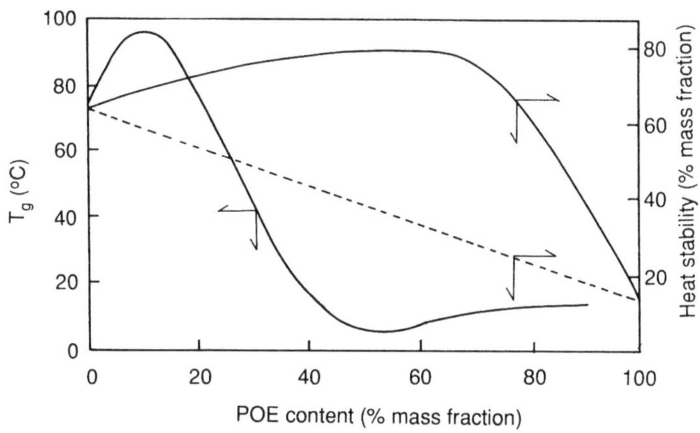

Fig. 12.11. Relationship between the glass transition temperature (T_g) and POE content in PAA–POE complexes. Heat stability of the PAA–POE complexes with various POE content retained at 300 °C for one hour

heated in air at 300 °C for one hour. A predicted weight constancy for mechanical mixtures of the polymer components is indicated by a dotted straight line, assuming that each polymer contributes proportionately (related to weight) to the heat stability of the mixtures. The upper curve shows that the actual heat stability of the complex rises rapidly with a slight increase of the PAA content, reaching a maximum for a PAA content ranging from 30 to 45%. It is interesting that the maximum value of 79% weight constancy is higher than that of each polymer component. This is due to the hydrogen bonding of the interpolymer complex.

Characteristic properties of PAA–POE (50:50 wt) complexes are summarized in Table 12.6 in comparison with polyoxyethylene [7]. Stiffness of the complex is significantly smaller than that of POE. That is because POE has a

Table 12.6. Characteristics of POE–PAA complex (50:50wt) (K.L. Smith et al., Ind Eng. Chem 51, 1361 (1959))

Property	Complex	POE
Stiffness 10^7 (dyne cm^{-2})	1.38	210–480
T_g	5	− 55
Heat stability[a]	77	5
Solubility in aq.[b]	7–13	100
Tensile strength 10^7 (dyne cm^{-2})	15	13–16
Ultimate elongation (%)	800	700–1200

[a] Retention (wt%) at 300 °C for 1 hr.
[b] Soluble portion (wt%) at room temperature

crystalline phase, while the complex is completely amorphous. POE also has an amorphous phase where the glass transition temperature is − 65 °C. In the case of the complex, T_g increases due to the complexation through hydrogen bonding. There are other characteristics of hydrogen bonding such as higher heat stability and lower water solubility.

12.7 Ion Conduction and Solid Polymer Electrolytes

12.7.1 Characteristics of Solid Polymer Electrolytes

Polymeric solid electrolytes are polymeric systems which enable efficient ion migration even in the solid state and are comparable to the solution state. Recently, it has obtained interest not only from pure scientific research fields but also from industrial application fields [65]. Since it was first recognized by Dr. Wright in the University of Sheffield, England in 1975 [66], it has been studied vigorously and has also been applied in several fields.

Compared with an inorganic ion conductor, the polymer systems have great advantages with regard to mechanical properties and a wide variety of chemical design is also available. The selective ion conduction (cation or anion) can be performed by the design of the functional polymer matrix. The kind of ions as carriers are easily changed by selecting additional salts. The advantages of a polymeric solid electrolyte system include flexibility, transparency, excellent processibility, light weight and a completely dry system. If polymeric solid electrolytes satisfying the above mentioned requirements are realized, lots of devices could be changed completely [67]. In this chapter, a brief introduction about solid polymer electrolytes is given.

12.7.2 Classification of Solid Polymer Electrolytes

In general, solid polymer electrolytes can be classified into four groups:

(1) porous polymer + solvent + salt
(2) porous polymer + oligomer + salt

(3) polar polymer + salt
(4) polar polymer with charged sites

The first group is the most simple system. An organic solvent and an inorganic salt are mixed and incorporated into the polymer matrix which has a micro porous structure [68–70]. In the second group, oligomers are used instead of an organic solvent [71–74]. This system contains no volatile solvent, and is expected to have a superior stability to the systems classified in the first group. These solid polymer electrolytes are developed from a solution system. Ionic conduction is realized through a solution column incorporated into the polymer matrix. The third group is composed of a polymer and inorganic salts, but without solvent or oligomer. This, therefore, can be called the exact solid polymer electrolyte. Ions migrate in polymer matrix and the molecular design plays an important part in enabling such a conduction. The last group is a rather advanced one. As the charged sites are fixed on the polymer structure, the mobile carriers are only counter ions of the fixed charges [75–89]. This system is expected to be a system which is extremely stable for devices driven under direct current.

12.7.3 Ion Conduction in Solid Polymer Electrolytes

The ion conduction is known to be given by

$$\sigma = ne\mu \tag{12.23}$$

where n, e, and μ are the total number of carrier ions, electric charge and carrier mobility, respectively. To design a solid polymer with a high ionic conductivity, it is necessary to increase the factors n and μ. n is related to the degree of ion dissociation of the salt incorporated into the polymer matrix.

12.7.3.1 Ion Dissociation

When a cation and an anion are transferred from the crystal lattice of the salt to their equilibrium position in a solvent (polymer), if the enthalpy and the entropy changes produce an overall reduction in the free energy of the system (Eq. (12.24)), the salt dissolves in the solvent [90].

$$\Delta G_{\mathrm{soln}} = \Delta H_{\mathrm{soln}} - T\Delta S_{\mathrm{soln}} < 0 \tag{12.24}$$

In polymer media, the loss of the entropy of the polymer solvent caused by ion solvation is much smaller than those of low molecular weight solvents. This is especially true if the ion is solvated by neighboring coordination groups or sites on the same polymer chain. The entropy change (ΔS_{soln}) in Eq. (12.24) is always positive. Therefore, the solubility of inorganic salts should be controlled primarily by the magnitude of the solvation energy of the dissociated ions in comparison with the lattice energy of the salt. Poly(oxyethylene) (POE) has been considered to be the best polymer material for ion dissociation. Ethers are quite strong donors and very poor acceptors, and the donicity of POE is estimated to be close to 20 [65, 91]. Thus, the solvation energy of the cation is high. Furthermore, the conformations of –C–O-(*trans*), –C–C-(*gauche*), and –O–C-(*trans*) are appropriate for the coordination of several ether oxygens to a cation [92–97]. Therefore, the salt can dissociate by pseudo-solvation of ether oxygens of POE.

12.7.3.2 Ion Conduction Coupled with Segmental Motion of POE

A cation can be transported by the segmental motion of POE and the exchange of the cation between ether oxygens. Therefore, the high segmental motion of the chain and the high exchange rate of the coordination are key points in the design of an ion conductor having high ionic conductivity [98].

To understand the importance of the glass transition temperature, the difference in the ion conduction mechanism is illustrated in Fig. 12.12. The potential field of the inorganic salt generates inevitably. In general, the ion does not migrate easily because of the high potential barrier. However, in the case of special salts such as AgI or β-Alumina, ion migration is explained by the ion hopping mechanism at an energy level low enough to jump the carrier potential. On the other hand, the ions in the solid polymer do not seem to migrate easily; however, it becomes easier when sufficient segmental motion is given to the system. Ions can be transferred along with the segmental motion of the matrix polymer, just like surfing. Higher segmental motion can be enhanced at room

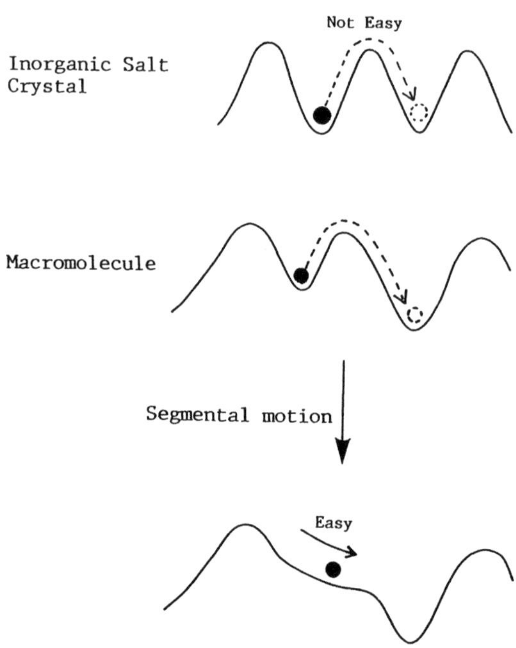

Fig. 12.12. Mechanism of the migration of ions in macromolecules

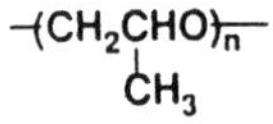

[98] $1.0x10^{-6}$ ($LiClO_4$, 25°C)

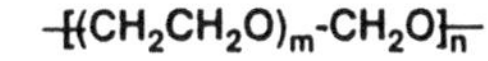

[99] $5.0x10^{-5}$ ($LiCF_3SO_3$, 25°C)

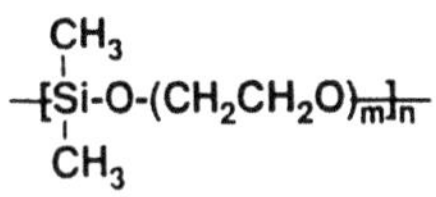

[100] $2.0x10^{-4}$ ($LiClO_4$, m=4, 25°C)

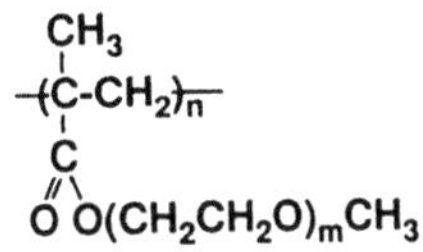

[101] $2.2x10^{-5}$ ($LiClO_4$, m=7, 25°C)

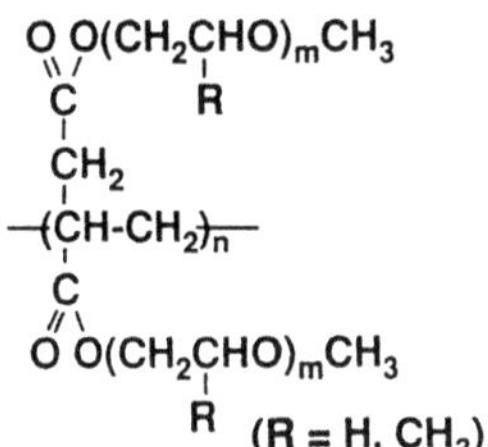

[102] $2.5x10^{-6}$ ($LiClO_4$, R=H, m=3, 25°C)

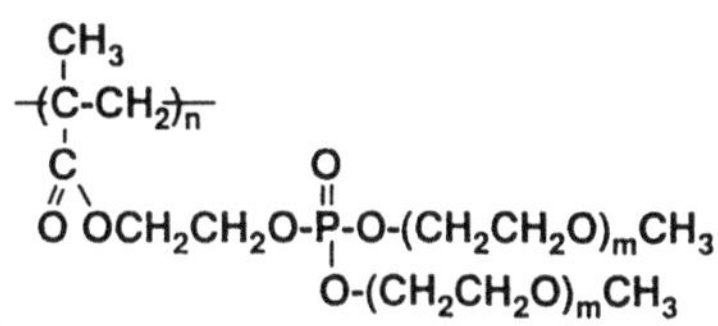

[103] $1.6x10^{-4}$ ($LiClO_4$, m=7, 25°C)

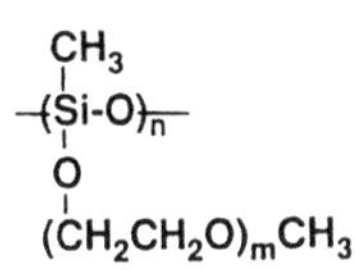

[104] $1.6x10^{-4}$ ($LiClO_4$, m=12, 25°C)

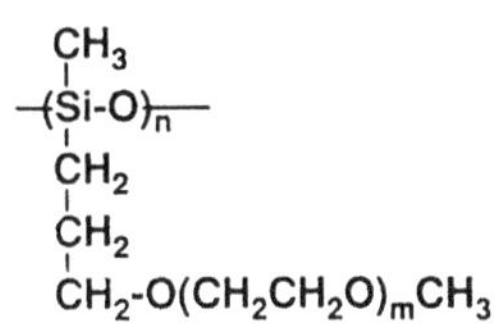

[105] $7.6x10^{-5}$ ($LiClO_4$, m=12, 25°C)

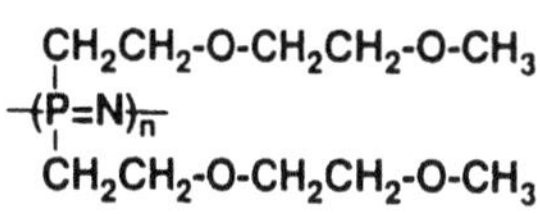

[106] $2.6x10^{-4}$ ($AgSO_3CF_3$, 25°C)

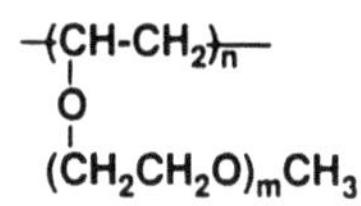

[107] $5.6x10^{-5}$ ($LiClO_4$, m=3, 25°C)

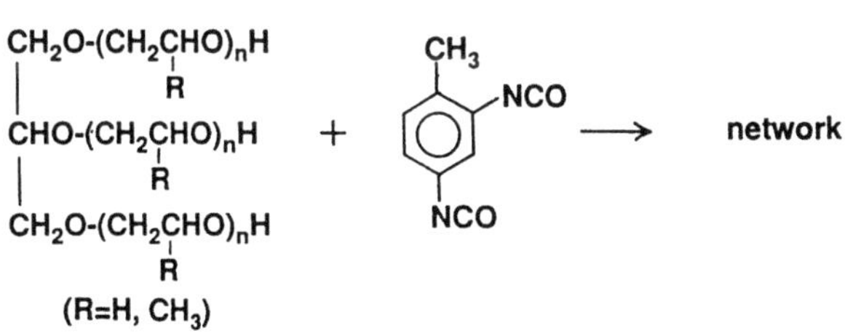

[108] $1.0x10^{-5}$ ($LiClO_4$, R=H, 30°C)

Fig. 12.13. POE derivatives synthesized for high ion conduction

temperature if the matrix polymer has a low glass transition point. The linear POE, however, isn't the best material for the design of an ion conductor because the regularity of the unit allows a high degree of crystallinity which involves about 70–85% of the polymer. Many amorphous POE derivatives have therefore been synthesized as shown in Fig. 12.13 [98–108]. This relation describes the transport properties in a viscous matrix. In general, an amorphous polymer shows weak mechanical stability; this can be improved by crosslinking or introducing polymer frames. Their ion conduction behavior often obey the Williams–Landel–Ferry (WLF) [109] (Eq. 12.25) or Vogel–Tammann–Fulcher (Eq. 12.26) equations [110]:

$$\log(\sigma(T)/\sigma(T_g)) = C_1(T - T_g)/(C_2 + (T - T_g)) \tag{12.25}$$

where T and T_g are temperature (K) and glass transition temperature (K), respectively. $\sigma(T)$ and $\sigma(T_g)$ are conductivities at T and T_g, respectively. C_1 and C_2 are WLF parameters.

$$\sigma = AT^{-/2}\exp(-B)/(T - T_0) \tag{12.26}$$

where A and B = empirical constants, T_0 corresponds to T_g.

12.8 Ion Conduction of Hydrogen-Bonding Complexes

12.8.1 Ion Conduction of PMAA/POE Complex Containing $LiClO_4$ [111]

The intermolecular complexes of PMAA–POE, to which the inorganic salt ($LiClO_4$) is added, act as solid polymer electrolytes. In this system, the POE part forms the ionic conduction column and the interpolymer complex of PMAA–POE supports this conduction column by the three dimensional polymer matrix. POE_{6000} has a sufficient chain length to form the intermolecular complex. The $PMAA/POE_{6000}/LiClO_4$ hybrid film containing 3.0 mol% of $LiClO_4$ showed an ionic conductivity of 4.0×10^{-9} S/cm, whereas, the system without the salt showed a conductivity of about 10^{-16} S/cm. Therefore, the proton conduction can be neglected. In the case of the oligomer POE_{400}, which is far shorter than the critical chain length for complexation in aqueous media, it might form the interpolymer complex with PMAA in the solid state. Indeed, the IR spectrum of $PMAA–POE_{400}$ mixture showed the different carbonyl stretching band due to hydrogen bonding. Thus, POE_{400} might behave not only as a polar liquid additive but also as a component of the intermolecular complex. Under the conditions of [PMAA]/[POE] = 1 and $[LiClO_4]$ = 20 mol%, the proportion of POE_{6000}/POE_{400} was changed and the ionic conductivities were measured as shown in Fig. 12.14. The ionic conductivity increases remarkably with increasing POE_{400} ratio. The ionic conductivity of the $PMAA/POE_{400}/LiClO_4$ system at 25 °C is 2.4×10^{-7} S/cm, being about 500 fold higher than that of the $PMAA–POE_{6000}/LiClO_4$ system. The high complexation ability of POE_{400} was considered from the constant ionic conductivity even in days. Further advantages of using POE_{400} as a component polymer are

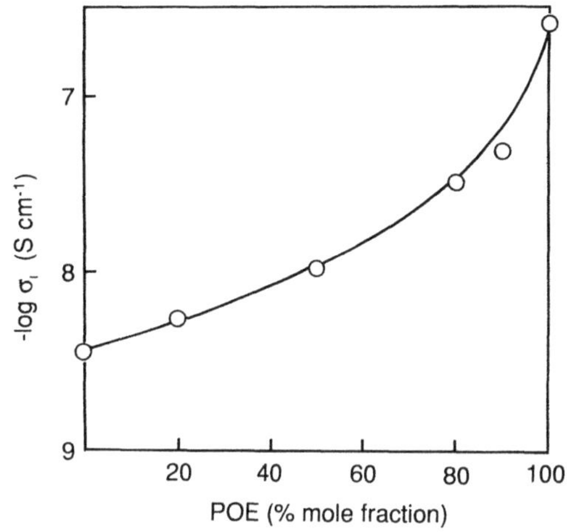

Fig. 12.14. Effect of POE_{400} mole fraction on ionic conductivity (σ_i) in the $PMAA/(POE_{400} + POE_{6000})/LiClO_4 =$ 40/40/20 in mol%

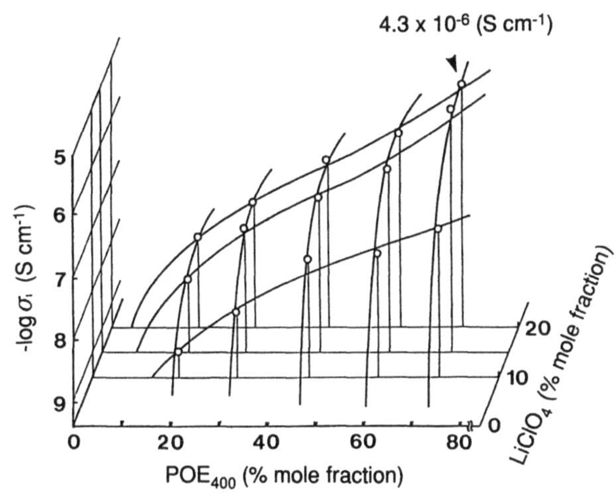

Fig. 12.15. Effect of POE_{400} and $LiClO_4$ content in $PMAA-POE_{400}/LiClO_4$ hybrid films on the ionic conductivity (σ_i) at 25 °C

good compatibility with $LiClO_4$ and the flexibility of resulting hybrid films. The $PMAA-POE_{400}/LiClO_4$ hybrid films were transparent and flexible when the concentration of $LiClO_4$ is less than 20 mol%, though the $PMAA-POE_{6000}/LiClO_4$ hybrid films containing more than 3 mol% of $[LiClO_4]$ are turbid and brittle.

Figure 12.15 shows the relationship between the composition of the $PMAA-POE_{400}/LiClO_4$ hybrid film and the ionic conductivity. The best conductivity in this system was observed to be 4.3×10^{-6} (S/cm) (25 °C) at the $PMAA/POE_{400}/LiClO_4$ unit molar ratio of 17.0/68.0/15.0. It was also noticed that an excess of POE_{400} was required to improve the ionic conductivity. This is due to the fact that a stoichiometrical amount of POE is used for interpolymer complexation and an excess of POE is necessary for solvation and migration of the ion. Therefore, this can be classified into the second group of solid polymer electrolytes.

12.8.2 *Single Ion Conduction of (PAANa–PAA)/POE Complex*

If PAANa and POE were miscible, single ion conduction would be possible because anions would be fixed to the polymer matrix. Since PAANa and POE, however, were not miscible in the solid state, acrylic acid moieties were introduced to PAANa for the purpose of the partial complexation through hydrogen bonding. In the experiment, the glass transition temperature and ionic conductivity were studied for PAANa–PAA–POE composite films having different neutralization of PAA and different incorporation of POE.

12.8.2.1 Glass Transition Temperature (T_g)

In the case of the mixture of POE_{400} and PAA ([POE]/[PAA] = 1), both the glass transition point ($T_g = -66.0\,°C$) and the melting point ($T_m = 4.7\,°C$) of pure POE_{400} completely disappeared and $T_g = -3.0\,°C$ was newly observed. This indicates that complex formation of POE_{400} with PAA occurs in spite of the considerably shorter chain length of POE_{400} – shorter than the critical chain length. When the [PAA]/[POE] ratios are 0.50, 0.33 and 0.25, T_g are −34.0, −42.5 and −50.5 °C, respectively as shown in Fig. 12.16. It shows that POE is miscible with PAA and takes an amorphous state in the mixture even at excess POE ratios. Therefore, T_g approaches that of pure POE with decreasing [PAA]/[POE] ratio.

In the partially neutralized system when the [PAA]/[POE] ratio is kept at 1.00 or 0.50, T_g increases with an increasing degree of neutralization because of

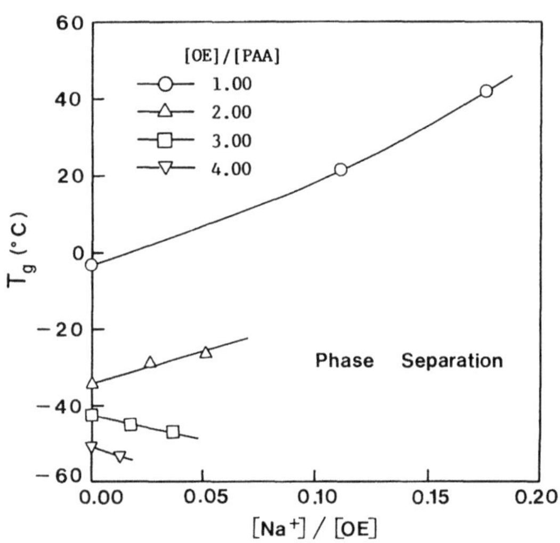

Fig. 12.16. Glass transition temperature (T_g) of the (PAA–PAANa)–POE_{400} hybrid films

the ionic crosslinking effect of the sodium ion. Moreover, phase separation occurs at a higher degree of neutralization, which can be confirmed from the appearances of T_g and T_m of the original POE domain. On the other hand, when the [PAA]/[POE] rations are 0.33 and 0.25, T_g decreases by increasing the neutralization point. This is due to an increase in intramolecular crosslinking of PAA–PAANa with increasing neutralization, resulting in the formation of a POE richer domain whose T_g could be detected from DSC measurement.

12.8.2.2 Na^+-Ionic Conductivity

Figure 12.17 shows the temperature dependence of Na^+-ionic conductivity for 10%-neutralized PAA/POE_{400} films with various POE incorporation. Each temperature dependence shows WLF type behavior. It indicates that ion-conduction significantly relates to the segmental motion of polymers. An increase in ionic conductivity (σ_i) with POE incorporation is due to the lower glass transition temperature of the film. The ionic conductivity of 4.9×10^{-7} (S/cm) was obtained at room temperature by incorporating four times the amount of POE_{400}.

PAA/POE_{400} hybrid films without neutralization also show ionic conduction, which is derived from the protons of carboxylic acids. The ionic conductivity is one or two orders lower than PAA–PAANa/POE_{400} systems. A.C. impedance measurements showed an increase in interface resistance when sodium amalgam was used as active electrodes. This indicates the reaction of carboxylic acids to active electrodes. Therefore, the PAAM/PAA/POE_{400} is not suitable for secondary battery systems constructed from active electrodes such as alkali metals.

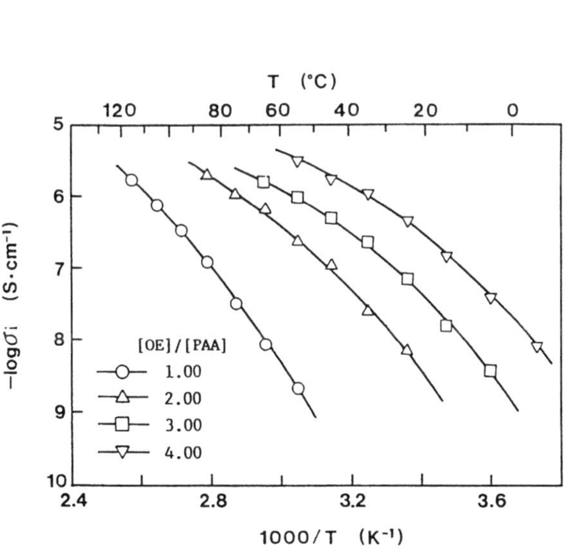

Fig. 12.17. Ionic conductivity $(Na)^+$ of (PAA–PAANa)–POE_{400} with different POE content

12.8.3 Application of Interpolymer Complexes to Electrochromic Devices

Electrochromic displays have some remarkable characteristics, such as memory effect, color variation, low mean power consumption etc. In most cases, however, electrochromic displays have been constructed as a solution containing system. The leakage of electrolyte solution and low stability are inevitable drawbacks for these electrochromic devices. Solid polymer electrolytes (SPEs) can be expected to improve these drawbacks [112, 113]. All solid state electrochromic displays based on SPEs should be developed.

All solid state WO_3/Prussian Blue (PB) based electrochromic devices (ECDs) were prepared by using SPEs of interpolymer complexes from PAA with POE which contains $LiCF_3SO_3$ [114]. Their electrochromic characteristics and redox behaviors were studied. The casting of the PAA/POE_{1000}/POE_{600000}/$LiCF_3SO_3$ methanol solution was carried out onto the WO_3- and PB-coated ITO electrodes and then the solvent was completely evaporated under dry argon over P_2O_5 at room temperature. After that, SPE film-coated WO_3 and PB ITO electrodes were contacted with each other by light pressing in a dry argon atmosphere as schematically shown in Fig. 12.18. The ECD cell was completely sealed with epoxy resin to prevent it from the influence of moisture. The conductivity of the SPE thus prepared was ca. 10^{-6} (S/cm) at 20 °C.

The oxidized and reduced forms of WO_3 films are transparent and blue, respectively, whereas PB films are Prussian blue and transparent in the oxidized and reduced states, respectively. These are shown in Eqs. (12.27) and (12.28).

$$\underset{\text{(transparency)}}{WO_3} + nK^+ + ne^- \rightleftharpoons \underset{\text{(blue)}}{K_nWO_3} \tag{12.27}$$

$$\underset{\text{(Prussian blue)}}{Fe_4^{3+}[Fe^{II}(CN)_6]_3} + 4K^+ + 4e^- \leftrightarrows \underset{\text{(transparency)}}{K_4Fe_4^{2+}[Fe^{II}(CN)_6]_3} \tag{12.28}$$

Consequently, the optical density change between coloring and bleaching can be expected to be large in this WO_3/PB system, compared with the case where either of them is employed as coloring materials. The visible spectra of a WO_3/SPE/PB cell are depicted in Fig. 12.19. A reversible color change between blue and colorless was observed when an appropriate voltage was applied

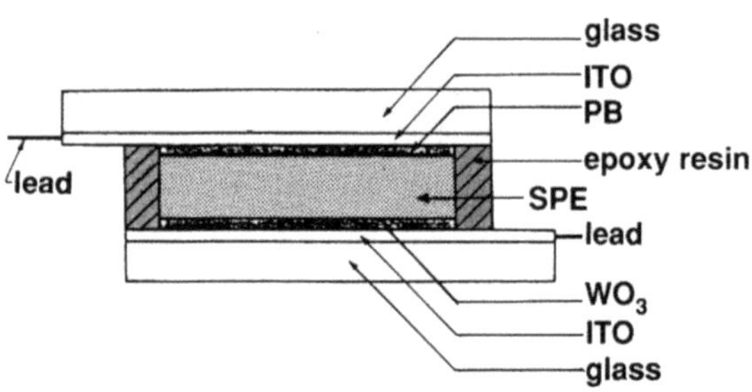

Fig. 12.18. Cell constitution of all solid-state WO_3/PB based ECD

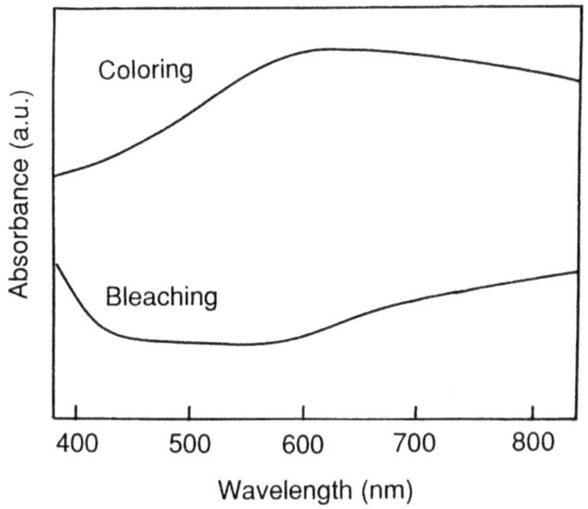

Fig. 12.19. UV/visible absorption spectrum of all solid-state WO_3/PB based ECD at the coloring and bleaching state. Applied voltage is 2.0 V (coloring) and − 1.0 V (bleaching) vs. the WO_3 electrode

repeatedly to the WO_3 and PB-coated electrodes. The coloration and bleaching of the ECD were found to be controlled with the electrochemical reaction of WO_3 electrode which depends on the injection (and disintercalation) of Li^+ ions to (and out of) the WO_3 layer.

12.9 References

1. Tsuchida E, Abe K (1982) In: Advances in Polymer Science 45, Interactions between macromolecules in solution and intermacromolecular complexes. Springer, Berlin Heidelberg New York
2. Michaels AS (1968) In: Encyclopedia of polymer technology vol 16. Wiley, New York
3. Rogachova B, Zezin AB, Kargin VA (1970) Vyscomol Soedin Ser B 12:826
4. Zezin AB, Rogachova B, Kavavov VA, Kargin VA (1972) Vyscomol Soedin Ser A14:772
5. Bailey Jr. FE, France HG (1961) J Polym Sci 49:397
6. Osada Y (1980) J Polym Sci, Polym Lett Ed 18:281
7. Smith KL, Winslow AE, Peterson DE (1959) Ind Eng Chem 51:1361
8. Osada Y, Sato M (1976) Nippon Kagaku Kaishi 1976:175
9. Papisov IM et al. (1974) Dokl Akad Nauk SSSR 214:861
10. Ikawa T, Abe K, Honda K, Tsuchida E (1975) J Polym Sci, Polym Chem Ed 13:1505
11. Osada Y et al. (1973) Vyscomol Soyed A14:2462
12. Papisov IM, Baranoviskii VYu, Kavanov VA (1975) Vsycomol Soyed A17:2428
13. Osada Y (1979) J Polym Sci, Polym Chem Ed 17:3485
14. Wilson RW, Bloomfield VA (1979) Biopolymers 18:1205
15. Osada Y, Sato M (1980) Polymer 21:1057
16. Ohno H, Matsuda H, Tsuchida E (1981) Markromol Chem 182:2267
17. Osada Y, Takeuchi Y (1981) J Polym Sci, Polym Lett Ed 19:303
18. Saltybaeva SS, Bimendia LA, Bekturov EA (1981) Izv Akad Nauk Kaz SSR, Ser Kim 1:24
19. Ohno H, Abe K, Tsuchida E (1978) Makromol Chem 179:755
20. Bimendia LA, Roganov VV, Bekturov EA (1974) J Polym Sci, Symp 44:65
21. Ohno H, Nii A, Tsuchida E (1980) Makromol Chem 181:1227
22. Morishima Y, Fujisawa K, Nozakura S (1976) J Polym Sci, Polym Lett Ed 14:467
23. Abe K et al. (1978) Makromol Chem 179:2043
24. Bekturov EA, Bimendia LA (1978) Izv Akad Nauk, Kaz SSR, Ser Kim 28(1):68
25. Litmanovich AA, Papisov IM, Kavanov VA (1980) Vysokomol Soyed A22:1180

26. Saltybaeva SS, Bimendina LA, Bekturov EA (1980) Izv Akad Nauk Kaz SSR, Ser Khim 3:35
27. Tleubaeva GS, Bimendina LA, Bekturov EA (1980) Izv Akad Nauk Kaz SSR, Ser Khim 6:69
28. Higashi F, Taguchi Y (1980) J Polym Sci, Polym Chem Ed 18:2875
29. Alberda van Ekenstein GOR, Koetsier DE, Tan YY (1981) Europ Polym J 17:845
30. Williamson GK, Wright B (1965) J Polym Sci, Part A3:3885
31. Okhrimenko IS, Dyakonova EB (1964) Vysokomol Soyed 6:1891
32. Dyakonova EB, Okhrimenko IS, Yefremov IF (1965) Vysokomol Soyed 7:1016
33. Kuhm W et al. (1950) Nature 165:514
34. Horiuchi H (1980) Koubunshi Ronbunshu 37:543
35. Murohashi S, Kuwabara A (1981) Nagaoka Kogyo Koto Senmon Gakko Kenkyu Kiyo 17:17
36. Horiuchi H, Ohshita T (1981) Koubunshi Ronbunshu 38:407
37. Gandurin LI, Smirnova EA, Bulysheva LK (1979) Izv Vyssh Uchebn Zaved, Tekhnol Tekst Prom-sti 6:68
38. Nikolaev AF et al. (1979) Vysokomol Soyed B21:723
39. Klenina OV, Fain EG (1981) Vysokomol Soyed A23:1298
40. Otocka EP, Eirich FR (1968) J Polym Sci, Part A-2 6:895
41. Otocka EP, Eirich FR (1968) J Polym Sci, Part A-2 6:913
42. Anufrieva E et al. (1979) Makromol Chem 180:1843
43. Anufrieva E et al. (1972) Vysokomol soyed A14:1430
44. Ohno H, Tsuchida E (1979) Polym Prep Jpn 28:1055
45. Ohno H, Nii A, Tsuchida E (1978) Polym Prep Jpn 27:964
46. Chatterjee SK, Prokopova E, Bohdanecky M (1979) J Macromol Sci, Phys B16:9
47. Bimendina LA, Frolova VA, Bekturov EA (1977) Izv Akad Nauk, Kaz SSR, Ser Khim 27:66
48. Jaycox GD, Smid J (1981) Makromol Chem, Rapid Commun 2:499
49. Matsubayashi K, Hirano Y (1960) Seni Gakkaishi 17:637
50. Staszewska D, Bohdanecky M (1981) Europ Polym J 17:245
51. Godovskii YuK et al. (1979) Vysokomol Soyed A21:127
52. Lohmeyer JHGM et al. (1975) J Polym Sci, Polym Lett Ed 13:725
53. Lohmeyer JHGM et al. (1978) Polymer 19:1171
54. Sulzberg T, Cotter RJ (1968) Macromolecules 1:554
55. Mulvaney JE, Brand RA (1980) Macromolecules 13:244
56. Geissler U, Schulz RC (1980) Makromol Chem 181:1483
57. Geissler U, Schulz RC (1980) Makromol Chem 181:1495
58. Leyte JC, Mandel M (1964) J Polym Sci A2:1879
59. Osada Y (1971) Kagaku no Ryouiki 25:625
60. Tsuchida E, Osada Y, Ohno H (1980) J Macromol Sci. Phys B17(4):683
61. Tsuchida E, Osada Y (1972) J Polym Sci, A-1 10:3397
62. Tsuchida E, Osada Y (1973) Koubunshi Kagaku 30:515
63. Abe K, Koide M, Tsuchida E (1977) Macromol 10:1259
64. Anufrieva YeV et al. (1977) Dokl Akad Nauk SSSR 232:1096
65. MacCallum JR, Vincent CA (1987, 1989) In: Polymer Electrolyte Reviews 1, 2. Elsevier Applied Science, London & New York
66. Wright PV (1975) Brit Polym J 7:319
67. Armand MB, Chabagno JM, Duclot M (1978) In: Second International Conference on Solid Electrolytes, 1978. St Andrews
68. Watanabe M, Kanba M, Matsuda H, Tsunemi K, Mizoguchi K, Tsuchida E, Shinohara I (1981) Makromol Chem Rapid Commun 2:741
69. Ohno H, Matsuda H, Mizoguchi K, Tsuchida E (1982) Polym Bull 3:271
70. Tsuchida E, Ohno H, Tsunemi K (1983) Electrochim Acta 28:591
71. Shigehara K, Kobayashi N, Tsuchida E (1985) Solid State Ionics 14:85
72. Tsuchida E, Shigehara K (1984) Mol Cryst Liq Cryst 106:361
73. Itoh T, Saeki K, Kohn K, Koseki K (1989) J Electrochem Soc 136:3551
74. Takeoka S, Sakai H, Shin H, Ohta T, Tsuchida E (1991) Polym Adv Tech 2:69
75. Hardy LF, Shriver DF (1985) J Am Chem Soc 107:3823
76. Bannister DJ, Davis GD, Ward IM, McIntyre JE (1984) Polymer 25:1291
77. Takeoka S, Sakai H, Tsuchida E (1990) Chem Lett :1539
78. Takeoka S, Horiuchi K, Yamagata S, Tsuchida E (1991) Macromolecules 24:2003
79. Tsuchida E, Kobayashi N, Ohno H (1988) Macromolecules 21:96
80. Ganapathiappan S, Chen K, Shriver DF (1988) Macromolecules 21:2301
81. Ganapathiappan S, Chen K, Shriver DF (1989) J Am Chem Soc 111:4091

82. Zhou G, Khan IM, Smid J (1989) In: 2nd International Symposium on Polymer Electrolytes. 14–16 June 1989. Siena, Italy
83. Tsuchida E, Ohno H, Kobayashi N, Isizaka H (1989) Macromolecules 22:1771
84. LeNest JF, Gandini A, Cheradame H, Cohen-Addad J (1987) Polym Commun 28:302
85. Watanabe M, Nagano S, Sanui K, Ogata N (1988) Solid State Ionics 28–30:911
86. Ogumi Z, Uchimoto Y, Takehara Z, Foulkes FR (1990) J Electrochem Soc 137:29, 35
87. Yeh TF, Okamoto Y, Skotheim TA (1990) Mol Cryst Liq Cryst 190:205
88. Liu H, Okamoto Y, Skotheim TA (1990) Mol Cryst Liq Cryst 190:213
90. Papke BL, Ratner MA, Shriver DF (1982) J Electrochem Soc 129:1694
91. Gutmann V (1978) In: The Donor Acceptor Approach to Molecular Interactions, Plenum, New York
92. Takahashi Y, Tadokoro H (1973) Macromolecules 6:672
93. Matsuura H, Miyazawa T (1969) J Polym Sci, A-2 7:1735
94. Maxfield J, Shepherd IW (1975) Polymer 16:505
95. Mark JE, Flory PJ (1966) J Am Chem Soc 88: 3702
96. Chatani Y, Okamura S (1987) Polymer 28:1815
97. Chatani Y, Okamura S, Fujii Y (1989) Polym Prepr, Polym Chem Div, Am Chem Soc 30:404
98. Armand MB, Chabagno JM, Duclot MJ (1979) In: Vashishita P, Mundy JN, Shenoy GK (eds) Fast Ion Transport in solids. Elsevier, North Holland, New York, p 131
99. Nicolas CV, Wilson DJ, Booth C, Giles JRM (1988) Br Polym J 20:289
100. Nagaoka K, Naruse H, Shinohara I, Watanabe M (1984) J Polym Sci, Polym Lett Ed 22:659
101. (a) Xia DW, Soltz D, Smid J (1984) Solid State Ionics 14:221; (b) Bannister DJ, Davis GR, Ward IM, McIntyre JM (1984) Polymer 25:1600; (c) Kobayashi N, Uchiyama M, Shigehara K, Tsuchida E (1985) J Phys Chem 89:987
102. Cowie JMG, Martin ACS (1987) Polymer 28:267
103. Yamada A, Shigehara K, Kurata Y (1985) Polym Prepr 34:903
104. (a) Hall PG, Davis GR, McIntyre JE, Ward IM, Bannister DJ, LeBrocq KMF Polym Commun 27:98; (b) Fish D, Khan IM, Smid J (1986) Makromol Chem Rapid Commun 7:115
105. Fish D, Khan IM, Smid J (1988) Br Polym J 20:281
106. Blonsky PM, Shriver DF, Austin P, Allcock HR (1984) J Am Chem Soc 106:6854
107. Cowie JMG, Martin ACS, Firth AM (1988) Br Polym J 20:247
108. Watanabe M, Nagano S, Sanui K, Ogata N (1986) Solid State Ionics 18–19:338
109. Williams ML, Landel RF, Ferry DJ (1955) J Am Chem Soc 77:3701
110. (a) Vogel H (1921) Z Phys 22:645; (b) Tammann VG, Hesse WZ (1926) Anorg Allg Chem 156m:245; (c) Fulcher GC (1925) J Am Chem Soc 8:339
111. Tsuchida E, Ohno H, Tsunemi K, Kobayashi N (1983) Solid State Ionics 11:227
112. Nishikawa M, Ohno H, Kobayashi N, Tsuchida E, Hirohashi R (1988) Nippon Shashin Gakukai-shi 51:184
113. Kobayashi N, Nishikawa M, Ohno H, Tsuchida E, Hirohashi R (1988) Nippon Shashin Gakukai-shi 51:375
114. Oyama N, Oshaka T, Menda M, Ohno H (1989) Denki Kagaku 12:1172

13 Fluorescence Probe Studies of Poly(acrylic acid) Interchain Complexation Induced by High Shear Flow and Influence of Cationic Surfactants on the Complexation

Oh-Kil Kim and Ling-Siu Choi

A dilute solution of poly(acrylic acid) was found to undergo a stable interchain complexation (through H-bonding) in a high shear flow. This requires the PAA to be ultra-high molecular weight, low concentration ($< 5 \times 10^{-4}$ M) and at a pH range of 6–8. The mode and strength of the PAA complexation was strongly influenced by the presence of salts and cationic surfactants. Addition of salts to the shearing PAA solution disrupted the complexation, while the modification of the PAA with the surfactant enhanced the complexation but only below a critical concentration that depends on the surfactant chain length and pH employed. The chain morphology developed by such shear-induced complexation is a local rigidity formation in the PAA chains. Conformational transition toward the PAA rigidity was monitored by fluorescence probe techniques using auramine O. Discussion is made with respect to conformational transition and chain morphology developed in the PAA by shear in relation to molecular environmental changes.

13.1 Introduction

Macromolecules in dilute solution undergo orientational and extensional chain transitions as a result of shear force by flow. The extent of the transition depends mainly on external conditions such as solvent, concentration and shear strength. In many flexible macromolecular systems, a significant viscosity increase [1] is observed with increasing shear and particularly with ultra-high molecular weight (UHMW) [2]. This is due to the elongational deformation of polymer chains induced by shear in that intermolecular association [3] is often accompanied and facilitated by a relatively high concentration at which molecular overlap becomes appreciable under shear. However, the viscosity decreases as a higher chain orientation and extension are attained in the macromolecules. These are favored by decreasing the concentration and increasing the shear rate. Under such molecular configurations, a multiple-site interchain interaction is expected to occur more effectively among the aligned macromolecules. Poly (acrylic acid), PAA, can provide an example of this behavior. In order for PAA to behave in this fashion, acidic pH is required so that H-bonding interaction prevails among the PAA chains. However, questions remain regarding the H-bonding stability under shear of such a highly hydrophilic polymer at an extremely low concentration in water.

From fluorescence probe studies using auramine O (AuO), a water-soluble fluorescence dye, we have observed [4] that UHMW PAA exhibits a highly

Macromolecular Complexes
in Chemistry and Biology
Eds.: Dubin/Bock/Davis/Schulz/Thies

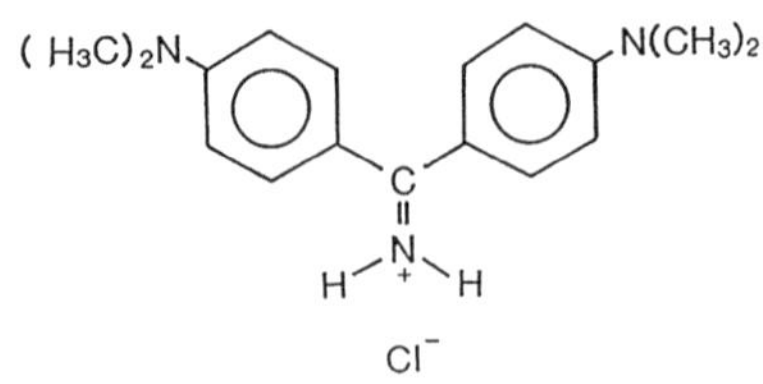

Auramine O

stable interchain complexation at around neutral pH in very dilute solution (1.9×10^{-4} M) under a high shear rotating disk flow (used for drag reduction measurements). Drag reduction (DR) of the PAA decreased sharply depending on the shear rate and time but without causing turbidity in the solution [5]. Such behavior is sensitive to the external conditions.

On the other hand, we thought it would also be of interest to study the mode and strength of interchain interactions of the PAA when modified with cationic surfactants, because such modified PAA molecules are capable of forming hydrophobic microdomains [6–9] which occur normally at a concentration far below their critical micelle concentration (CMC). This unique environmental development along the PAA chains can exert a form of resistant force against the shear applied, and subsequently, changes in the PAA complexation behavior may be expected.

In the first part of this report, we discuss a fundamental aspect of shear-induced conformational transition of the PAA resulting from the interchain association that was monitored using the AuO fluorescence probe. Further, we discuss the influence of molecular environment on PAA conformation in relation to the DR behavior of the PAA. In the second part, we discuss a particular molecular environmental effect of cationic surfactants on the interchain association of the PAA, in that the chain conformation is dictated by the extent of the surfactant binding and the surfactant structure.

13.2 Experimental

13.2.1 Materials

PAA was prepared by a conventional free radical polymerization using ammonium persulfate as the initiator. The PAA used for this study has the viscosity, $[\eta]_{1M\ NaCl} = 9.9$, corresponding to $\bar{M}_n \simeq 8 \times 10^6$. Alkyltrimethylammonium bromides (C_nTMAB) with different alkyl chains (C_n) were purchased from Aldrich Chemical Co., and hexadecyl tributylammonium bromide (C_{16}TBAB) was prepared by reacting hexadecyl bromide with tributylamine. These surfactants were recrystallized from ethanol. Fluorescence probes, AuO and pyrene, were commercial products and were purified by recrystallization from ethanol.

13.2.2 *Flow Processing*

The shear field was generated by a rotating disk flow apparatus as described in a previous report [5]. For the flow studies, an appropriate amount of PAA Na salt was dissolved in deionized water to give the concentration, 1.9×10^{-4} M and the pH around 8 which was then adjusted by adding acid or base when necessary. The polymer solution was sheared in the rotating disk flow usually at 2800 rpm (Reynolds number, $Re \simeq 1 \times 10^6$) up to 4 min. DR was calculated from the torque changes during shear in the presence and absence of polymer in solution.

13.2.3 *Fluorescence Measurements*

The resulting solution was mixed with AuO to give the probe concentration, 7.64×10^{-7} M. Emission spectra of the sheared (or unsheared) solution were recorded with a Spex Fluorolog II fluorimeter. The same experiment was also carried out in the presence of cationic surfactants (C_{10}-, C_{12}-, C_{14}-, and C_{16}TMAB and C_{16}TBAB) at a concentration in the range of 2×10^{-6}-10^{-5} M, in which a mixture of the PAA and the surfactant was sheared as described above. The resulting solution was mixed with pyrene [10] (or AuO) in water to give a probe concentration of 1.31×10^{-8} M (or 7.64×10^{-7} M). Emission spectra of pyrene ($\lambda_{ex} = 313$ nm) and AuO ($\lambda_{ex} = 366$ nm) of the sheared (or unsheared) solution were recorded as described above. The fluorescence intensity ratios of pyrene (I_1/I_3) and AuO (I/I_0) are used as a measure of local polarity and rigidity, respectively.

13.3 Results and Discussion

13.3.1 *Drag Reduction (DR) and PAA Conformation*

Conformational transitions in the PAA under shear were observed from the change in DR with shearing time. As shown in Fig. 13.1, %DR decreases depending on the shear rate applied; the higher the shear rate, the faster the decrease in DR. Polymer DR commonly increases with increasing shear rate up to a certain point, after which DR may decrease mainly due to shear degradation. In this regard, such behavior is unusual, unless there is a chain degradation. The variation in the onset of DR (Fig. 13.1) is most likely to be due to differences in the conformational states of the PAA molecules at the onset of shear. Accordingly, the sharp decrease in DR with shearing is to be related to a transition of the extended chain conformation, suggesting that extended chains are more susceptible to a subsequent conformational transition. This brings about a decrease in viscosity [5] with further shearing. This implies the occurrence of an interchain association.

Such decreases in DR are sharply pH-dependent as shown in Fig. 13.2. Interestingly enough, the sharpest decrease in DR (at a fixed shear rate) occurs at a pH of 6–8 but not at more strongly acidic or basic values. The implication of this is that such unique time dependent chain deformation requires that the PAA initial conformational state be a loosely coiled state but not in a tight coil or an extended coil. Although the extended coil state (pH 9, for example) could be more susceptible to shear to undergo a chain elongation, in comparison with a tight coil state (pH 4, for example), no such drastic deformation occurs,

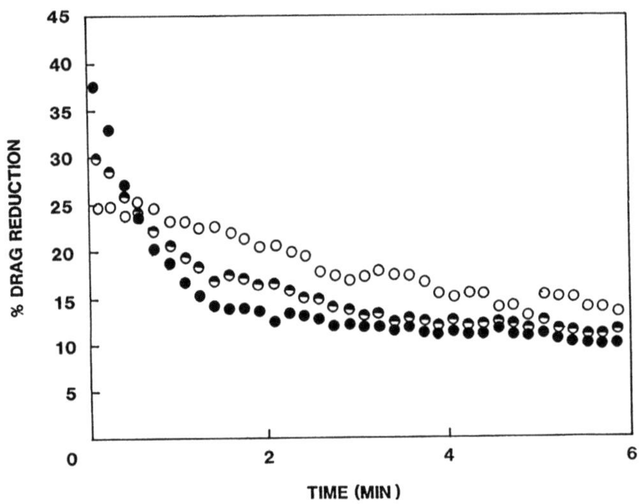

Fig. 13.1. Influence of shear rate on % DR of PAA solution (1.9×10^{-4}) at pH = 8.0; shear rates are 1600 (○), 2000 (◓), and 2800 (●) rpm

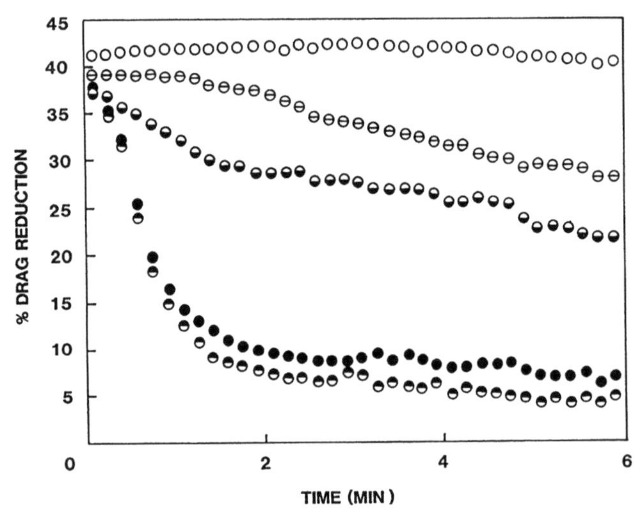

Fig. 13.2. Influence of pH on % DR of PAA solution (1.9×10^{-4} M) at shear rate, 2800 rpm; pHs are 4.0 (◒), 6.0 (◓), 8.0 (●), 9.0 (⊖) and 9.5 (○)

suggesting that the deformation is sensitive to the electro-static repulsion. This seems to further imply that the change involves interchain H-bonding. If this is the case, the change should be favored by a low pH. However, it was difficult to bring about such a drastic chain transformation, probably because the conformational state (of the PAA) at that pH is highly resistant (due probably to intramolecular H-bonding) to the shear applied. Consequently, the pH range, 6–8, must be the best condition for the PAA association. In order for this association to occur in the PAA, the interchain H-bonding under the pH condition must be strong enough to overcome the repulsive force of polyions and the concomitant shear force. An immediate question is whether a small number of free acid groups (about 30 % at pH 8.0) in the PAA can sustain their H-bonding against these disruptive forces.

One plausible answer to this is the role of counter-ion binding to the poyions. ^{23}Na NMR study of the PAA solution at pH = 8.0 (Na^+ as counter ion) indicated that the total number of bound Na^+ ions increased with shearing, reaching a saturation. The implication is that with shearing, more Na^+ ions tend to remain in a bound state with polyions, thereby leading to intermolecular bridging. This is probably the supplemental force to stabilize the H-bonding.

One can argue that such a decrease in DR might be the result of chain degradation. If that were so, a high pH condition should be more vulnerable to the degradation. It is also inconsistent with the observation [5, 11] that the once decreased DR is fully recovered instantaneously by the addition of a minute amount of salts (0.1 g/L NaCl, for example) to the shearing solution of the PAA.

Another aspect of conformational susceptibility of PAA to shear is the concentration dependence [12]. At PAA concentration in the range of 1.0–3.5 $\times 10^{-4}$ M (at pH 8.0), its DR exhibits a sharp decrease (as shown in Fig. 13.2) while at PAA concentrations beyond that, DR of PAA (at the same pH) remains almost unchanged or somewhat increased over the prolonged shearing. This seems to suggest that PAA molecules are more susceptible to shear force at low concentrations, but with increasing concentrations, shear resistance of PAA chain elongation increases as a result of a growing chain entanglement.

The interpretation of pH dependence and concentration dependence of PAA on the DR change is that the initial conformational state (before shearing) of the PAA dictates the conformational transition of the polymer under shear. In other words, a sharp decrease in %DR requires that the PAA have conformational flexibility for chain elongation and free acid groups for interchain interactions. PAA molecules at high pH might be conformationally flexible but do not have enough free acid groups to form intermolecular H-bonding. At a low pH, PAA molecules are intramolecularly aggregated and at a high concentration, intermolecularly entangled. One additional requirement for the transition of PAA conformation is UHMW; PAA with a molecular weight below a few millions does not exhibit such a transitional behavior in the DR [12]. This seems to suggest that multiple-site H-bondings between PAA chains are essential for the stability of the PAA complexation.

13.3.2 Local Chain Rigidity

We found that the chain morphology developing in the PAA through the interchain association contributes to a growing local rigidity. AuO is a unique fluorescence probe for monitoring the rigid local environment [13]. Without shearing, no AuO fluorescence is observable from the PAA solution, whereas with shearing it appears and its intensity increases with shearing time up to 2 minutes, levelling off after that (Fig. 13.3). The AuO fluorescence disappears, however, when the decreased DR is fully recovered upon the addition of salts, suggesting that the locally rigid microenvironment developed is completely disrupted and returned to the initial unsheared state. Thus, AuO is such a sensitive probe that its fluorescence intensity (or the PAA rigidity) was found to correspond inversely to the DR change of the PAA [4]. It is most likely that the growing rigidity by shearing occurs through the interchain H-bonding among the PAA molecules. This is evidenced by FTIR spectra and also by transmission electromicrographs [14]. The unsheared sample of the PAA solution shows a texture of a randomly expanded coil, while the sheared sample exhibits a bundle of locally stiff chains.

Since the shear induces an unusual conformational transition in the PAA to form a self-complexation, it is worthwhile to investigate how the complexation is affected by the interaction with competitive species such as poly(ethylene oxide) [15] and cationic surfactants. With the latter, it is particularly interesting to see how the PAA interchain complexation is interfered with by the presence of a hydrophobic environment due to the bound surfactant molecules, and how their microdomain formation is altered by the presence or absence of shear stress. To monitor the conformational state changes of the PAA, we chose two fluorescence probes, pyrene and AuO; the former as a probe for microenvironmental polarity in the absence of shear and the latter as a probe for local rigidity along the PAA chain under shear.

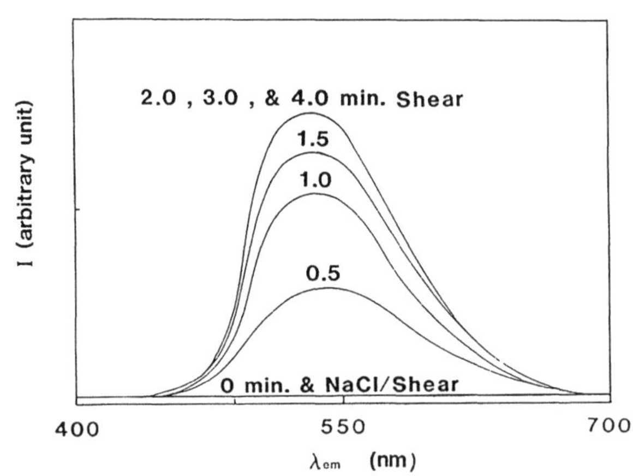

Fig. 13.3. Changes in fluorescence intensity of AuO in the PAA solution (1.9×10^{-4} M) at pH = 8.0 upon shearing (up to 4 min.) and addition of NaCl (0.10 g/l)

13.3.3 Hydrophobic Association

When the PAA interacts with a cationic surfactant in solution without shear, the hydrophobicity increases along the chains with increasing surfactant concentration, and at a certain concentration a micelle-like structure begins to form that will be reflected in a sharp decrease in the fluorescence parameter (I_1/I_3) of pyrene. Figure 13.4 shows the decrease in the micropolarity of the modified PAA depending on the alkyl tail length of the surfactants; the polarity ($I_1/I_3 = 1.6$) of the PAA alone in solution decreases to 1.4 in the presence of 5×10^{-6} M C_{16}TMAB, for example, while the same decrease requires the C_{12}-homolog to be 3.5×10^{-5} M. The sharp transitions (corresponding to the critical aggregation concentration (CAC)) in the fluorescence parameter for C_{16}- and C_2TMAB systems appear at surfactant concentrations, 5.7×10^{-6} M and 2.2×10^{-5} M, respectively, which are far below their CMCs (8.0×10^{-4} M and 1.3×10^{-2} M, respectively). Similar observations have been seen in other complex systems [8, 9]. Turro and coworkers [7] reported about a PAA (M.W. $= 4 \times 10^4$) system modified with C_{12}TMAB, in which a CAC was observed at around 7×10^{-4} M, that is one order of magnitude higher than ours (2.2×10^{-5} M). This is probably due to the low PAA concentration (1.9×10^{-4} M) in our system compared to the reported system (1.06×10^{-2}), suggesting that CAC cannot be determined independently but depends on the host polymer concentration. This implies further that the association of the polymer-bound surfactants can be favored by a low concentration and a high molecular weight of the host polymer [16].

As illustrated in Fig. 13.4, regardless of shearing, there is a general tendency for the hydrophobic domains to grow along the PAA chains with increasing surfactant concentrations and with increasing alkyl chain length (C_{12}, C_{14}, and C_{16}) of the surfactant tail. However, when a comparison is made with respect to the hydrophobicity change in the individual surfactant systems due to shearing, a marked difference appears depending on the surfactant tail length (C_{12}, C_{14},

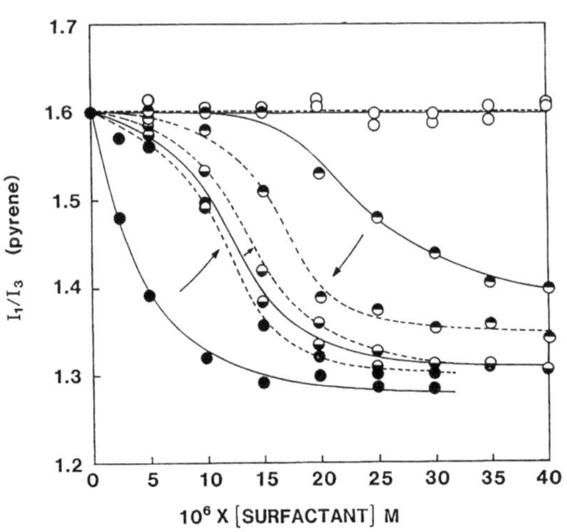

Fig. 13.4. Relative intensity (I_1/I_3) of pyrene fluorescence as a function of concentration of various lengths of cationic surfactants (C_nTMAB) in PAA solution (1.9×10^{-4} M) at pH = 8.0 with shear (---) and without shear (——). The lengths of surfactant tails are C_{16}(●), C_{14} (◒), C_{12} (◓) and C_{10} (○). I_1/I_3 is a measure of polarity in the local environment of the PAA, where I_1 and I_3 are the intensities of the first (372 nm) and the third (383 nm) peaks of pyrene fluorescence

and C_{16}); the microenvironment formed by the C_{16}- and C_{14}TMAB systems becomes less hydrophobic, more noticeably with the C_{16}, while the C_{12}-homolog system becomes more strongly hydrophobic. On the other hand, the C_{10}TMAB system is not affected at all by shearing and increasing concentration, suggesting that the hydrophobicity of its alkyl chain is not strong enough to form a micellar association even at high concentrations. These results can be interpreted to mean that the hydrophobic association of the C_{12}TMAB residues is strongly intensified by shear while that of the longer chain counterparts (C_{14}- and C_{16}-homologs) is disrupted under the shear condition. In view of the fact that, with the longer chain surfactants, the onset of the hydrophobic association occurs at a very low concentration, their microdomains must be loosely formed (compared to the C_{12}-homolog), becoming more vulnerable to the shear stress because shear is an elongational force. Thus, the conformational states of the surfactant-modified macromolecules in a high shear flow are determined by the balance of two competing forces, namely, shear stress and molecular cohesive force.

13.3.4 Hydrophobe-Assisted Rigidity

As shown by relative intensity of AuO fluorescence (Fig. 13.5), upon shearing, the interchain complexation (based on H-bonding) of the PAA (pH 8.0) is facilitated at low surfactant concentrations; the local rigidity along the PAA chain is about 50 % greater for 5×10^{-6} M C_{16}TMAB than in the absence of the surfactant, as indicated by the fluorescence intensity ratios, I/I_0 with and without a surfactant. Such enhancement in the local rigidity in the PAA is higher with longer alkyl chains of the surfactants. Since the PAA self-complexation is based on the interchain interaction, it can be said that the PAA complexation is promoted (or stabilized) by the interchain hydrophobic interaction among the elongated PAA molecules modified with the surfactants, suggesting an important role of the bound surfactant residues. With a further increase beyond a critical concentration, referred to as the critical shear deformation concentration

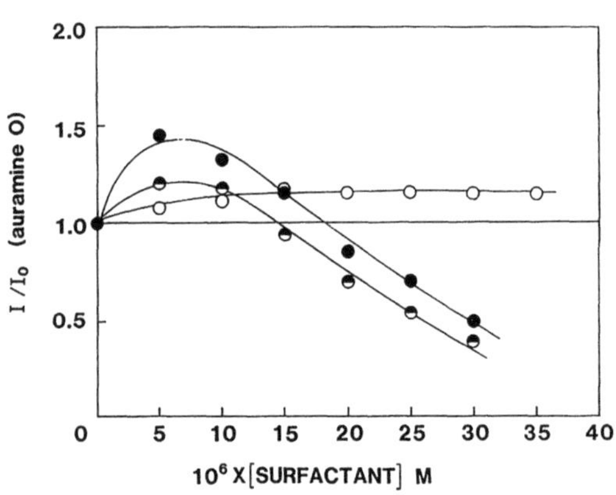

Fig. 13.5. Relative intensity (I/I_0) of auramine O fluorescence as a function of C_{10}- (○), C_{12}- (◓) and C_{16}TMAB (●) concentrations in PAA solution (1.9×10^{-4}) at pH = 8.0 under shear. I/I_0 is a measure of the local rigidity along the PAA, where I and I_0 are the intensities in the presence and absence of the surfactants, respectively

(CSDC), (1.8×10^{-5} M for C_{16}- and 1.4×10^{-5} M for C_{12}TMAB systems (Fig. 13.5)), the local rigidity of the PAA sharply decreases. It is, therefore, most likely that the morphological state of the surfactant-modified PAA below the CSDC is domain-free under the shear, but at concentrations above CSDC, microdomains develop through the intramolecular association of the bound surfactants, which can override the shear stress. however the C_{10}-homolog has insufficient associative force to influence the PAA self-complexation. Incidentally, this demonstrates that AuO fluorescence is a probe for microenvironmental rigidity but not for hydrophobicity (Fig. 13.5).

The pH effect on the hydrophobic domain formation and subsequent micropolarity changes in the PAA chains are associated with the charge-density (depending on α) of the polymer. At a highly charged state, an efficient electrostatic binding of surfactants and subsequent hydrophobic association occur in the PAA, but due to the repulsive interaction among the remaining charged segments, the hydrophobic association is not favored, resulting in somewhat increased micropolarity that leads to some increase in CAC. On the other hand, a low pH favors the hydrophobic association of the surfactant residues on the PAA, which is a potential resistant force against shear, rendering the PAA complexation (through interchain H-bonding) difficult. In the presence of salts added, the surfactant binding interaction of the bound surfactant residues is greatly retarded, leading to an increase in CAC [7]. This situation becomes worse with increasing pH, resulting in a highly polar microenvironment in the PAA.

A marked environmental effect of hydrophobicity on the PAA complexation appears at a relatively low pH. This is quite unexpected, because the shear resistance of the micelle-like domain formed by the modified PAA is presumed to be stronger at that pH, leading to the suppression of the interchain interaction. With the modification of C_{16}TMAB, the PAA attained the highest rigidity at pH 5.8 (Fig. 13.6): nearly twice as large as that at pH 8.0 and about three times as large as that of the unmodified PAA. However, the rigidity increase appears always within the limit of a critical surfactant concentration, CSDC, that depends on the surfactant chain length. The maximum rigidity at each pH (5.8 and 8.0) appears at surfactant concentrations around 1.1×10^{-5} M and 6×10^{-6} M, respectively. A similar trend was also observed with the C_{12}-homolog. Such a high rigidity formation of the surfactant-modified PAA relative to that of the unmodified PAA at those pHs is attributable to the hydrophobic environment, in that interchain interaction is promoted by shear, resulting in the stabilization of interchain H-bonding.

It is interesting to notice (Fig. 13.6) that with C_{16}TMAB-modified PAA at pH 5.8, the I/I_0 increases almost linearly until reaching a maximum at a high surfactant concentration (ca. 1.1×10^{-5} M), while at pH 8.0, it comes to a saturation at a somewhat low surfactant concentration (ca. 6×10^{-6} M). At pH 8.0 ($\alpha \simeq 0.7$) there are more binding sites (anionically charged) available for surfactants, and it seems to be likely that with increasing addition of surfactants (to PAA), the distribution of the bound surfactants onto the polymer tends to be

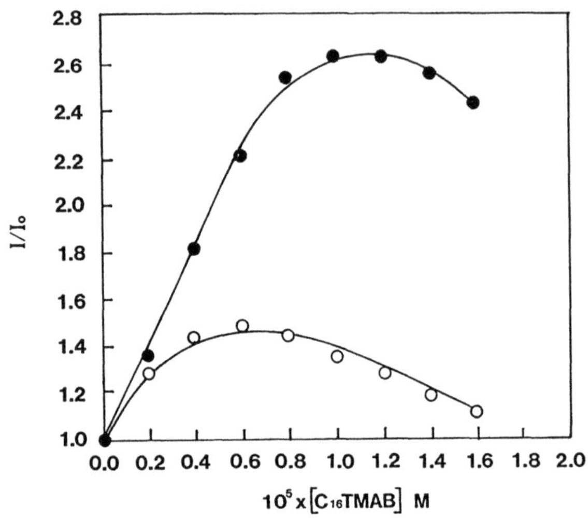

Fig. 13.6. Influence of pH on the relative intensity (I/I_0) of AuO fluorescence in PAA solution (1.9 $\times 10^{-4}$ M) under shear as a function of C_{16}TMAB concentrations; pHs are 5.8 (●) and 8.0 (○)

disproportionate (forming a block) due to the affinity interaction, leading to an increase in the micelle-like association at a relatively low surfactant concentration (a low CAC). On the other hand, since at low pH such as 5.8 ($\alpha \simeq 0.2$), there are less binding sites available in PAA, and the binding sites are rather uniformly distributed along the chain, relatively even distribution of the surfactant binding (onto PAA) is attainable. This brings in the microenvironment a steady growth of hydrophobicity upon increasing the surfactant concentration. A similar tendency in the pH dependence of the local rigidity was also observed with C_{14}- and C_{12}-homolog surfactants with the same head group.

With surfactants having a bulkier head group such as C_{16}- tributyl-ammonium bromide (C_{16}TBAB), the situation becomes more complex. Although the pH-dependent binding behavior of C_{16}TBAB should be about the same as that of C_{16}TMAB, due to the steric factor some distinction is perceivable with the former. One noticeable distinction is that the I/I_0 maximum and CSDC of C_{16}TBAB at pH 5.8 are unexpectedly low compared to those of C_{16}TMAB (Figs. 13.6 and 13.7). This can be interpreted as being due to the steric repulsion of the bulky head groups. That destabilizes interchain H-bonding formation in the modified PAA even at such a low concentration of CSDC as 8×10^{-6} M (Fig. 13.7). It is also plausible that the binding of C_{16}TBAB to the PAA and the intramolecular association among the same surfactant residues (bound to the polymer) could be loose, despite its hydrophobic strength. However, at pH = 8.0 the anticipated steric factor of C_{16}TBAB (bound to the PAA) does not look significant compared to that of C_{16}TMAB with respect to I/I_0 maximum and CSDC. This seems to imply that the influence of the bulky head groups on the H-bonding is diminished by the strong repulsive

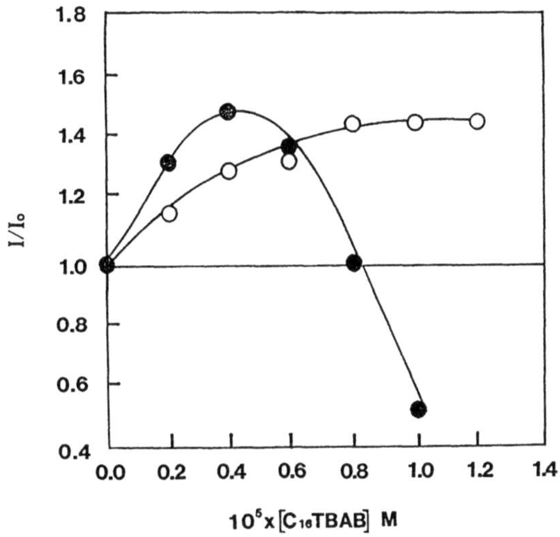

Fig. 13.7. Influence of pH on the relative intensity (I/I_0) of AuO fluorescence in PAA solution (1.9×10^{-4} M) under shear as a function of C_{16}TBAB concentrations; pHs are 5.8 (●) and 8.0 (○)

interaction among polyions, leading to a lowered rigidity in the PAA. In any case, regardless of the surfactant chain lengths and the head groups, the enhancement of local rigidity in the PAA occurs only within the limit of surfactant concentration (CSDS) that is dependent on the surfactant structure and external conditions employed.

To conclude, a dilute solution of PAA forms a stable self-complexation through the interchain H-bonding induced by shear at pH 6–8, resulting in the development of a microenvironmental rigidity in the PAA. Such time-dependent transition in the PAA conformation is closely related to the decreases in the viscosity and thus in DR of the polymer. Addition of a small amount (10^{-4}–10^{-5} M) of salts disrupted the complexation, resulting in a full recovery of DR and also the viscosity. For the complexation, the PAA is required to be of ultra-high molecular weight and at very low concentration ($< 5 \times 10^{-4}$). Such conformational transition was monitored by fluorescence probe techniques using AuO. In the presence of cationic surfactants (below CMC), the modified PAA chains undergo a marked polarity change. The subsequent interchain complexation is assisted by hydrophobic interaction depending on the alkyl chain length of the surfactants. However, it occurs only below a critical concentration (CSDC) of the surfactants, which depends on the surfactant structure and pH of the system, and decreases rapidly above the CSDC, suggesting the formation of intramolecular micelle-like association of the polymer-bound surfactant. There is a pronounced pH effect on the extent of rigidity developed in the surfactant-bound PAA, resulting mainly from the environmental polarity and total number of binding sites and free acids available for the complexation. A relatively low pH (5.8, for example) always gives a marked enhancement in the rigidity. There is a significant steric effect on the PAA complexation due to a bulky head group of the bound surfactant, and that is reflected in the lowered rigidity.

Acknowledgement. Partial funding support from the Office of Naval Research is gratefully acknowledged.

13.4 References

1. Witten TA Jr., Cohen MH (1985) Macromolecules 18:1915
2. McHugh AJ, Blunk RH (1986) Macromolecules 19:1249
3. Peiffer DG, Kim MW, Lunberg RD (1986) Polymer 27:493
4. Kim O-K, Choi LS, Long T, Yoon TH (1988) Polym Comm 29:168
5. Kim O-K, Long T, Brown F (1986) Polym Comm 27:71
6. Binana-Limbele W, Zana R (1987) Macromolecules 20:1331
7. Chandler P, Somasundaran P, Turro NJ (1988) Macromolecules 21:950
8. Abwin EB, Scaiano JC (1984) J Am Chem Soc 106:6274
9. Dubin PL, The SS, Gan LM, Chew CH (1990) Macromolecules 23:2500
10. A saturated solution of pyrene in hot water was cooled to room temperature (25 °C) and then filtered. The concentration of pyrene in the solution is determined to be 6.53×10^{-7} M. See: May WE, Wasik SP, Freeman DH (1978) Anal Chem 50:997
11. Kim O-K, Choi LS (1990) Makromol Chem, Makromol Symp 39:203
12. Kim O-K, Long T, Yoon TH, Choi LS, unpublished results
13. Oster G (1955) J Polym Sci 16:235
14. Kim O-K, Choi LS, Long T, McGrath K, Armstead JP, Yoon TH (1993) Macromolecules 26:379
15. Kim OK, Choi LS, Long TD, Yoon TH (1987) Polym Preprints 28(2):68 and unpublished results
16. Choi LS, Kim O-K (1994) Langmuir, in press

Part III

Biopolymer Systems

14 Water-Soluble Biospecific Polymers for New Affinity Purification Techniques

Patrick Hubert and Edith Dellacherie

New affinity purification techniques involving water-soluble biospecific polymers have been reported during the past 15 years, with the ambition of providing an alternative to affinity chromatography. This approach was prompted owing to certain limitations among which the difficulty to properly control the synthesis of the affinity matrix and the risk of non-specific binding of contaminants are directly related to the use of insoluble supports.

The new concept consists of the preparation of water-soluble biospecific polymers, the macroligands, so that a reversible biospecific water-soluble complex is formed with the desired target molecule. Provided that the parent polymer possesses any distinctive feature allowing an easy discrimination of this complex from the rest of contaminants – high molecular weight, high density of charges, precipitability, etc – all kinds of new, versatile techniques then become available to fractionate the mixture.

Examples illustrating the basic principles of these techniques are described, and their advantages and drawbacks compared to those of affinity chromatography are discussed.

14.1 Introduction

Since the initial work by Cuatrecasas and Anfinsen in 1971 [1], affinity chromatography has been the object of theoretical studies and successful purifications [2–4] have been reported at such a rate that this technique is nowadays almost compulsorily included in purification protocols, as routinely as classical nonspecific techniques like, for instance, ammonium sulfate precipitation and gel filtration or ion-exchange chromatographies.

However, irrespective of all these impressing practical achievements, the affinity chromatography concept – i.e. the highly specific binding of the target molecule to a ligand covalently immobilized onto a solid support, followed by washing of the contaminants prior to elution with an appropriate buffer – theoretically has limitations inherent to the use of a water-insoluble matrix:

1. As the synthesis of the affinity support is carried out in a heterogeneous medium, the amount of immobilized ligand on the solid support may turn out to be rather low and difficult to appreciate. Moreover, due to steric hindrance and diffusion limitations, the accessibility of the target molecule to these binding sites may be somewhat limited. The binding capacity of the affinity matrix as well as the conditions for the release of the target molecule

Macromolecular Complexes
in Chemistry and Biology
Eds.: Dubin/Bock/Davis/Schulz/Thies

from the adsorbent must therefore be preliminarily and empirically established.

2. The corollary of a low binding capacity is the necessity for using relatively large amounts of the affinity matrix. Since the occurrence of nonspecific interactions of all kinds – electrostatic, hydrophobic, hydrogen bonding, etc. – can hardly be totally avoided, the risk of contamination of the target molecule during the elution step obviously increases as the amount of matrix required for an efficient binding is higher.
3. Some severe problems concerning downstream processing are encountered when affinity chromatography is used on a large scale, as well as clogging when crude extracts are treated [5, 6].

To overcome these difficulties, the idea of adapting the affinity chromatography concept in new purification techniques involving water-soluble biospecific polymers emerged as early as 1974 [7–10] and is presently the object of recrudescent interest. The general principle on which all these techniques are based is schematically presented in Fig. 14.1.

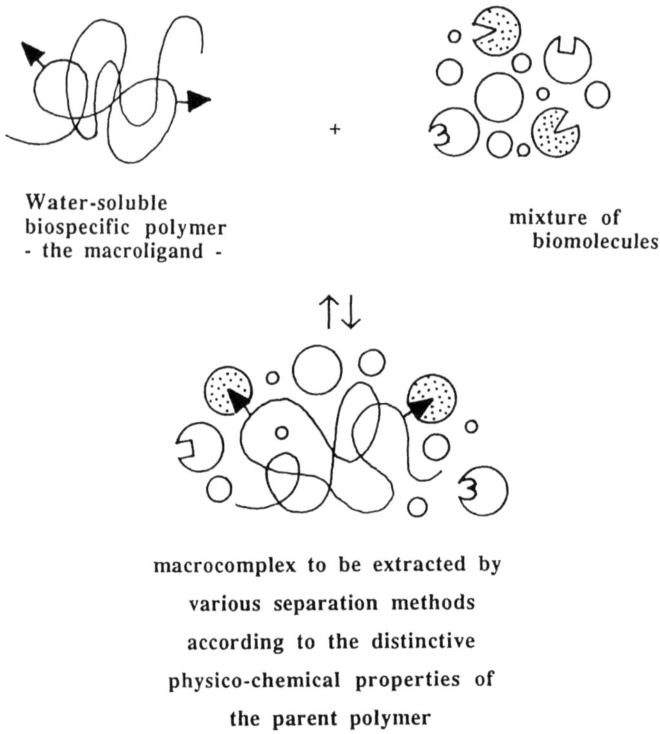

Fig. 14.1. Basic principle of the separation methods involving the formation of a complex between the target molecule and a water-soluble biospecific polymer

A water-soluble polymer, made biospecific after covalent coupling to an adequate ligand – resulting all together in a so-called macroligand – forms a water-soluble specific and reversible complex with the target molecule, when dissolved in the starting mixture.

At this stage, the only problem to be solved is to selectively extract the resulting macroligand-target molecule complex from the mixture. This can easily be obtained if the starting affinity polymer possesses a distinctive physico-chemical property, allowing its discrimination from the contaminants. So far, four different types of discrimination features have been exploited: high molecular weight, high density of charges, pH- or salinity-sensitive solubility and surface tension properties. The emergence of new techniques based on other distinctive features is only limited by the imagination capacity of the scientific community.

In the present article, examples are taken from our work and others, to illustrate how these water-soluble biospecific polymers can be used in various techniques and in certain situations as an alternative approach to affinity chromatography.

14.2 Discrimination on the Basis of High Molecular Weight

When the water-soluble affinity material is a high molecular-weight polymer, the reversible, biospecific complex formed with the desired target molecule is characterized by a high hydrodynamic volume, compared to that of contaminants in the starting mixture. In other words, the molecular weight of the target molecule is artificially but reversibly increased so that it becomes readily differentiable from other species in solution. Then, all techniques routinely used to fractionate mixtures on the basis of hydrodynamic volume differences – ultracentrifugation, gel permeation chromatography, membrane ultrafiltration, etc. – are therefore eligible to carry out the selective extraction of this complex.

14.2.1 Biospecific Ultrafiltration

Membrane ultrafiltration has been selected in a few instances for this purpose [11–14]. With this technique, contaminants can be eliminated in the filtrate while the macrocomplex concentration in the retentate is kept constant (diafiltration) or even increased (ultrafiltration) during the operation. The latter procedure results in enhanced protection of the target molecule against denaturation, well-known to occur as purification advances and dilution increases.

Adamski-Medda et al. [11] attempted to fractionate an artificial mixture of trypsin and chymotrypsin – whose separation is somewhat delicate owing to the similarity of properties and molecular weights – by ultrafiltration (membrane molecular weight cut-off 300 000), using a macroinhibitor of trypsin prepared by covalent coupling of *p*-aminobenzamidine (PAB) onto dextran ($\bar{M}_w$: 2×10^6). The resulting dextran-PAB exhibited some affinity towards trypsin, characterized by an inhibition constant of 0.5 μM (i.e. 1 g/l) on the basis of its competition

Table 14.1. Ultrafiltration of mixtures of trypsin and chymotrypsin

	Trypsin	Chymotrypsin	Trypsin[a]	Chymotrypsin[a]
R %	15	17	76	35

0.05 M Tris buffer pH 8.3; total concentration of proteins 1 g/l; molecular weight cut-off of the membrane: 300 000; pressure: 0.5 bars; 20 °C; *R*: rejection ratio.

[a] Ultrafiltration of proteins was carrried out in the presence of dextran-PAB (10 g/l)

Table 14.2. Ultrafiltration of a crude extract of *Pseudomonas testosteroni*

	Isomerase	Contaminants	Isomerase[a]	Contaminants[a]
R %	64	50	92	50

Molecular weight cut-off of the membrane: 10^6; pressure: 0.2 bars; *R*: rejection ratio.

[a] Ultrafiltration carried out in the presence of dex-E_{II} (3 g/l) ($\bar{M}_w$ around 2×10^6, 0.7×10^{-4} mol estradiol/g polymer)

with DL-benzoylarginine-*p*-nitroanilide. The main characteristics of this experiment are collected in Table 14.1

These preliminary results indicate that there is a rather specific interaction between dextran-PAB and trypsin, though relatively weak compared to that observed in other experiments, in which a dextran-estradiol was used to separate $\Delta_{5\to4}$ 3-oxosteroid isomerase (isomerase, in the following) from a crude extract of *Pseudomonas testosteroni* (Table 14.2) [15].

Owing to limitations inherent to the ultrafiltration technique – difficulty to find membranes entirely permeable to lower molecular weight contaminants, due to the large distribution of the pore sizes and to problems of fouling and concentration polarization – the efficiency of separation highly depends on the relative rejections of the complex and the contaminants. In the present example, the addition of the dextran-PAB macroinhibitor not only affects the rejection of trypsin (from 15 to 76%), as expected, but also and in a nonneglectable extent that of chymotrypsin (from 17 to 35%). The authors suggest that to obtain a better fractionation would require an increase in the macroinhibitor concentration (this is limited, from a practical point of view, by the correlative viscosity increase), or better, the synthesis of a macroinhibitor exhibiting a more clear-cut specificity towards trypsin. Similar conclusions were drawn by Choe et al. on the same topic [14].

14.2.2 Biospecific Gel Filtration

The use of gel filtration chromatography to extract the macrocomplex from the rest of the contaminants has received only little attention so far, although the quality of the fractionations achieved by this technique might have prompted

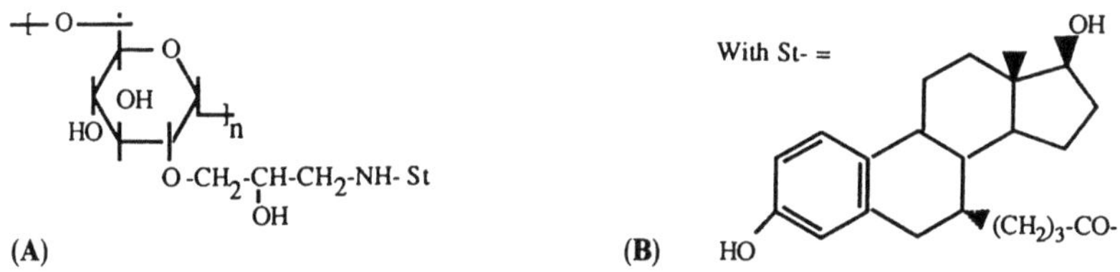

Fig. 14.2. **A** dex-E_{II}: 3×10^{-2} mmol estradiol/g of polymer; **B** Chemical structure of the dextran-estradiol (dex-E_{II}) conjugate

further studies. Besides the purification of certain proteins (e.g. pyruvate kinase and some blood coagulation factors) which bind nonspecifically but preferentially to the chromophore of blue dextran [16–18], the only examples reporting the fractionation of water-soluble biospecific macrocomplexes by gel filtration, concern the purification of the placental insulin receptor [19] and that of the cytoplasmic calf uterus receptor of estradiol [20].

Concerning the steroid hormone receptor, it was, because of its sensitivity towards denaturation and its low concentration in target tissue extracts – about 10^{-5} that of most other proteins –, a very good model to demonstrate the ability of the technique involving the use of a water-soluble biospecific polymer, to solve extremely difficult problems.

A soluble macroligand was prepared by covalent attachment of estradiol (E_{II}) at its 7α position via an eight-atom spacer chain, to high molecular weight dextran ($\bar{M}_W$: 500 000) (Fig. 14.2). This macroligand exhibited binding capacity for calf uterine cytosol receptor, characterized by an inhibition constant K_i around 0.3 μM (calculated on the basis of its competition with estradiol), compared to a dissociation constant K_d around 0.2 nM between the receptor and estradiol, under the same conditions. This binding was a reversible process, as demonstrated by the fact that when excess [^{3}H]-estradiol was added to the dextran-estradiol-receptor complex, a complete exchange of polymer for radioactive estradiol occurred within 1 h, at 25 °C. The incubation of this dextran-E_{II}, at a concentration corresponding to 250 K_i (2.5 g/l), with a prepurified "4S-trypsin" estrogen receptor preparation [21], free of aggregates, afforded a dextran-estradiol-receptor complex, which was separated from the bulk of proteins by filtration on a gel whose porosity was selected so that the complex could be eluted in the void volume of the column (Fig. 14.3A). As a control, the incubation of the estrogen receptor preparation with the same amount of a modified dextran, identical to the dextran-E_{II} derivative, except for omission of the last step of synthesis, i.e. fixation of the estradiol residue, led to the subsequent elution of the receptor at its usual 4S-form position (Fig. 14.3B).

Once the fractions corresponding to the void volume had been pooled, the solution was treated with a large excess of radioactive estradiol, in order to release the receptor from the dextran-E_{II} macroligand. The filtration of the mixture on a gel, identical to that previously used in the first step, allowed the recovery of the purified receptor at its usual elution volume (Fig. 14.4).

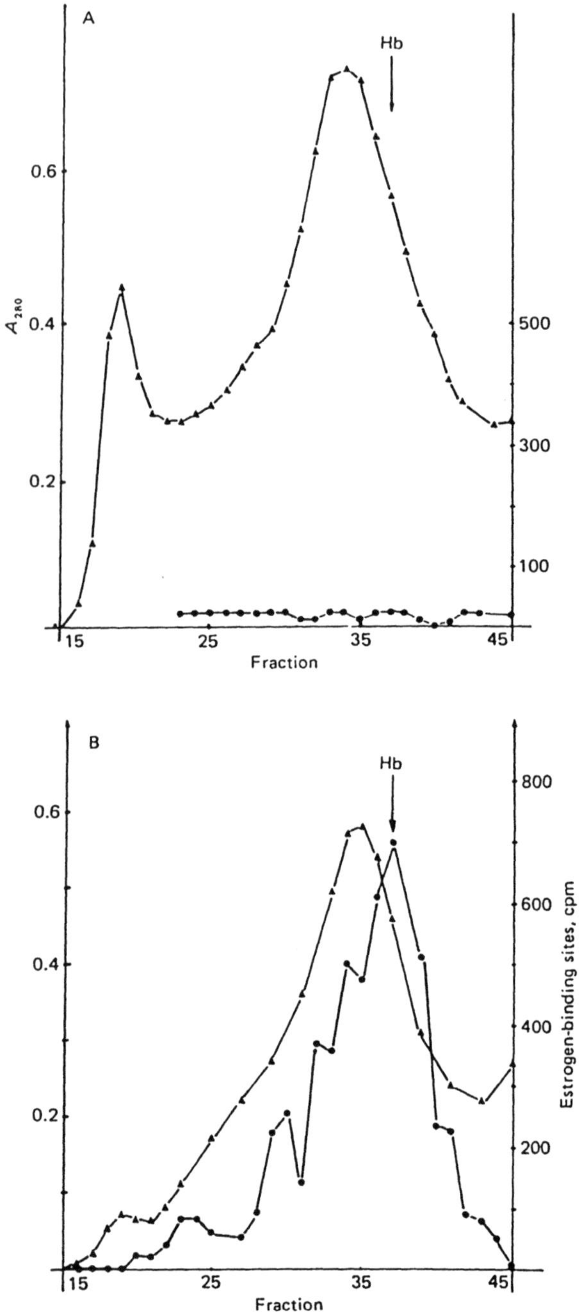

Fig. 14.3. Gel filtration on AcA 34 Ultrogel of uterine cytosol after treatment with dextran-estradiol (dex-E_{II}) conjugate (**A**) or with dextran-3-amino-2-hydroxy-propylether (**B**). (●) concentration of estrogen-binding sites; (▲) protein concentration (A_{280}). The arrow corresponds to the normal elution volume of haemoglobin, used as a molecular weight marker

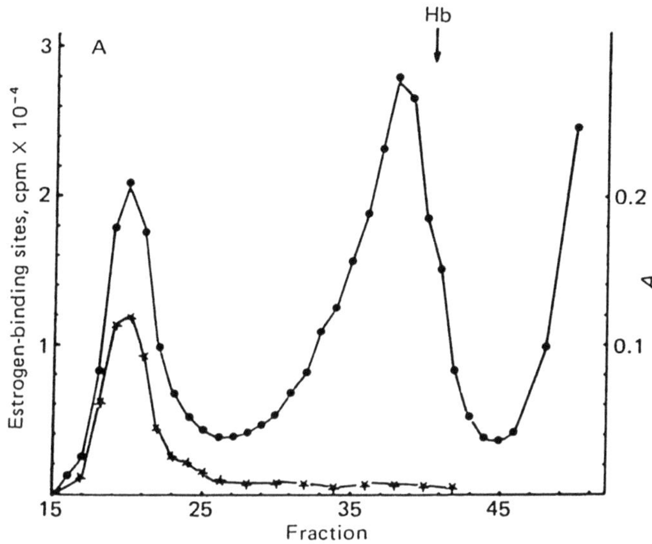

Fig. 14.4. Isolation of purified receptor-[^{3}H] estradiol complex, by gel filtration on AcA 34, after exchange of the dex-E_{II}-receptor complex collected in the void volume of the preceding chromatography, by an excess of [^{3}H] estradiol. (●) concentration of estrogen-binding sites; (*) protein concentration (A_{280})

The recovery was 50–75%, with respect to the amount of receptor present in the aggregate-free preparation (depending on the initial quality of the calf uterus tissue), and the purity was estimated to about 80%. These results were quite competitive with those afforded by traditional affinity chromatography [22, 23], if not better, and were obtained in a short time without previous experience in the field of estrogen receptor purification. The quantity of material that could be treated by this technique was only limited by the size of the gel filtration columns. On the other hand, the water-soluble macroligand could be recovered in the void volume of the column, at the second filtration step, and recycled for further purification experiments, after removal of the non-specifically bound contaminants.

The very same principle was applied by Suzuki et al. [19] for the purification of the placental insulin receptor, starting with a prepurified sample, obtained from a plasma membrane fraction of human placenta by a fractionation protocol including affinity chromatography on Sepharose-concanavalin A as the last step. The following purification step was carried out, using insulin bound to dextran ($\bar{M}_w$: 40000) as the biospecific, water-soluble carrier. After filtration of the dextran-insulin-receptor complex on an adequate gel and then dissociation by mild acidification, the insulin receptor was obtained in a purified form. This method permitted a 8700 fold purification, from crude plasma membrane and, even more remarkable, a 60-fold purification after the affinity chromatography step.

The main inconvenience of this technique concerns the effect of dilution, occurring during the second gel filtration, on the stability of the purified target molecule. This drawback, irrelevant to proteins not too sensitive to denaturation, can be partly overcome, simply by starting with larger quantities, owing to the protective role played by the increased concentration of purified material, against denaturation.

14.3 Discrimination on the Basis of High Density of Charges: Affinophoresis

A second distinctive feature, selected so as to allow an easy discrimination between contaminants and a target molecule, after its biospecific complexation with a water-soluble macroligand, is the high density of charges. The technique, first described and subsequently developed by the group of Shimura was named affinophoresis [24] (not to be mistaken with more conventional affinity electrophoresis [25–27], where the affinity ligand is immobilized on the insoluble electrophoresis gel) and exploits the large electrophoretic mobility of a polyionic water-soluble polymer, made biospecific after coupling to a desired ligand (Fig. 14.5).

The affinophoresis is applicable to substances having little or even no intrinsic mobility, such as for instance proteins at their isoelectric point, and also, since affinophoresis does not necessarily require an insoluble support, to suspended species like cells.

Affinophoresis can be carried out with either polycationic [24] (e.g. diethylaminoethyldextran) or polyanionic [28] (e.g. polyacryl-β-alanyl-β-alanine or succinyl polylysine) affinophores. However, the use of polyanionic derivatives is recommended, since positively charged affinophores can interact with dyes such

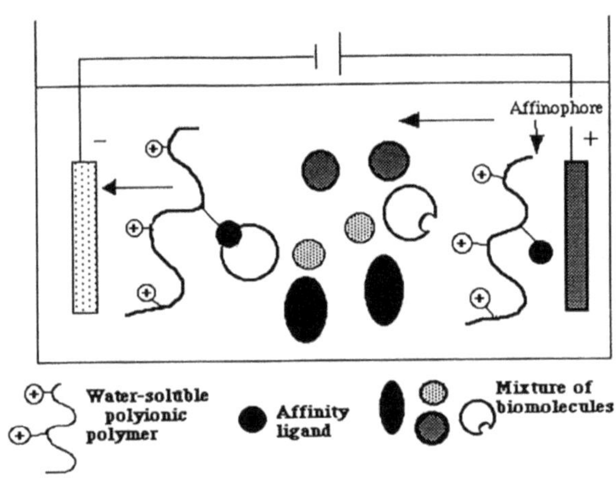

Fig. 14.5. Principle of affinophoresis

as Coomassie blue and hence interfere with the detection of proteins in the gel, and on the other hand are adsorbed on agarose, one of the most frequently used electrophoresis matrix, owing to electrostatic attractions with carboxylate or sulfate groups carried by this gel.

One-dimensional affinophoresis gave satisfactory results for samples of simple composition, e.g. an artificial mixture of trypsins from 3 different sources, with *m*-aminobenzamidine-DEAE dextran as the affinophore [24]. This macroligand consisted of dextran of about 60 glucose residues, carrying about 10 affinity ligands and about 25 positive charges, on average. When this affinophore was added to the electrophoresis gel (pH 7), the migration of cationic trypsins from bovine and *Streptomyces griseus* was increased and that of anionic trypsin from *Streptomyces erythreus* was decreased. The biospecific character of this experiment was further demonstrated by the facts that the mobility of chymotrypsin, a protein for which *m*-aminobenzamidine is not a specific ligand, was not affected and the effect of the affinophore on the migration of the various trypsins was completely suppressed by adding a competitive inhibitor, leupeptine, to the gel or when trypsins were irreversibly inhibited by tosyllysine chloromethyl ketone.

For more complex mixtures, e.g. separation of mannose-binding lectins from extracts of legume seeds [29] or extraction of anti-hapten antibody in rabbit serum [30], the principle of two-dimensional electrophoresis can be applied, the second step being carried out at right angles to the first electrophoresis and in the presence of the adequate affinophore. Nonspecific proteins should lie on a diagonal line and the specific protein should be found away from the line, as its mobility is changed in the second electrophoresis by the affinophore.

The idea of using water-soluble biospecific polymers is to afford an alternative to affinity chromatography. One of the reasons for this, as pointed out in the introduction of this article, is the risk of nonspecific binding to the insoluble matrix of the chromatographic procedure. From this point of view, the use of polyionic polymers in the affinophoresis techniques seems, a priori, a real wager, since one can hardly think of better traps than ionic polymers, to attract all kinds of substances of opposite charge, precisely by nonspecific interaction. Nevertheless, provided that the experiments are carried out at an ionic strength sufficiently high to suppress these undesirable side-effects, the affinophoresis techniques has proved successful and may find further applications in various situations (for a review see [31]).

14.4 Discrimination on the Basis of Surface Tension Properties: Affinity Partition

Surface tension of polymers, related to their electrostatic and hydrophobic properties, is such that certain couples – among which poly(ethylene glycol)/dextran, poly(vinyl alcohol)/dextran, poly(ethylene glycol)/$MgSO_4$, etc. – exhibit mutual incompatibility in water and the mixing of their concentrated aqueous solutions results in the formation of a biphasic system, in which the top phase segregates almost completely one polymer and the bottom phase the other [32–35]. Biomolecules may partition between the two phases with more or less attraction for one phase, according to their surface properties. Mixtures can therefore be fractionated in such aqueous biphasic systems, on the basis of their partition coefficient differences. This technique of partition between aqueous two-phase systems has been widely used for the fractionation of proteins, nucleic acids, membranes and cell particles, although certain drawbacks – high viscosity, slow settling of the two phases due to their low interfacial tension, frequent requirement of multistep operations, etc. – have certainly prevented more extensive use. Among various improvements which are proposed, for example, the use of magnetic polymer derivatives to increase the rate of phase separation [36] and the extrapolation of the partition technique principles in chromatographic processing [37], the idea of binding an affinity ligand onto one of the phase-forming polymers was definitely an important milestone in this purification technique [7–10, 38]. In fact, this modification allows one to attract specifically and almost quantitatively the target molecule in one phase – the one where the affinity macroligand is segregated – while the contaminants are directed as much as possible to the other phase, by varying pH, ionic strength, salt nature or other experimental conditions.

In one of the earliest studies [39], we carried out the purification of $\Delta_{5\rightarrow4}$ 3-oxosteroid isomerase from a crude extract of *Pseudomonas testosteroni* by affinity partioning in a poly(ethylene glycol)/dextran system. The macroligand was prepared by covalent coupling on poly(ethylene glycol) (PEG) of the same steroid derivative as previously described in Fig. 14.2. The resulting macroligand PEG-E_{II} had an inhibition constant K_i around 5 μM towards the isomerization of Δ_5-androstene-3,17-dione, by isomerase.

Figure 14.6 shows the variation of the partition coefficients (the partition coefficient K is the ratio of the concentration of the considered species in the top phase to that in the bottom phase) of isomerase and contaminants, vs the concentration of PEG-E_{II} added to the system. While the contaminants are only very slightly affected, the partition coefficient of isomerase increases, to finally reach – when the concentration of affinity ligand added to the system, i.e. 0.5 mM corresponding to 100 K_i, is sufficient to provide a complete complexation of the desired protein – an upper limit corresponding to the theoretical partition coefficient of poly(ethylene glycol) itself [the amount of poly(ethylene glycol) in the top phase is around 95%, hence corresponding to a partition coefficient around 15].

Salt concentration or pH highly influenced the partition coefficients (Fig. 14.7) and both effects could be exploited to improve the separation. Using a multistep extraction procedure (i.e. collecting the phase of interest and mixing it, at each step, with the desired amount of fresh opposite phase) isomerase was obtained in five operations with a 50% yield and a purification ratio of 170.

Poly(ethylene glycol) is the most widely used polymer to prepare macroligands for affinity partition. This is probably partly due to its solubility both in

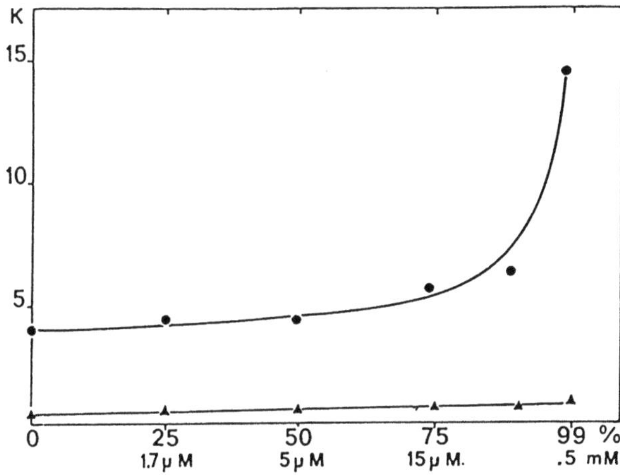

Fig. 14.6. Partition coefficient (*K*) of isomerase (●) and total proteins (▲), vs the theoretical percentage of complex formed by affinity between isomerase and PEG-E_{II}, or the molar concentration of PEG-E_{II} represent in the system

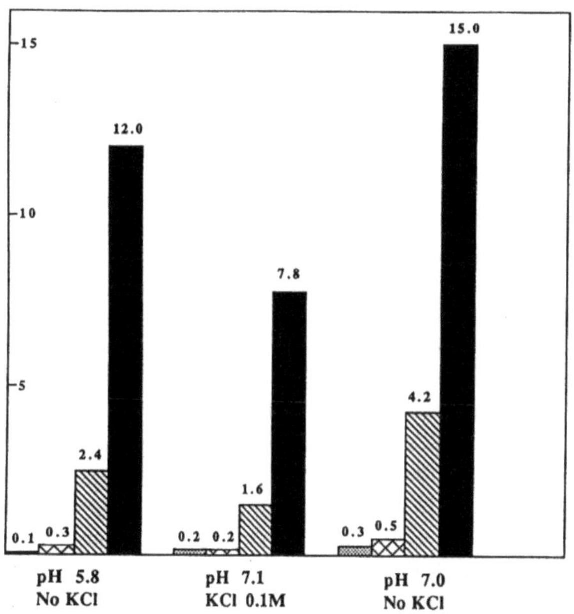

Fig. 14.7. Effect of PEG-E_{II} (0.5 mM) on the partition coefficient of isomerase and total proteins under various pH and KCl conditions. Partition coefficients of proteins without PEG-E_{II} (dotted columns) and with PEG-E_{II} (cross-hatched columns). Partition coefficients of isomerase without PEG-E_{II} (single-hatched columns) and with PEG-E_{II} (solid columns)

water and common organic solvents, allowing the use of a wide scope of chemical modifications for its subsequent coupling to a large variety of ligands [40]. Among these, one may quote the use of fatty acids to produce hydrophobic affinity macroligands for cell separation [41, 42] or the immobilization of various dyes to purify protein [43] and nucleic acid fractions [44]. More recently, the modifications of IgG with poly(ethylene glycol) to produce affinity ligands to separate cells based on antigenic determinants [45], and the use of metal-chelate pseudo-affinity partition [46] – a method combining metal-protein interaction and partition in aqueous two-phase systems – were reported.

If the success of affinity partition is largely demonstrated by the accumulation of articles reporting the use of this technique for analytical purposes as well as a purification tool, certain important considerations must, nevertheless, be pointed out. First, the recovery of the target molecule is limited by the partition coefficient between the two phases of the macroligand itself. Since the mutual incompatibility of the two phase-forming polymers leads to incomplete, though extensive, segregation, the unavoidable presence of a macroligand in both phases obviously leads to losses, especially important as multistep operations are carried out. The second point concerns the elimination of contaminants. When performing multistep extractions, one observes that the partition coefficient of contaminants does not remain constant but increases as the steps are successively performed. This means that a fractionation of these contaminants takes place and that the system is progressively enriched in species which naturally favor the "wrong" phase. This consequently leads to a situation where further extraction steps are almost totally inefficient, unless pH and salt conditions are varied during the procedure, to take into account the evolution, step after step, of the physicochemical properties of the contaminants remaining to be eliminated.

Among hampering factors to the development of large scale applications of partition techniques, the cost of polymers involved in the biphasic system has played an important role. For instance, if poly(ethylene glycol)/dextran systems are traditionally employed at the laboratory scale, it is financially unrealistic to consider the use of highly purified dextran as the bottom phase component. Various new and inexpensive polymers, e.g. hydroxypropylstarch, crude dextran or gum acacia [47–50] have been introduced recently to find a solution to this problem. Another practical problem concerns the recovery of the desired biomolecule at the end of the partition procedure, or in other words, the elimination of the considerable amount of polymers involved in the phase system. The traditional way to operate is to collect the top phase containing the affinity complex, change the conditions in order to obtain dissociation, then mix it with a fresh bottom phase, generally a salt aqueous solution. Upon a new partition experiment, the biomolecule of interest is recovered in the bottom phase and is subsequently submitted to desalination by diafiltration or dialysis. Obviously, all these final purification steps on a purified, hence rather unstable biomolecule, are additional risks of denaturation.

14.5 Discrimination on the Basis of Reversible Solubility: Affinity Precipitation

The last distinctive feature that has been exploited to discriminate between contaminants and the complex formed between a target molecule and a water-soluble biospecific macroligand, is the reversible soluble-insoluble character of the parent polymer.

Synthetic polymers [5, 51] as well as derivatives of natural macromolecules, such as hydroxypropylcellulose acetate succinate [52, 53] or chitosan [54, 55] have been used. Precipitation can be induced by pH shift, salinity modification or temperature. Certain polysaccharides, such as sodium alginate whose solubility is sensitive both to acidic conditions and to the presence of divalent cations (e.g. Ca^{++}) or carrageenan which precipitates upon decreasing temperature, have not yet been exploited but are eligible candidates for a future use in affinity precipitation. Various works can be selected from the literature to illustrate the basic principles of this technique.

As a first example, the flow diagram of the affinity precipitation procedure, used by Taniguchi et al. [53], for the purification of staphylococcal protein A, is represented in Fig. 14.8. The polymer used for the macroligand preparation is hydroxypropylmethylcellulose acetate succinate (HPMCAS), an enteric coating polymer, soluble at pH higher than 5.2 and insoluble below due to the neutralization of the carboxylate functions at acidic pH. The soluble complex between protein A and the affinity carrier (IgG bound to HPMCAS) is formed at pH 7.2, and then precipitated by a pH shift from 7.2 to 4.2. The dissociation step of the isolated solid complex can be carried out either by washing with a low pH eluent (pH 2.5) or by treatment of the redissolved complex (pH 7.2) with a high concentration of a chaotropic agent (3.5 M KSCN). Both approaches afford protein A with a purity around 25% and a recovery in the 50–70% range.

In one of the earliest reports, Schneider et al. [5] followed a similar approach by using a reversibly water-soluble-insoluble copolymer of *N*-acryloyl-*m*-aminobenzamidine, acrylamide and *N*-acryloyl-*p*-aminobenzoic acid where benzamidine was the biospecific ligand, to purify trypsin from beef pancreas by affinity precipitation.

Another group [51] reported on the purification of various proteins, using a precipitable carrier obtained by copolymerization of *N*-isopropylacrylamide (NIPAM) with *N*-acryloxysuccinimide (NASI) or glycidyl methacrylate (GMA) (Fig. 14.9). NASI and GMA are the reactive groups on the copolymer, necessary for the immobilization of the desired ligands, while the hydrophobic isopropyl groups of the NIPAM moiety confer to the copolymers their character of precipitability, upon addition of salt at high concentration e.g. around 2 M ammonium sulfate. Such copolymers, after coupling to *p*-aminobenzamidine or to IgG, were used for the purification by affinity precipitation of trypsin and protein A, respectively.

Since the availability of large scale technology for harvesting of precipitates is good, it seems that affinity precipitation is the easiest and more suitable

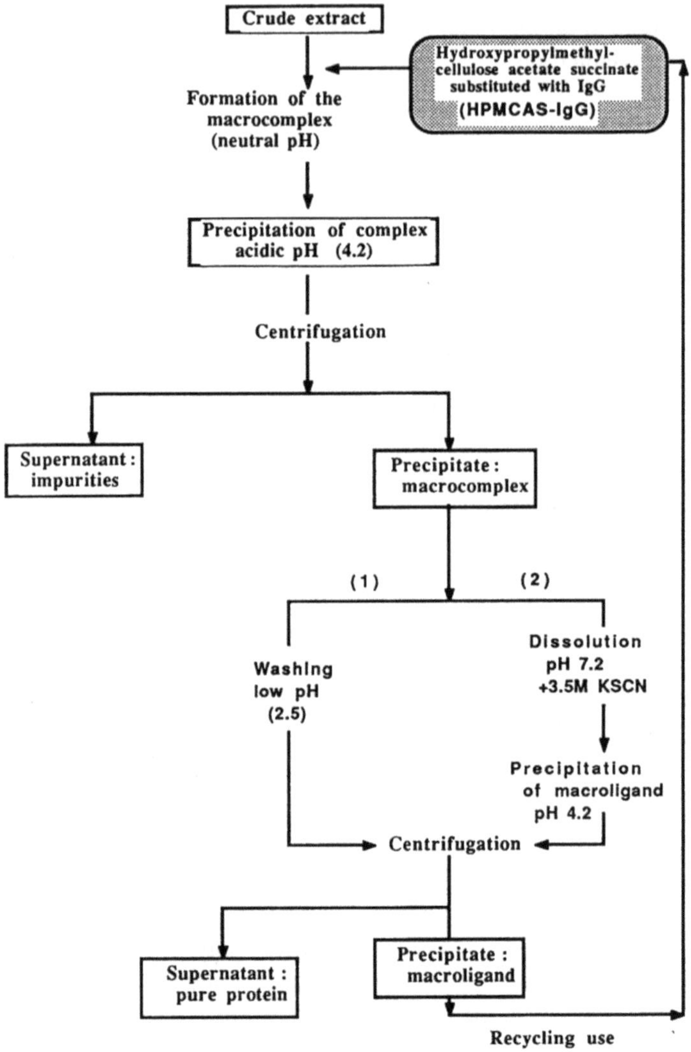

Fig. 14.8. Flow diagram of the two affinity precipitation pathways for the purification of staphylococcal protein A

technique when large amounts are to be purified. This is obviously of considerable importance, with respect to the development of affinity techniques on an industrial scale. However, it seems that the Achille's heel of this method is located at the precipitation step itself. In fact, since a modification, sometimes rather drastic, of the pH or the salinity must be applied to obtain the precipitation of the complex, this concept is only applicable to target molecules able to withstand this modification without loosing most of their activity. The affinity precipitation technique is therefore not of universal scope and requires the

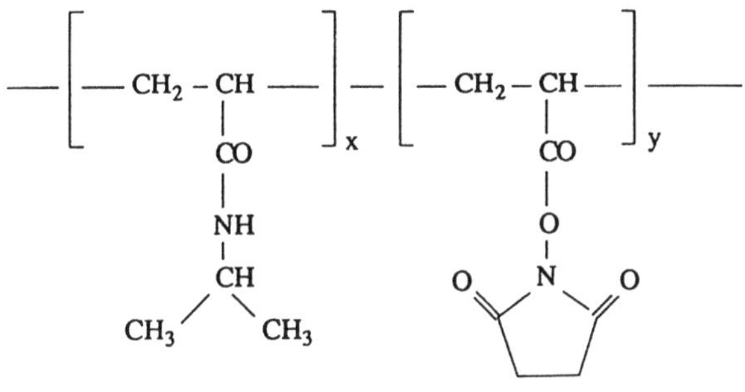

NIPAM-NASI copolymer

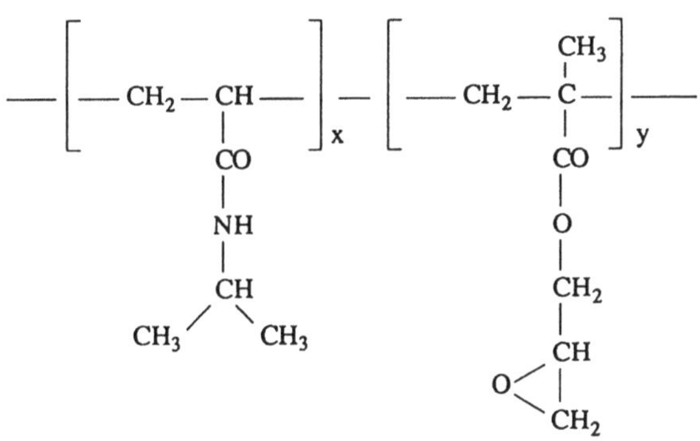

NIPAM-GMA copolymer

Fig. 14.9. Chemical structure of the *N*-isopropylacrylamide, *N*-acryloxysuccinimide (NIPAM-NASI) and N-isopropylacrylamide, glycidylmethacrylate (NIPAM-GMA) copolymers

preliminary selection of an adequate polymer whose physicochemical properties are compatible not only with the stability domain of the target molecule but as well with that of the chemical bond between the polymer and the affinity ligand, if one considers to properly recover the macroligand for subsequent cycles.

14.6 Advantages and Drawbacks of Techniques Involving Water-Soluble Biospecific Polymers

The advantages of the techniques involving water-soluble biospecific polymers can be found by merely reversing the arguments, developed in the introduction of this article, concerning the drawbacks of affinity chromatography.

The solubility in water and, for some polymers like poly(ethylene glycol), in certain commonly used organic solvents, first allows the use of a wide scope of chemical reactions, to modify or activate the parent polymer for subsequent coupling to desired ligands. This point, of course, holds as well to a certain extent for insoluble supports, but the advantage of working in a homogeneous solution, obviously leads to higher substitution yields without the necessity of reacting a large excess of sometimes quite expensive materials.

Second, the solubility in water allows one to control the reactions on the carrier better. This concerns first the activation procedures and subsequently the coupling step. The amount of immobilized ligand can therefore be precisely determined using, for instance, all routine spectroscopic methods (UV-visible, NMR spectroscopies, etc.) which can otherwise be hardly considered for insoluble compounds.

Once the macroligand is prepared, it can be studied in solution for analytical purposes. Its binding capacity towards the target molecule can be determined straightforward by competition studies with substrates or inhibitors, using classical Dixon [56] or double reciprocal Lineweaver-Burk [57] plots. The knowledge of the association constant has the immediate consequence of allowing the calculation of the amount of macroligand required to afford a complete formation of the complex with the target molecule. On the other hand the conditions for a complete dissociation of this complex, at the second stage, can as well be readily appreciated.

This possibility of calculating the minimum, though sufficient, amount of macroligand to complex the desired biomolecule, affects the quality of the subsequent purification. In fact, the use of minimum quantities of affinity macroligand – for instance, 25 mg of dextran-E_{II} for the purification of 10 ml of the 4S-trypsin calf uterus estradiol receptor preparation [20] – enables one to reduce to a very low level the nonspecific adsorption of contaminants on the affinity material. Compared to affinity chromatography, where gram amounts of the matrix are frequently required, this low contamination at the association stage must obviously lead to better purification ratios.

Besides the inconvenience specifically related to technological limitations of the various methods discussed in the preceding chapters – denaturation risks related to dilution during gel filtration, limited recovery during partition owing to incomplete segregation of the macroligand in one phase, lack of selectivity of the membranes cut-off, etc. – the main drawback that all techniques share, concerns the solubility limit of the macroligand. In fact, when the ligand to be immobilized has a marked hydrophobic character, e.g. steroids, fatty acids, etc., the water-soluble polymer has a substitution limit beyond which its solubility properties may be substantially altered. This point, almost irrelevant to very highly soluble polymers like for instance poly(ethylene glycol), may turn out critical in situations where polymers of limited solubility, e.g. poly(vinyl alcohol) or polysaccharides are involved. For instance, during the synthesis of macroligands based on dextran and steroids, the attempts to anchor estradiol-7αbutyric acid in amounts greater than 2×10^{-4} moles per gram of polymer, led to the insolubilization of the affinity material.

This phenomenon obviously results in a major limitation when the target molecule exhibits a weak affinity towards the immobilized ligand, since the amount of macroligand thus necessary to achieve a complete formation of the complex, must be increased to an extent where nonspecific interactions may become unacceptable.

14.7 References

1. Cuatrecasas P, Anfinsen CB, (1971) Annu Rev Biochem 41:259
2. Lowe CR, Dean PDG (1974) Affinity chromatography. John Wiley and Sons Ltd, New York
3. Jacoby WB, Wilchek M (eds) (1974) Methods in enzymology. Academic Press, New York, Vol. 34, Part B
4. Turkova J (1978) Affinity chromatography. Elsevier, Amsterdam
5. Schneider M, Guillot C, Lamy B (1981) Ann NY Acad Sci 369:257
6. Arnold HF, Chalmers JJ, Saunders MS, Croughan MS, Blanch HW, Wilke CR (1985) ACS Symp Ser 271:113
7. Takerkart G, Segard E, Monsigny M (1974) FEBS Lett 61:223
8. Shanbhag VP, Johansson G (1974) Biochem Biophys Res Commun 61:1141
9. Flanagan SD, Barondes SH (1975) J Biol Chem 250:1484
10. Flanagan SD, Taylor P, Barondes SH (1975) Nature 254:441
11. Adamski-Medda D, Nguyen QT, Dellacherie E (1981) J Membrane Sci 9:337
12. Mattiasson B, Ramstorp M (1984) J Chromatogr 283:322
13. Choe JB, Massé P, Verdier A (1986) Biotechnol Lett 8:163
14. Luong JHT, Male KB, Nguyen AL (1988) Biotechnol Bioeng 31:516
15. Unpublished results
16. Haeckel R, Hess B, Lanterborn W, Waster KH (1968) Hoppe-Seyler's Z Physiol Chem 349:699
17. Swart ACW, Hemker HC (1971) Biochim Biophys Acta 222:364
18. Staal GE, Koster JE, Kamp H, Van Milligen-Boersma L, Veeger C (1971) Biochim Biophys Acta 227:86
19. Suzuki S, Sakamoto Y, Kuzuya T, Fujita-Yamaguchi Y (1981) Biomedical Res 2:457
20. Hubert P, Mester J, Dellacherie E, Néel J, Baulieu EE (1978) Proc Natl Acad Sci USA 75:3143
21. Fox LL, Redeuilh G, Baskevitch P, Baulieu EE, Richard-Foy H (1976) FEBS Lett 63:71
22. Truong H, Geynet C, Millet C, Soulignac O, Bucourt R, Vignau M, Torelli V, Baulieu EE (1973) FEBS Lett 35:569
23. Truong H, Baulieu EE (1974) FEBS Lett 46:321
24. Shimura K, Kasai K (1982) J Biochem 92:1615
25. Takeo K (1987) In: Crambach A, Dunn MJ, Radola BJ (eds) Advances in electrophoresis. UCH, Weinheim, New York, Vol 1, p 229
26. Horejsi V, Ticha M, Kocourek J (1977) Biochim Biophys Acta 499:290
27. Takeo K, Suzuno R, Tanaka T, Nakamura K, Kuwahara A (1986) In: Dunn MJ (ed) Electrophoresis 86. UCH, Weinheim, p 233
28. Shimura K, Kasai K (1984) Biochim Biophys Acta 802:135
29. Shimura K, Kasai K (1987) J Chromatogr 400:353
30. Shimura K, Kasai K (1987) Electrophoresis 8:135
31. Shimura K (1990) J Chromatogr 510:251
32. Albertsson PÅ (1971) Partition of cells, particles, macromolecules, 2nd ed. Wiley-Interscience, New York.
33. Walter H (1982) In: Pretlow TG, Pretlow TP (eds) Cell separation: Methods and selected applications. Academic Press, New York, Vol 1, p 261
34. Albertsson PÅ (1978) J Chromatogr 159:111
35. Fisher D (1981) Biochem J 196:1
36. Wikström P, Flygare S, Gröndalen A, Larsson PO (1987) Anal Biochem 167:331
37. Hubert P, Mathis R, Dellacherie E (1991) J Chromatogr 539:297

38. Cordes A, Flossdorf J, Kula MR (1987) Biotechnol Bioeng 30:514
39. Hubert P, Dellacherie E, Néel J, Baulieu EE (1976) FEBS Lett 65:169
40. Harris JM, Struck EC, Case MG, Paley MS (1984) J Polym Sci, Polym Chem Ed 22:341
41. Eriksson E, Albertsson PÅ, Johansson G (1976) Mol Cell Biochem 10:123
42. Van Alstine JM, Brooks DE (1983) Clin Chem 30:441
43. Kopperschlager G, Johansson G (1981) Anal Biochem 124:117
44. Müller W, Hattesohl I, Schuetz HJ, Meyer G (1981) Nucleic Acids Res 9:95
45. Sharp KA, Yalpani M, Howard SJ, Brooks DE (1986) Anal Biochem 154:110
46. Birkenmeier G, Vijayalakshmi MA, Stigbrand T, Kopperschläger G (1991) J Chromatogr 539:267
47. Kroner KH, Husted H, Kula MR (1982) Biotechnol Bioeng 24:1015
48. Ling TGI, Nilsson H, Mattiasson B (1989) Carbohydr Res 11:43
49. Tjerneld F, Berner S, Cajarville A, Johansson G (1986) Enzyme Microb Technol 8:417
50. Rastongi A, Chand S (1989) Biotechnol Techniques 3:33
51. Nguyen AL, Luong JHT (1989) Biotechnol Bioeng 34:1186
52. Taniguchi M, Kobayashi M, Natsui K, Fujii M (1989) J Ferment Bioeng 68:32
53. Taniguchi M, Tanahashi S, Fujii M (1990) J Ferment Bioeng 69:362
54. Senstad C, Mattiasson B (1988) Biotechnol Bioeng 33:216
55. Senstad C, Mattiasson B (1989) Biotechnol Bioeng 34:387
56. Dixon M (1953) Biochem J 55:170
57. Lineweaver H, Burk D (1934) J Am Chem Soc 56:658

15 Protein-Polyelectrolyte Complexes

Jiulin Xia and Paul L. Dubin

15.1 Introduction

Proteins interact strongly with both synthetic and natural polyelectrolytes. Ample evidence exists for the binding of polyanions and polycations to proteins below and above their isoelectric points, respectively. These interactions may result in soluble complexes [1, 2], complex coacervation [3–6], or the formation of amorphous precipitates [7–9]. The practical consequences of these phase changes may include the use of polyelectrolytes for protein separation [10–16] and immobilization or stabilization of enzymes in polyelectrolyte complexes [17–18]. In these two applications, the optimal physical states of the system are different. In the case of enzyme immobilization, highly deaggregated states may be less active. In purification or separation processes involving settling or filtration, aggregation is desirable. For efficient settling, close-packed aggregates are preferred, whereas in filtration processes more open-textured aggregates are needed to allow adequate solvent penetration. However, in both cases the aggregation should be essentially reversible.

All of the states of protein-polyelectrolyte complexes referred to above may be achieved by the selection of the polyelectrolyte (PE), choice of ionic strength and pH, and control of the concentration of the macromolecular components. Most reports on the practical applications of protein-polyelectrolyte complexes concern the effects of these factors. This chapter focuses on the effects of polymer charge density, solution pH and ionic strength on protein-polyelectrolyte complexation discussed in recent and current research. Techniques employed in the study of protein-polyelectrolyte complexes and applications of the complex to protein separation and enzyme immobilization are also described.

15.2 Investigation Methods

15.2.1 Turbidimetric Titration

The simplicity and sensitivity of the turbidimetric titration method as applied to protein-polyelectrolyte systems is based on the fact that turbidity increases with particle size [4, 16]. One example is the complexation of bovine serum albumin

Macromolecular Complexes
in Chemistry and Biology
Eds.: Dubin/Bock/Davis/Schulz/Thies

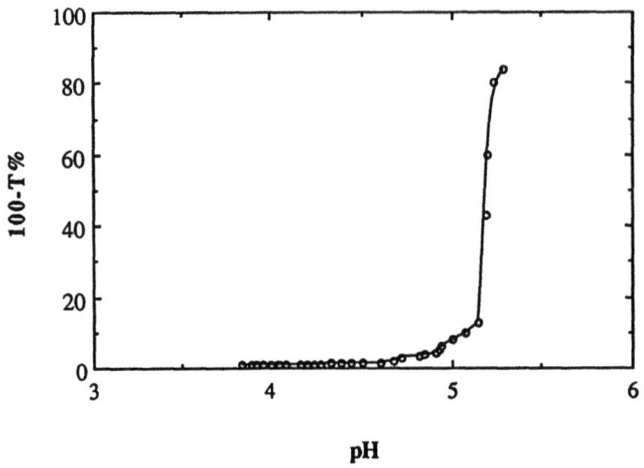

Fig. 15.1. Type 1 titrations of 0.10 g/L PDMDAAC in 0.60 g/L BSA at $I = 0.01$ M NaCl

(BSA) with polydimethyldiallylammonium chloride) (PDMDAAC) [19], as shown in Fig. 15.1. In this "type I" [19] titration (performed by adding HCl or NaOH to a mixture of protein and PE), the turbidity shows a gradual increase followed by an abrupt increase. The former is identified as the range of initial soluble complex formation, which usually corresponds to a turbidity change of $100 - T\% < 10$. The latter corresponds to phase separation. Both the initial complexation and phase separation are characterized by well-defined pH (or net protein charge) values, i.e. pH_c and pH_ϕ (Z_c and Z_ϕ), respectively. Such turbidimetric titrations usually show reversibility within a pH range where the protein is not denatured [19]. However, a hysteresis was found in the titration of human hemoglobin and dextran sulfate by Nguyen [7].

Employing "type III" turbidimetric titration (in which PE is added to protein at constant pH and ionic strength), Kokufuta [9, 20] studied complexation between human serum albumin (HSA) and PDMDAAC and potassium poly(vinyl alcohol sulfate) (KPVS). He found a turbidimetric "end point" at which the mole numbers of quaternary ammonium groups in PDMDAAC and sulfate groups in KPVS are approximately identical, respectively, to the contents of the acidic and basic groups in HSA. Therefore, the complexation between HSA and PDMDAAC and KPVS was described as stoichiometric. A similar result was also found in the BSA-PDMDAAC system [21].

15.2.2 Viscometry

Kabanov reported a considerable decrease in the reduced viscosity (η_{sp}/C) of quaternized poly-4-vinylpyridines (PVP) in pH 7 aqueous solution, upon addition of BSA [22]. The decrease in viscosity was explained by the formation of polycation-BSA soluble complexes. In a more recent study of human hemoglobin-dextran sulfate mixtures, Nguyen [7] found that η_{sp}/C decreases with the polymer concentration, as shown in Fig. 15.2. In contrast to the behavior of

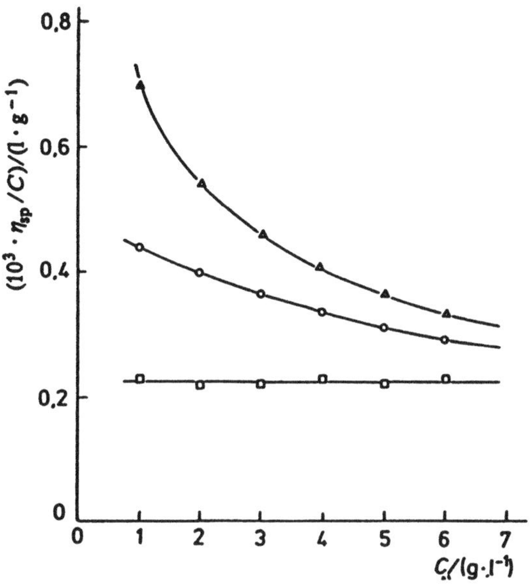

Fig. 15.2. The reduced viscosity of diethylaminoethyl (DEAE) dextran and DEAE dextranhemoglobin solutions at pH 7.2 as a function of DEAE dextran concentration. DEAE dextran alone (△); DEAE dextran/hemoglobin weight ratio 0.2 (○); DEAE dextran/hemoglobin weight ratio 0.2 (□). (from Ref. [7])

polyelectrolytes in pure water, as embodied by the relationship $\eta_{sp}/C \propto C^{-1/2}$, the ionic interactions between particles seems to be reduced through complexation, which is a reflection of the process of charge neutralization. Similar behavior was also described by Kuramoto [23] for BSA in the presence of poly(acrylic acid).

15.2.3 Ultrafiltration

An ultrafilter works like a sieve on a macromolecular scale. Theoretically, ultrafiltration would be an ideal technique for studying macromolecular binding. Unfortunately, two main drawbacks generally limit its appication: (1) the lack of membranes of suitable cut-off, and (2) concentration polarization and membrane fouling during the experiments. The latter phenomenon modifies the separative properties of the membrane, i.e. the sieving characteristic is not constant during the experiments.

One of the few experimental reports on ultrafiltration of protein-polyelectrolyte complexes was the study of hemoglobin-dextran sulfate complexation by Nguyen [7]. As shown in Fig. 15.3, the relative concentration of hemoglobin in the filtrate decreases when the ratio of polymer to hemoglobin increases. The linear decrease in the amount of hemoglobin permeating through the membrane as a function of the amount of dextran sulfate suggests that the complexation is stoichiometric. The extrapolation of the linear part of the plot to zero concentration in the permeate gives a binding weight ratio of 63 hemoglobin/dextran molecule.

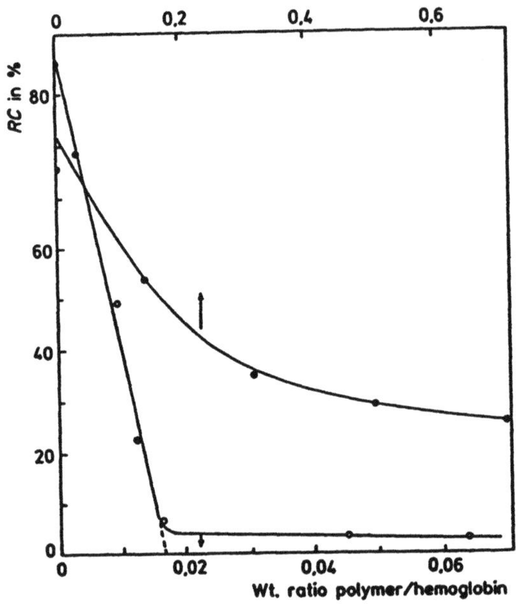

Fig. 15.3. The relative concentration (RC) of hemoglobin in the ultrafiltrate to that in the feed as a function of weight ratio of polymer to hemoglobin. Dextran sulfate (○); DEAE dextran (●). Operating conditions: hemoglobin concentration: 1 g/L; pH 7.2; temp.: 25 °C; pressure: 1.38×10^4 Pa. (from Ref. [7])

15.2.4 Sedimentation

Ultracentrifugation of the homogeneous BSA-PVP system at various ratios of the components, as shown in Fig. 15.4, was reported by Kabanov in 1977 [22]. An increase in the BSA concentration at a constant PVP concentration leads to a decrease in the area of the peak corresponding to the free polyelectrolyte (peak I), while the area of the presumed peak of the complex (peak II) increases. At a 3 : 1 ratio of the number of macromolecules of BSA/PVP, the peak of free PVP disappers, and only one peak remains on the sedimentogram. It was also shown that the BSA/PE ratio at which the free PVP disappears, at a fixed weight concentration of PVP, depends on the PVP molecular weight, i.e. the higher the PVP molecular weight the more BSA are needed to make the free PVP peak disappears. It was suggested that a portion of the PVP chain, of approximately constant average length l, is consumed for the binding of each protein. The average value of l was about 80 nm. A similar sedimentation pattern was also found for the complexation of BSA with a copolymer of 2-methyl-5-vinyl-pyridine and acrylic acid [24].

15.2.5 Size-Exclusion Chromatography

The Hummel–Dreyer method [25] of size-exclusion chromatography (SEC) has been used for the determination of the stoichiometries of protein-polyelectrolyte complexes. Samsonov [26] in 1969 showed that the strength of the complexes formed by ribonuclease (RNAse) and copolymers of sodium vinyl pyrrolidone and vinyl sulfonate diminishes with reduction in the content of sodium vinyl

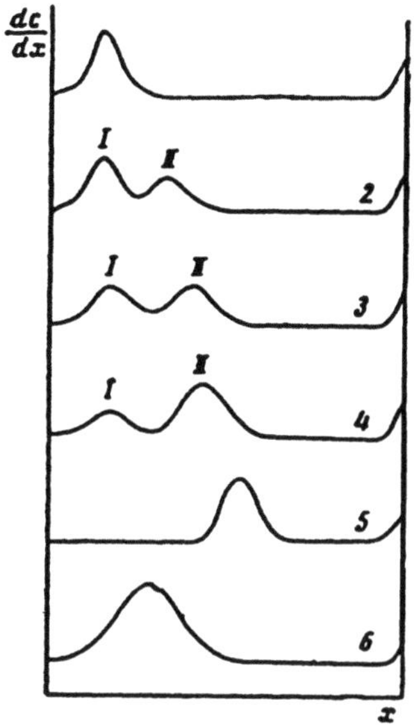

Fig. 15.4. Sedimentograms for PE (*1*) and its mixtures with BSA at various number concentration ratio n_{BSA}/n_{PE}: (*2*) 0.75; (*3*) 1.25; (*4*) 2; (*5*) 3; time of centrifugation 60 min; (*6*) 6; 20 min; $\omega = 56\,000$ rpm. Along *y* axis: concentration gradient (dc/dx); along *x* axis: distance from the center of rotation of the rotor (*x*) in arbitrary units. (from Ref. 22)

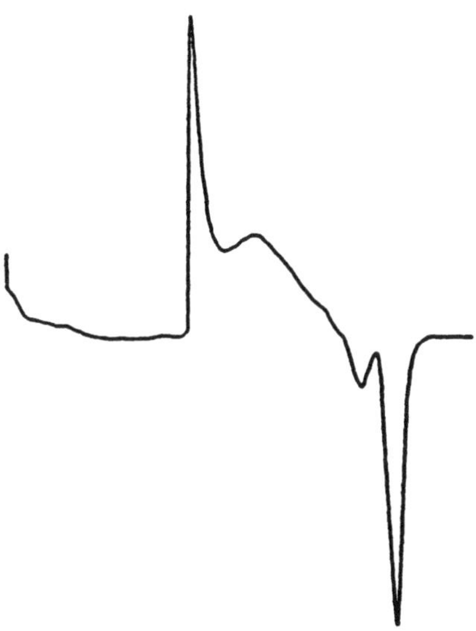

Fig. 15.5. UV (254 nm) chromatograms (Superose 6) of PDMDAAC ($MW = 3.2 \times 10^5$) using, as mobile phase, 0.25 M, pH 9.0, NaOAc buffer containing 5.0 mg/mL BSA. (from Ref. [16])

sulfonate in the copolymer and with increase in the content of the Na^+ ions in the solution. In 1986, Barberousse [27] reported SEC of soluble complexes formed by hemoglobin and dextran sulfate in 0.05 M and pH 7.0 Tris buffer. It was concluded that about 13 hemoglobin molecules were bound per polymer chain. More recently, Dubin [16] studied PDMDAAC-BSA in 0.25 M and pH 9.0 NaOAc buffer as shown in Fig. 15.5. From data at different BSA concentrations

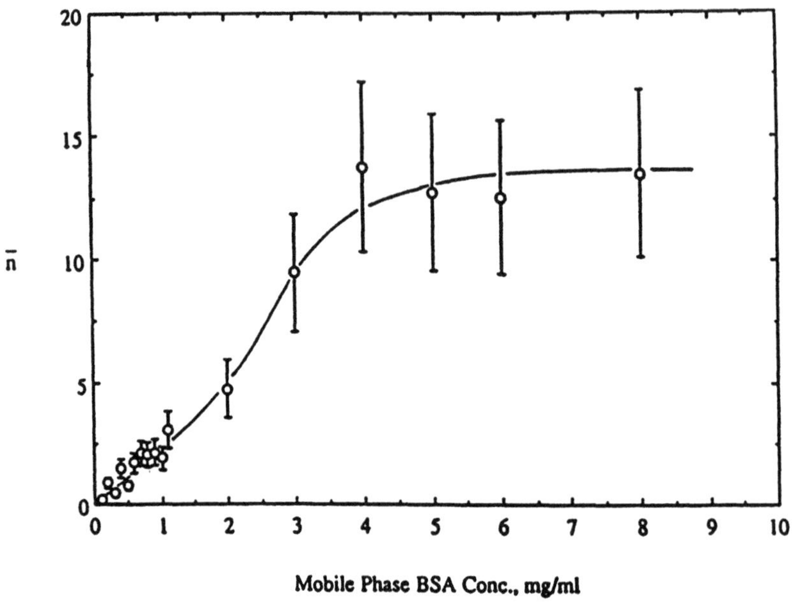

Fig. 15.6. Degree of binding n versus mobile phase BSA concentration ($I = 0.25$, pH 9.0, PDMDAAC concentration = 2.5 mg/mL). (from Ref. [16])

it was possible to construct a binding isotherm (see Fig. 15.6). The number of proteins bound per polymer molecule was found to be a function of pH and free protein concentration.

15.2.6 Polarized Luminescence

The polarized luminescence method is sensitive to intermolecular interactions between different components in solution because of a change in the mobility parameters (θ) of each of the components (the mobility of the protein macromolecules as a whole or the intramolecular mobility of the polymer chains). Pautov [28] used luminescence-labeled lysozyme and methacrylic acid-ethylene dimethyl acrylamide copolymer (PMAA) to study the mobility of PE-bound lysozyme as a function of the relative protein concentration. Figure 15.7 shows soluble complex formation at low and high protein relative concentrations (P/PEL); precipitation occurs at $1 < P/PEL < 12$. Based on the mobility change of the bound lysozyme in complexes with relative high protein concentration, Pautov concluded that the bound protein molecules are in a compact array. For the complexes formed at low relative protein concentrations, the compactness of the complex was diminished. Under such conditions the protein associates on the PMAA become less compact and the complex itself more stable.

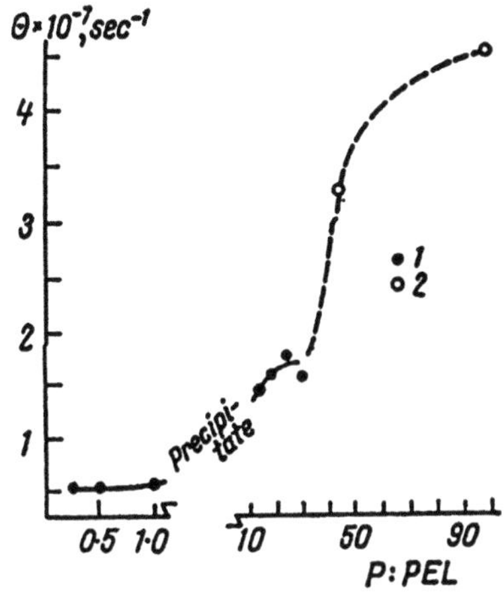

Fig. 15.7. Change in mobility ($\theta_{\text{bound}} = 1/\tau_\omega$) of PMMA-bound lysozyme molecules as a function of the relative contents of protein and PMAA (P: PEL) in aqueous solution (*1*); (*2*) change in the observed mobility coefficient $\theta_{\text{bound+unbound}}$ with the appearance in solution of free lysozyme molecules (298 K, $\alpha_{\text{PMMA}} = 0.13$, $c_{\text{lysozyme}} = 0.25$ g/L). (from Ref. [28])

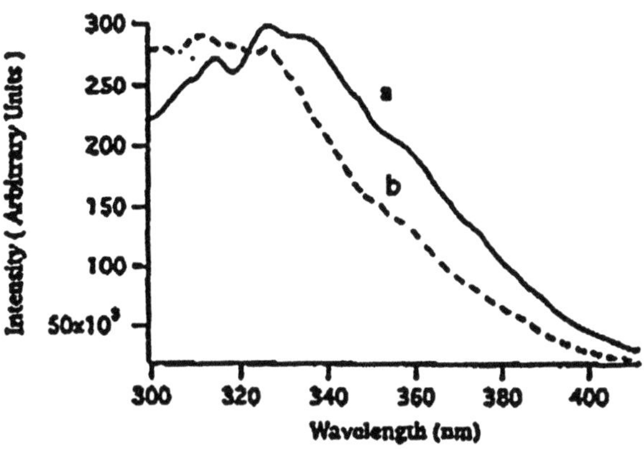

Fig. 15.8. Fluorescence emission spectra of papain: KPVS complexes excited at 280 nm. (*a*: papain, and *b*: papain : KPVS = 1.5 : 0.8). (from Ref. [29])

15.2.7 Spectroscopy

In recent fluorescence studies of papain complexation with potassium poly(vinyl alcohol sulfate) (KPVS), Frank et al. [29, 30] found a blue shift in the emission spectra of papain in the presence of KPVS, as shown in Fig. 15.8. These blue shifts were interpreted as the result of shielding of tryptophan residues from aqueous media by the complexed KPVS chains. It was suggested that Classes II (Trp 69) and III (Try 7, 177, and 181) tryptophans are initially exposed to the aqueous environment, but subsequently bind with KPVS. At increased papain concentration, the shielding effect was found to decrease (see Fig. 15.9). Thus, a high degree of complexation at low papain content was suggested.

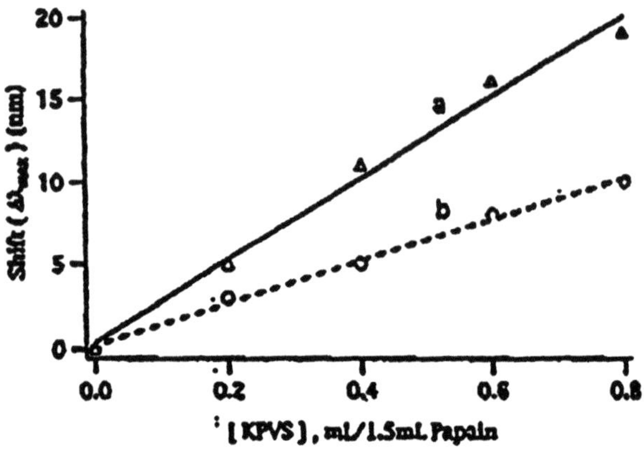

Fig. 15.9. Effects of KPVS concentration on papain: KPVS complexation. ([papain] = a: 3×10^{-6} M, b: 1×10^{-5} M). (from Ref. [30])

15.2.8 Static Light Scattering (SLS)

The molecular parameters of a macromolecule are usually obtained from static light scattering results by using Zimm-diagrams corresponding to Eq. (15.1) [31]:

$$\frac{Kc}{R_\theta} = \frac{1}{M_w}\left(1 + \frac{16\pi^2}{3\lambda^2} R_g^2 \sin^2\frac{\theta}{2}\right) + 2A_2 c \tag{15.1}$$

where c is the mass concentration of polymer, K is a constant which contains the optical parameters of the system; M_w and R_g are the weight average molecular weight and the root mean square radius of gyration of the macromolecule, respectively; and A_2 is the second virial coefficient. For typical protein-polyelectrolyte complexes, the low concentrations make it possible to ignore the particle interaction term [32, 34], i.e. the second virial coefficient in Eq. (15.1). Therefore, Eq. (15.1) can be simplified to

$$\frac{Kc_x}{R_\theta} \cong \frac{1}{M_x}\left(1 + \frac{16\pi^2}{3\lambda^2} R_g^2 \sin^2\frac{\theta}{2}\right) \tag{15.2}$$

where c_x and M_x are the concentration and molecular weight of the complex, respectively. c_x and M_x are related to the polymer concentration and polymer molecular weight M_p by

$$c_x = c_p(1 + \beta) \tag{15.3}$$

$$M_x = \alpha M_p(1 + \beta) \tag{15.4}$$

where α is the degree of aggregation, i.e. the number of polymer chains within one complex. β is defined as the weight ratio of protein to PE. Substitution of

Eqs. (15.3) and (15.4) into (15.2) yields

$$\frac{Kc_p}{R_\theta} \cong \frac{1}{\alpha M_p(1+\beta)^2}\left(1+\frac{16\pi^2}{3\lambda^2}R_g^2\sin^2\frac{\theta}{2}\right) \tag{15.5}$$

Therefore, we may use the known polymer concentration c_p to evaluate the molecular weight of the complex.

SLS of the bovine serum albumin (BSA)-polysaccharide (PSa) conjugate was studied by Chu in 1982 [33]. In this study the molecular weight of the conjugate (3.5×10^6) was found to be about 7 times larger than that of the pure polymer (5.3×10^5), yet their sizes are comparable: $Rg \sim 54$ nm for the conjugate and ~ 50 nm for the polymer. Dautzenberg et al. [32] reported on the static light scattering of complexes formed from the enzyme invertase and a polycation, poly(*N*-methyl-*N*,*N*-diethylamino-ethylacrylate). In this complex, the mass ratio of protein to polymer was found about 24. Recently, our SLS results [34] for BSA-PDMDAAC complexes showed that the structure of the BSA-PDMDAAC complex in solutions of $C_{pr} = 0.6$, $I = 0.01$ M and pH 7.88 depends on the polymer concentration. At low polymer concentration, an intrapolymer complex is saturated with BSA. The mass ratio of the bound BSA to polymer in the saturation limit is about 40. This intrapolymer complex aggregates to interpolymer species upon increase in the polymer concentration.

15.2.9 Quasi-Elastic Light Scattering (QELS)

In QELS measurements we obtain the homodyne intensity-intensity correlation function $G(q, t)$, with q, the amplitude of the scattering vector, given by $q = (4\pi n/\lambda)\sin(\theta/2)$, where n is the refractive index of the medium, λ is the wavelength of the excitation light in a vacuum, and θ is the scattering angle. For a Gaussian distribution of intensity profile of the scattered light, $G(q, t)$ is related to the electric field correlation function $g(q, t)$ by [35–36]:

$$G(q,t) = A(1 + bg(q,t)^2) \tag{15.6}$$

where A is the experimental baseline and b is a constant, which depends on the number of coherence areas that generates the signal ($0 < b < 1$). The electric field correlation function depends on the Fourier transform of the fluctuating number density of particles or molecules. For the center of mass diffusion of identical particles the following simple relation holds:

$$g(q,t) = e^{-t/\tau}$$
$$\frac{1}{\tau} = Dq^2 \tag{15.7}$$

where τ is decay time and D is diffusion coefficient. For polydisperse systems such as mixtures of protein and polyelectrolyte the correlation function $g(q, t)$ can be expressed as an integral of the exponential decays weighted over the distribution of relaxation times $\rho(t)$:

$$g(t) = \int_0^\infty e^{-t/\tau} \rho(\tau)\, d\tau \tag{15.8}$$

Experimentally the distribution function $\rho(\tau)$ is obtained from numerical analysis of $g(t)$. Therefore, the mean relaxation time $\langle \tau \rangle$ is resolved from each of the distribution modes of $\rho(\tau)$, as the first moment of the normalized relaxation spectrum. The diffusion coefficient, which corresponds to each value of $\langle \tau \rangle$, can be calculated using

$$D = \frac{\lambda^2}{16\pi^2 \sin^2(\theta/2) \langle \tau \rangle} \tag{15.9}$$

From each D value Stokes' radius, R_s, is obtained by the Einstein equation

$$R_s = \frac{kT}{6\pi\eta D} \tag{15.10}$$

where k is Boltzmann's constant, T is the absolute temperature and η is the viscosity of the solvent.

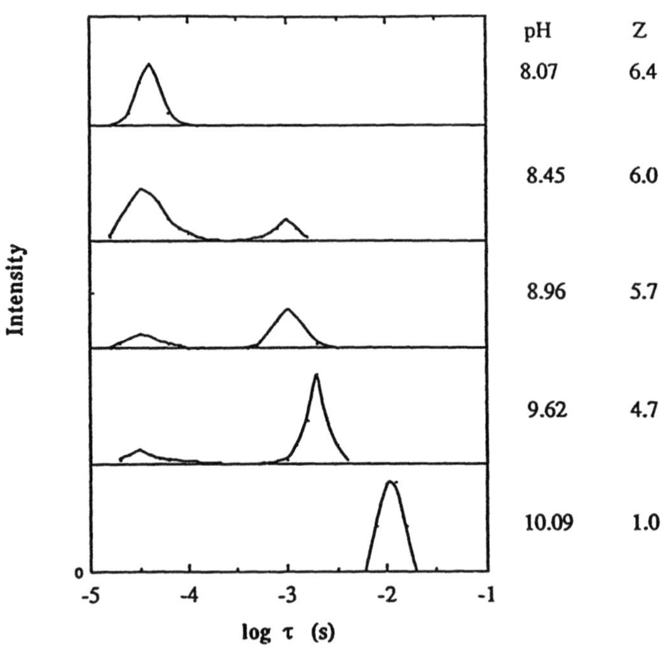

Fig. 15.10. Distribution functions of LBN 66-lysozyme complex in 0.10 M NaCl obtained by CONTIN program from the autocorrelation functions at different pH (from top to bottom) 8.07, 8.45, 8.96, 9.62 and 10.09. (from Ref. [38])

Few QELS studies have been reported on protein-polyelectrolyte complexes. From light scattering of the BSA-PSa system [33], Chu obtained both R_s and R_g. By comparing the values to that of the pure polymer, it was concluded that the complex has a rather compact structure compared to the polymer. In 1988, Dubin [37] reported on QELS of BSA-PDMDAAC and showed a wide size distribution, sensitive to pH. Recently, more systemic studies [38] on QELS of protein-polyelectrolyte complexes have been carried out. Figure 15.10 shows results for the complexation between lysozyme and a polycation. The distribution function, $\rho(\tau)$, of the relaxation time changes with pH from a single mode to a bimodal distribution. At pH < 8.07, one diffusion mode is resolved with a relaxation time constant identical to that of the pure protein. A bimodal distribution is first resolved at pH 8.50, with a fast mode of the same relaxation time as the single distribution observed at pH < 8.07. While the fast mode decreases in intensity with increasing pH, the slow mode increases with respect to both intensity and relaxation time. The slow modes have slower diffusion than that of the polymer. These results suggest that the bimodal distribution corresponds to the diffusion of the pure protein and the complex formed by the protein and the polymer. The onset of slow modes corresponding to complex formation is defined as pH_c (or Z_c in terms of net protein charge). With increasing pH, binding and soluble complex formation are enhanced. At pH 10.02, the fast mode disappears and the solution becomes cloudy, corresponding to phase separation. This pH is defined as pH_ϕ (or Z_ϕ).

15.2.10 Electrophoretic Light Scattering (ELS)

In ELS, the photon-counting heterodyne correlation function for a solution with an electrophoretically monodisperse solute can be written as [39]:

$$C(\tau) = \beta_0\delta(\tau) + \alpha_0 + \alpha_1 \exp(-2Dq^2\tau) + \alpha_2 \exp(-Dq^2\tau)\cos(\Delta\omega\tau) \tag{15.11}$$

where $\beta_0, \alpha_0, \alpha_1$, and α_2 are constants independent of correlation time, τ, and $\delta(\tau)$ is the delta function. D and q have the same definitions as in QELS. The cosine term is due to simultaneous electrophoresis and diffusion.

The Fourier transform of Eq. (15.11) respect to time, as stipulated by the Weiner–Khinchine theorem [40], gives the power spectrum:

$$S(\omega) = \beta_0 + \alpha\delta(\omega) + \frac{2(\alpha_1/\pi)Dq^2}{\omega^2 + (2Dq^2)^2} + \frac{\alpha_2 Dq^2}{2\pi}\left[\frac{1}{(\omega + \Delta\omega)^2 + (Dq^2)^2} + \frac{1}{(\omega - \Delta\omega)^2 + (Dq^2)^2}\right] \tag{15.12}$$

where a is a constant independent of ω.

In Eqs. (15.11) and (15.12), $\Delta\omega$ is the difference between the angular frequency of the scattered light, ω_s, and that of the reference beam, ω_r, which is the same as that of the incident beam. Since the frequency of the incident beam is modulated in the scattered light by the amount of the so called Doppler shift frequency, $\Delta\omega$ is given by:

$$\Delta\omega = \frac{2\pi n}{\lambda} E u \sin\theta \tag{15.13}$$

where E (volts/cm) and u ((μm s^{-1})/(V cm^{-1})) are the applied electric field strength and electrophoretic mobility, respectively. Therefore, u can be directly evaluated from frequencies of the power spectrum. Detailed discussion on ELS measurements can been found in several reviews [39, 41].

ELS is a sensitive tool for the study of charge molecules in solution. Most recently we applied this technique to the study of protein-polyelectrolyte complexation [38], as shown by the example in Fig. 15.11, which is the mobility obtained at different pH for RNAse-LBN 66. One notes that the mobility of the complex starts decreasing around 9.10 as the pH progresses toward phase separation. When pH $\leq$ 9.10 the measured u values are identical to the mobility of LBN 66 polymer. The onset of the mobility change at pH 9.10 can be easily understood as a result of initial protein-polymer complexation by considering the motion of a complex in a electric field.

The motion of the complex at steady state in a field of strength E (V cm^{-1}) can be described by [38]

$$u = \frac{v}{E} = \frac{q_p + n q_{pro}}{f_p + n f_{pro}} \tag{15.14}$$

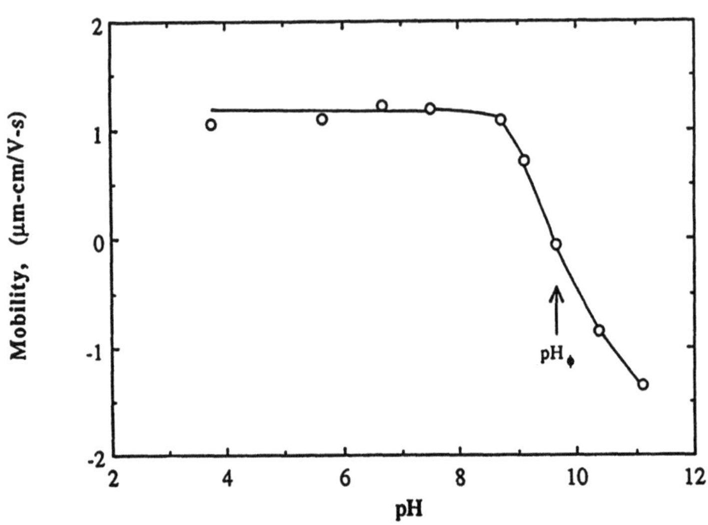

Fig. 15.11. Electrophoretic mobility of RNAse-LBN 66 in 0.10 M NaCl as a function of pH. Note the charge neutrality at the phase separation point. The mobility at pH > pH$_\phi$ were made over a short time period on stirred suspensions and are stable and reproducible

where q_{pro} and q_p are the net charge of a bound protein and total polymer charge, respectively; f_{pro} and f_p are the friction coefficients of the bound protein and the polymer, respectively, which are assumable to combine additively; E is the external field strength and v is the center of mass velocity of the complex; and n is the number of bound proteins per polymer chain. Clearly, the pH dependence of the complex mobility is expected from Eq. (15.14) since the net protein charge q_{pro} is a function of pH, as is n. Equation (15.14) is also reduced to the pure polymer mobility at $n = 0$. Therefore, the mobility change with pH may be used to characterize the protein-polyelectrolyte complexation via Eq. (15.14).

15.2.11 Miscellaneous Techniques

15.2.11.1 Small Angle X-Ray Scattering

Ruckpaul and co-workers [42] have obtained small angle X-ray scattering curves of heparin-hemoglobin complex. Their data show the existence of a heparin-hemoglobin complex containing 2 molecules of hemoglobin (Hb) per heparin chain.

15.2.11.2 Electron Spin Resonance (ESR)

ESR spectra of hemoglobin in the absence and presence of heparin were observed by Ruckpaul [42]. It was found that the heparin-hemoglobin complex exhibits a hyperfine structure with a line interval of 16.5 gauss because of the interaction between heparin and hemoglobin.

15.2.11.3 Electron Microscopy

Kabanov and co-workers [22, 24] have obtained electron micrographs of the soluble complex formed by BSA and quaternized poly-4-vinylpyridine. These micrographs show that the complexes are rod-like particles, assembled from proteins linked by the polymer chain. The thickness of the rod complex is ca. 100 Å.

15.2.11.4 *Circular Dichroism Spectroscopy*

Strelzowa et al. [43] have observed CD spectral changes in the range of 250-330 nm for α-chymotrypsin upon mixing the protein with dextran sulfate. However, the CD spectrum of δ-chymotrypsin is not affected by adding dextran sulfate to the protein solution. Recently, we [44] studied the complexation of PDMDAAC with papain, insulin and hemoglobin by CD spectroscopy. The CD spectra for papain and insulin show a α-helix structure change upon complexation, but the structure of hemoglobin is not affected by complexation.

15.3 Factors Influencing Protein-Polyelectrolyte Complexation and Structures of the Protein-Polyelectrolyte Complexes

15.3.1 Salt

The addition of inorganic electrolytes generally decreases the electrostatic interactions of polyions [45] due to charge shielding effect. For complexation between protein and polyelectrolyte, the salt effect produces a decrease in the degree of initial binding, as shown in Fig. 15.12 for sodium poly(stryenesulfonate) (NaPSS)-RNAse system. Clearly, the higher the ionic strength, the lower the critical pH, since RNAse needs more positive charge to overcome the charge shielding at high ionic strength. Reduced protein binding with increasing ionic strength was also found for the lysozyme-PMAA system by Anufrieva [28]. The addition of salt can also affect the stability of the protein-polyelectrolyte complexes. For example, Morawetz [46] found that complexes formed by catalase and polyacrylic acid more readily precipitate upon increase in BaCl concentration. However, an opposite effect is shown for the complexation of BSA and PDMDAAC [4], where the phase separation pH goes higher at high ionic strength. It seems that the reduction of the electrostatic interaction by adding salt can stabilize complexes or may also make complexes aggregate. The effect of ionic strength on the mechanism of protein-polyelectrolyte binding was recently studied for the BSA-PDMDAAC system by Dubin et al. [21, 34]. It was found that the BSA-PDMDAAC complexes are stoichiometrically formed in low ionic strength solution. At high ionic strength the complexes are formed with a broad distribution in structure.

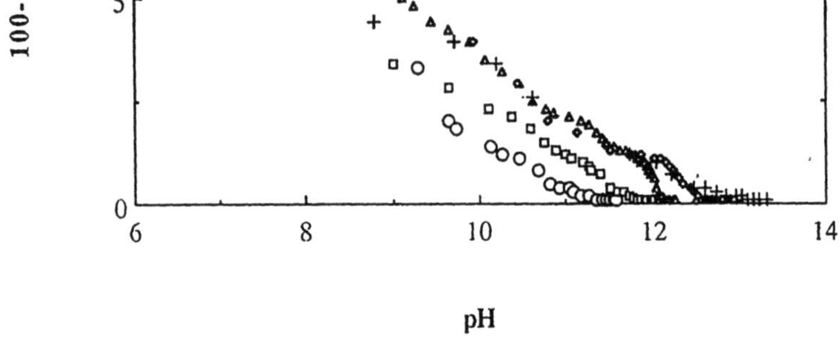

Fig. 15.12. Turbidimetric titration of NaPSS-RNAse system in NaCl solution with various ionic strength. Note the inverse dependence of pH_c on I

15.3.2 pH or Protein Charge

Since the complexation between proteins and polyelectrolytes is caused by Coulomb interaction, it has been evident that increasing pH promotes the formation of protein-polycation complexes, and decreasing pH enhances complexation of proteins and polyanions [19]. From the fluorescence study of papain-PVS complex, Cha et al. [29] found the the emission shift for the complex strongly depends on the pH, as shown in Fig. 15.13.

From studies of the complexation of HSA with both anionic PVS and cationic PDMDAAC, Kokufuta [9, 20] concluded that protein-polyelectrolyte complexes are stoichiometrically formed through salt linkages. However, we recently showed that the initial complexation between polyelectrolytes and proteins can even occur under conditions where the net protein charge is the same as that of the PE. This result may be attributed to non-uniform protein charge patches. These may be observed by computer modeling [47].

Regarding pH-induced phase separation, all the reported results show that pH_{ϕ}, the critical pH for phase separation, corresponds to a net protein charge opposite to that of the polyelectrolyte. If charge neutralization takes place at the point of phase separation, the following equation holds [19] for the charge on the complex:

$$Z_T = Z_p + n_{\phi} Z_{\phi} = 0 \qquad (15.15)$$

where Z_p is the formal charge of the polyion, n_{ϕ} is the average number of protein molecules bound per polyion chain at phase separation, and Z_{ϕ} is the net charge of a bound protein at phase separation. The validity of Eq. (15) has been observed for some protein-polyelectrolyte pairs [38]. But non-zero mobility was also detected for other complexes at the phase separation point. This may suggest that some other factors are also involved in the processes of the protein-polyelectrolyte phase separation.

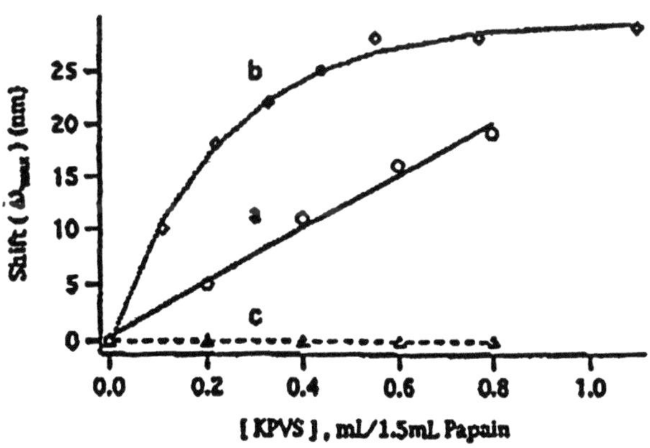

Fig. 15.13. Effects of pH on the emission shift of papain-KPVS complexes. (*a*: pH 2, *b*: pH 7, and *c*: pH 10). (from Ref. [29])

15.3.3 Concentration

In the study of BSA-PDMDAAC complex [34], it was found that the initial complexation pH is insensitive to the concentration of BSA and PDMDAAC. However, the coacervate of the complex formed at phase separation is not reversible at low BSA : PDMDAAC ratio. For several protein-polyelectrolyte complexes, Kokufuta [9, 20] found that the amount of polyelectrolyte needed to precipitate a protein is linearly proportional to the amount of the protein in solution. The concentration dependence for the efficiency of protein-polyelectrolyte phase separation was also reported by Morawetz [46] for several other systems.

15.3.4 Charge Density of Polyelectrolytes

There is little systematic study of the effect of polyelectrolyte charge density on protein-polyelectrolyte complexation. One exception [19] is the observation of strong interaction between BSA, RNAse or lysozyme and PVS, a polyanion with a large geometric charge density [48]. It was found that the critical initial binding pH increases with the polyion negative charge density even when the protein binding is ocurring at pH > IEP. This was explained by recognizing that the increased electrostatic attraction between the more densely charged polyions and a local protein positive region overcomes possible repulsion between the polyanion and the global protein charge.

15.3.5 Molecular Weight of Polyelectrolytes

The effect of polymer molecular weight on protein-polyelectrolyte complexes was recently reported. Zaitsev [49] found that the MW of the soluble BSA-PEVP complex is linearly dependent on the degree of the polymerization of the PEVP polycation, as shown by Fig. 15.14. Based on this proportionality, it was concluded that the complex has an intrapolymer structure. For the same BSA-PEVA system, Kabanov [22] found that the reduced viscosity is linearly proportional to the contour length of the polycation. From fluorescence studies on papain-NaPSS system, Cha et al. [30] found that the emission shift depends on the NaPSS MW. This MW dependence was interpreted by the complexation mechanism involving the bridging of papain by NaPSS. We recently studied the RNAse-NaPSS complexes formed with varying MW of NaPSS by using QELS. It was found that the diffusion coefficients of the complex follow a scaling law with the polymer MW, as shown by Fig. 15.15. This may also suggest an intrapolymer structure of the RNAse-NaPSS complex.

15.3.6 Models of the Structure of Protein-Polyelectrolyte Complexes

Based on viscometry and sedimentation data, Kabanov [22] proposed a rod-shape intrapolymer structure for the soluble BSA-PVP complex at pH 7. This is shown in Fig. 15.16, in which the polycation chain encircles the butt-joined BSA globules. The thickness of each complex chain was believed to be ca. 100 Å. The solubility of the complex was attributed to the hydrophilic contribution from the loops of the bound polymer chain. Zaitsev [50] proposed a different model

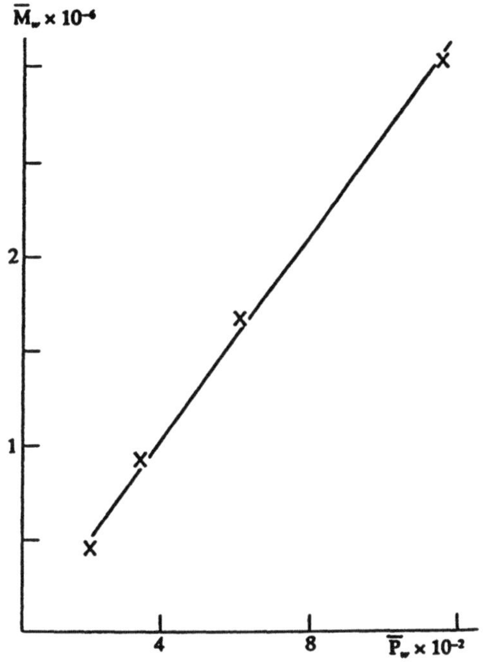

Fig. 15.14. Dependence of the molecular weight of water-soluble BSA-PEVP complex on the degree of polymerization of the PEVP polycation. (from Ref. [49])

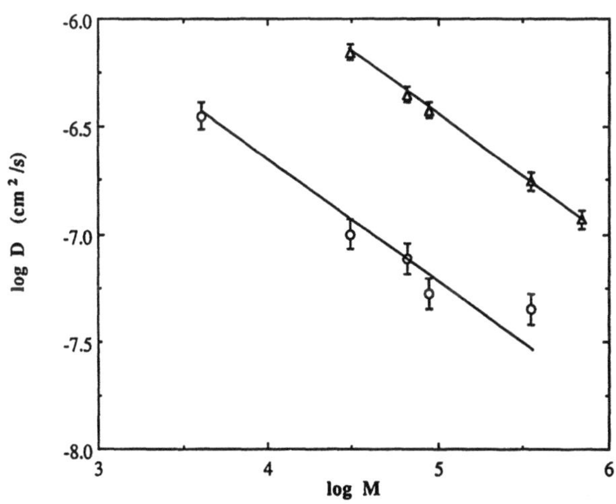

Fig. 15.15. Diffusion coefficients of NaPSS-RNAse complexes in 25 mM $Na_2B_4O_7$ pH 11.0 buffer as a function of NaPSS MW

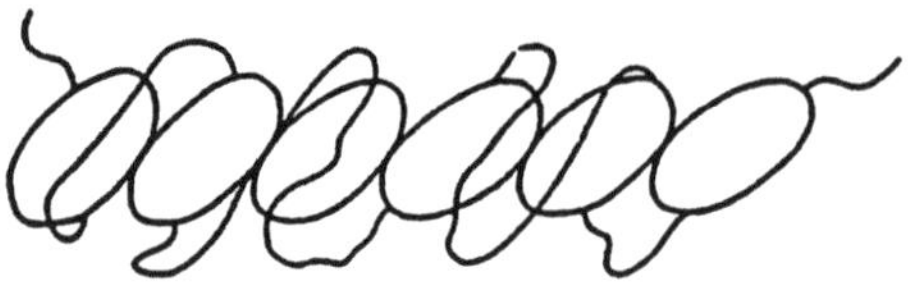

Fig. 15.16. Proposed structure of the soluble BSA-PVP complexes by Kabanov. (from Ref. [22])

Fig. 15.17. Proposed structure of the soluble BSA-PEVP complexes by Zaitsev. (from Ref. [50])

(Fig. 15.17) for soluble BSA-PEVP complex. In this model the bound PEVP polycation chains are localized in a hydrophobic center wherein the positive charges from PEVP are neutralized by the negative BSA charge; the hydrophobic center is also surrounded by negative charged BSA, which provides a net complex surface charge and is thus responsible for the solubility of the complex. For the lysozyme-PMAA complex, Anufrieva [51] recently suggested that some PMAA domains are filled with protein, while other domains remain unoccupied (Fig. 15.18).

Soluble protein-polyelectrolyte complexes are usually formed at a pH close to the protein IEP, and the soluble complexes aggregate to form coacervates toward phase separation by adjusting pH. Dubin [16, 34] suggested the existence of the primary intrapolymer complexes (in which a single polymer chain is bound by several protein molecules) for the excess protein case. These primary complexes could aggregate to form interpolymer complexes, in which several polymer chains are involved, as shown in Fig. 15.19. The interpolymer complex could also be soluble up to the point of the large scale aggregation.

Fig. 15.18. Arrangement (disposition) of structured parts of the lysozyme-PMMA polycomplex in the aqueous solution when the relative amount of protein in solution is 1 and $\alpha_{pMAA} = 0.4$). (from Ref. [50])

Fig. 15.19. Intrapolymer and interpolymer structures of protein-polyelectrolyte complexes

15.3.7 Mechanism of the Formation of Protein-Polyelectrolyte Complexes

It is generally believed that the electrostatic interactions constitute the primary driving force for protein-polyelectrolyte association. The formation of protein-polyelectrolyte complex at pH > IEP for polycations, and at pH < IEP for polyanions has been attributed to salt linkage [9, 20]. Recently it was found that some polyanions bind proteins at pH > IEP [19]. Protein heterogeneous charge distribution [19, 38] can help to interpret this observation. As shown by the results of computer modeling for proteins, in Fig. 15.20, RNAse can simultaneously have both positive and negative potential regions in aqueous solution. This local non-uniform potential region or portein charge patch is believed to provide an attraction force that overcomes the repulsion between the global protein charge and the polyelectrolyte. Therefore, the complexes formed by polyanions (polycations) with global negatively (positively) charged proteins are suggested to be a manifestation of local interaction between the protein charge patch and polyelectrolytes.

15.4 Protein Separation by Polyelectrolytes

Precipitation by polyelectrolytes is an important method in the processing of protein products because of the advantage of low additive consumption, high

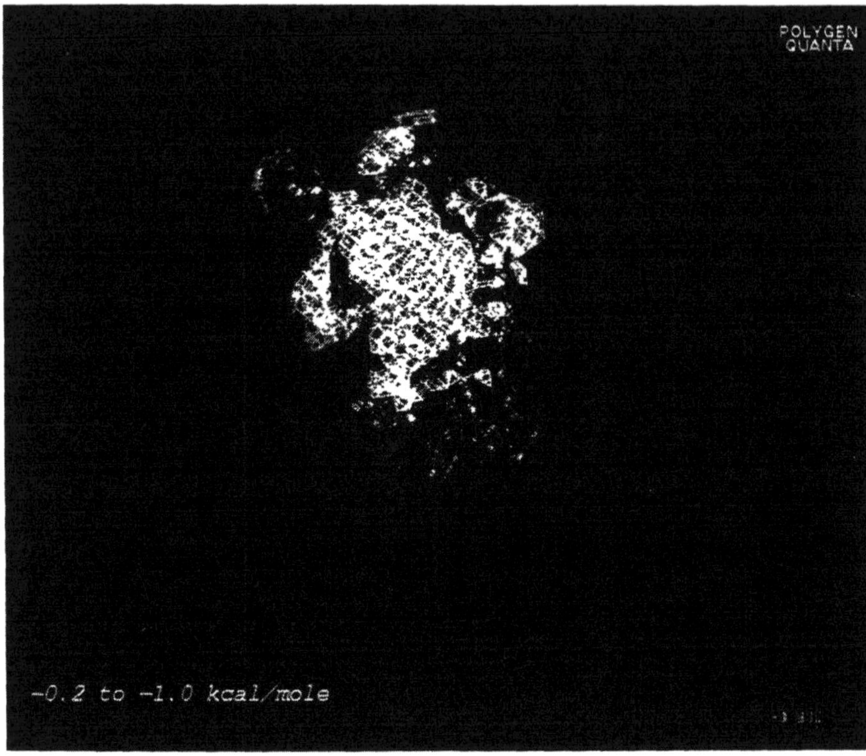

Fig. 15.20. Electrostatic potential of RNAse calculated by using UHBD program. Contours at + 1.0 and − 1.0 Kcal/mol are shown in blue and red respectively. The protein backbone is highlighted by a yellow ribbon. The view is looking-down on to the protein active site

protein recovery and high degree of fractionation [14]. A schematic representation of protein precipitation by polyelectrolytes, as proposed by Glatz [13], is shown in Fig. 15.21. The first stage is the formation of protein-polyelectrolyte complexes. These complexes then quickly aggregate to form primary particles, in the size range between 0.1 and 1.0 μm [14]. The final, and rate-determining stage is the aggregation of the primary particles into flocs.

Polyelectrolyte precipitation has been successfully used to recover some proteins as summarized in Table 15.1. Beside these examples, separation of nucleic acids and related proteins by PEI has been reviewed by Jendrisak [53]. Protein purification by using biospecific polyelectrolytes is not included in Table 1 since it has been discussed in Chapter 14.

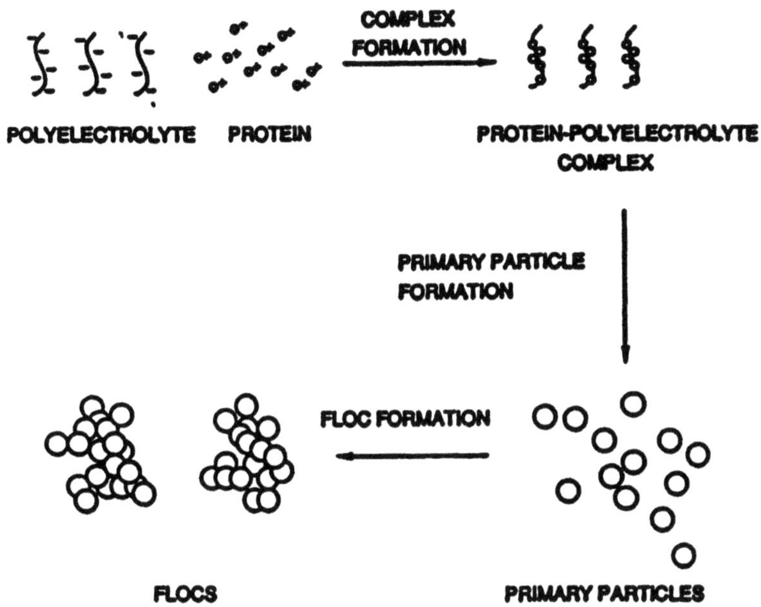

Fig. 15.21. Schematic diagram of floc formation in protein precipitation by polyelectrolytes. (from Ref. [13])

Table 15.1. Examples of Protein Separation by Polyelectrolytes

Proteins	Polyelectrolytes	Conditions	Best Recovery
BSA [10, 12]	PMAA, MAVP, PVAm, MAS, DEMMA	pH 3.95–8.70 I: H_2O, NaCl solution	~93%
Catalase and BSA [46]	PAA	pH ~ 5.2	~100%
Catalase-Ferrihemoglobin [46]	PAA	pH ~ 5.8	~92%
Lysozyme from egg-white [8, 52]	PAA	pH 5.45	~92.3
Lysozyme from egg-white [13]	CMC	pH 4.5, I = 0.07	~96%
RNA polymerase [53]	PEI	pH 7.0 I = 0.2 NaCl	~90%
Whey proteins [54]	CMC	pH 2.5–4.0	68%
Amyloglucosidase [55]	PAA	pH 3.2–4.0	95%
Lactase [55]	PAA	pH 3.2	100%
Microbial [55]	PAA	pH 3.2	85–96%
α amylase [55]	PAA	pH 4.0	93%
β amylase [55]	PAA	pH 4.0–4.5	77%
Lipoxygenase [55]	PAA	pH 4.8	76%
Calf Rennet [55]	PAA	pH 3.5	73%

15.5 Enzymes in Polyelectrolyte Complexes

The existence of soluble protein-polyelectrolyte complexes makes polyelectrolytes potentially important enzyme carriers, in which the enzyme activity could be controlled. For example, the tryptic digestion of hemoglobin with PAA as a carrier was studied by Morawetz [56] in 1955. The trypsin activity was varied by adjusting pH and polymer concentration. It was found that the carrier could even activate the enzyme at low pH, as shown by Fig. 15.22. As shown in Fig. 15.23, the complex of trypsin with KPVS was found to retain activity in a wide pH range by Kokufuta [57]. Based on the study of trypsin-carboxymethylcellulose complexes, Larionova et al. [58] concluded that some protein-polyelectrolyte complexes can provide extra thermal stability at the preservation of high biological activity.

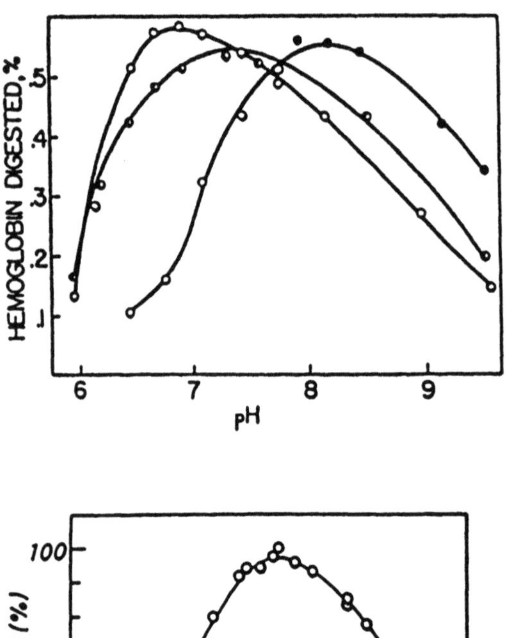

Fig. 15.22. Effect of PAA on tryptic digestion of hemoglobin as a function of pH. ● No PAA added; ◐ 0.25 g/L PAA; ○ 0.5 g/L PAA. (from Ref. [56])

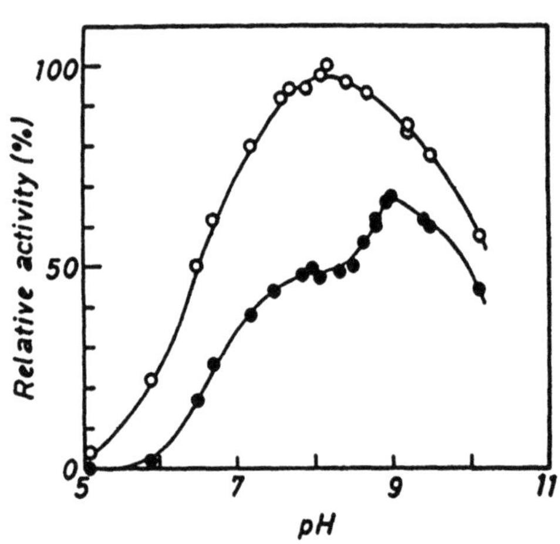

Fig. 15.23. pH activity curves of native (○) and complexed trypsin (●) determined at 30 °C by using a 2.66 mM BANA solution. Relative activity was calculated from the initial rate of BANA hydrolysis, and 100% activity refers to the hydrolysis rate (4.61×10^{-2} mM min^{-1}) by the native enzyme at pH 8.18. (from Ref. [57])

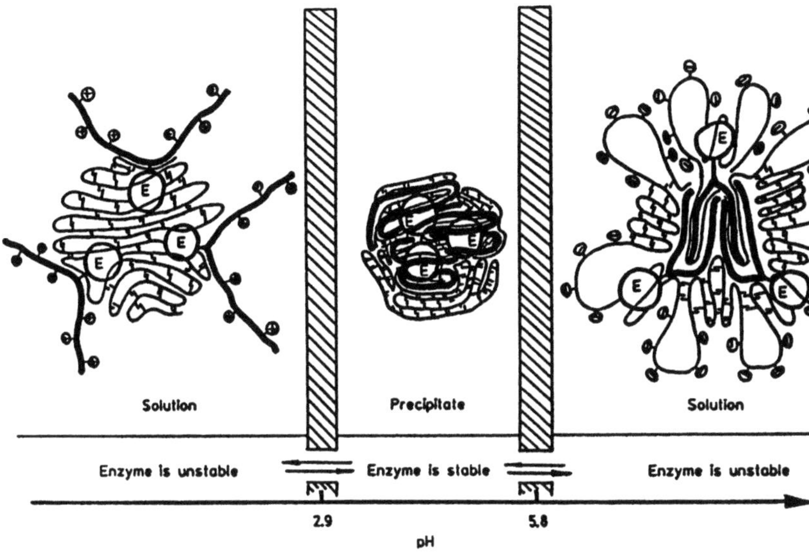

Fig. 15.24. Schematic representation of pH induced phase transitions in polyelectrolyte complex solutions. (from Ref. [18])

Because of the potential of Dextran polymers as blood substitutes, the oxygen affinity of hemoglobin coupled by the polymers are extensively studied. Reduced oxygen affinity of hemoglobin by Dextran polymers was reported by Nguyen [7] and Sacco [1]. Recently, Dubin et al. [44] found that the ligand-binding cooperativity of hemoglobin is affected by the complexation with PDMDAAC.

Kabanov [18] summarized two major reasons for a polyelectrolyte to affect a bound enzyme. The first is redistribution of charged particles ($H+$ ions, substrate molecules, products etc.) between the carrier and bulk solution, which results in (a) a shift or alteration of the pH dependencies of the kinetic and equilibrium parameters; (b) a change in the values of binding constants for substrates, products and effectors; and (c) the appearance of negative or positive cooperativity. These effects are observed at low ionic strength and disappear at a salt concentration above 0.5 M. Second, variation of pH and ionic strength can cause strong conformational changes in a polyelectrolyte carrier, which involve alteration of the size and shape of the particles or in their aggregation state. Such rearrangements and reversible phase transition of the gel$\rightleftharpoons$liquid type are inherent in enzyme stabilization, as shown in Fig. 15.24 where enzyme stabilization occurs only when precipitate is formed.

15.6 Conclusion

Because of the growing interest in the experimental and theoretical problems posed by complex systems, future progress in understanding the behavior of these systems seems assured. The practical importance of protein-polyelectrolyte complexes, as in purification via phase separation and enzyme immobilization, will also drive developments in this field. The diversity of experimental techniques used to study of protein-polyelectrolyte systems seems likely to continue to grow. In many cases methods currently applied to pure protein or pure polymer systems will need to be adapted. The elucidation of the molecular structure of protein-polyelectrolyte complexes will rest on this development of experimental methods, and also on the insights gained from computational molecular modeling.

15.7 References

1. Sacco D, Bonneaux F, Dellacherie E (1988) Int J Biol Macromol 10:305
2. Dellacherie E (1991) Am Chem Soc, Div Polym Chem Prepr 32(1):602 and references therein
3. Lenk T, Thies C (1987) In: Eisenberg A, Bailey FE (eds) Coulombic Interactions in Macromolecular Systems. American Chemical Society, Washington, DC, (1987) chap 8
4. Dubin P, Ross TD, Sharma I, Yegerlehner B (1987) In: Hinze WL, Armstrong DW (eds) Ordered media in chemical separations. American Chemical Society: Washington, DC, chap 8
5. Veis A (1991) Am Chem Soc, Div Polym Chem Prepr 32(1):596, and references therein
6. Burgess DJ, Carless JE (1984) J Colloid Interface Sci 98:1
7. Nguyen TQ (1986) Makromol Chem 187:2567
8. Sternberg M, Hershberger C (1974) Biochim Biophys Acta 342:195
9. Kokufuta E, Shimizu H, Nakamura I (1981) Macromolecules 14:1178
10. Morawetz H, Hughes WL (1952) J Phys Chem 56:64
11. Berdick M, Morawetz H (1954) J Biol Chem 206:959
12. Bozzano AG, Andrea G, Glatz CE (1991) J Membr Sci 55:181
13. Clark KM, Glatz CE (1987) Biotechnol Prog 3:241
14. Fisher RR, Glatz CE (1988) Biotechnol Bioeng 32:777
15. Shieh J, Glatz CE (1991) Am Chem Soc, Div Polym Chem Prepr 32(1):606
16. Strege MA, Dubin PL, West JS, Daniel Flinta CD (1990) In: Ladisch M, Willson RC, Painton CC, Builder SE (eds) Protein purification: from molecular mechanisms to large-scale processes. American Chemical Society: Washington, DC, chap 5
17. Burgess RR, Jendrisak JJ (1975) Biochemistry 14:4634
18. Margolin A, Sheratyuk SF, Izumrudov VA, Zezin AB, Kabanov VA (1985) Eur J Biochem 146:625
19. Park JM, Muhoberac BB, Dubin P, Xia J (1992) Macromolecules 25:290
20. Kokufuta E, Shimizu H, Nakamura I (1982) Macromolecules 15:1618
21. Dubin P, Ahmed L, Xia J (1994) J Macromol Sci A31:17
22. Kabanov VA, Evdakov VP, Mustafaev MI, Antipina AD (1977) Molekulyarnaya Biologiya 11:582
23. Kuramoto N, Sakamoto M, Komiyama J (1984) Makromol Chem 185:1419
24. Kabanov VA, Mustafaev MI, Belova VV, Evdakov VP (1978) Molekulyarnaya Biologiya 12:1264
25. Hummel JP, Dreyer WJ (1962) Biochim Biophys Acta 63:530
26. Samsonov GV, Ponomareva RB, Luchko RG (1969) Biofizika 14:634
27. Barberousse V, Sacco D, Dellacherie E (1986) J Chromatography 369:244
28. Pautov VD, Kuznetsova NP, Mishayeva RN, Anufriyeva YV (1983) Vysokomol soyed 25:1599
29. Cha HJ, Izumi T, Kokufuta E, Frank CW (1992) Am Chem Soc, Div Polym Chem Prepr 33:872
30. Cha HJ, Izumi T, Kokufuta E, Frank CW (1992)
31. Zimm BH (1948) J Chem Phys 16:1093
32. Dauzenberg H, Rother G (1991) J Appl Polym Sci Appl Polym Symp 48:351
33. Patkowski A, Bujalowski W, Chu B (1982) Biopolymers 21:1503
34. Xia J, Dubin P, Dauzenberg H (1993) Langmuir 9:2015

35. Pecora R (1976) Dynamic light scattering: Application of photon correlation spectroscopy. Plenum, New York
36. Schmitz KS (1990) An Introduction to Dynamic Light Scattering by Macromolecules. Academic Press, New York
37. Dubin PL, Murrell JL (1988) Macromolecules 21:2291
38. Xia J, Dubin P, Muhoberac BB, Kim YS, Klimkowski V (1993) 97:4528.
39. Ware BR, Haas DD (1983) Fast Methods in Physical Biochemistry and Cell Biology; Shaafi RI, Fernandez SM (eds) Elsevier, Amsterdam
40. McQuarrie DA (1976) Statistical Mechanics. Harper and Row, New York
41. Ware BR (1974) Adv Colloid Interface Sci 4:1
42. Ruckpaul L, Rein H, Janig GR, Pfeil W, Ristau O, Damaschun I, Damaschun H, Muller JJ, Purschel HV, Behlke V, Scheler W (1972) Studia Biophysica 34:81
43. Strelzowa SA, Tolstogusow WB (1977) Colloid & Polymer Sci 255:1054
44. Xia J, Dubin P, unpublished results
45. Hiemenz PC (1986) Principles of Colloid and Surface Chemistry. Marcel Dekker, New York
46. Berdick M, Morawetz H (1953) J Phys Chem 57:959
47. Davis ME, Madura JD, Luty BA, McCammon JA (1991) Comp Phys Comm 62:187
48. Davis RM, Russell WB (1986) J Polym Sci Polym Phys Ed 24:51
49. Zaitsev VS, Izumrudov VA, Zezin AB (1992) Polymer Science USSR 34:54
50. Zaitsev VS, Izumrudov VA, Zezin AB, Kabanov BA (1992) Doke Akad Nauk USSR 332:319
51. Anufrieva YV, Pautov VD, Kuznetsova NP, Lushchik VB, Krakovyak MG (1987) Polymer Science USSR 29:663
52. Chen W, Walker S, Berg JC, (1992) Chemical Engineering Science 47:1039
53. Jendrisak J (1987) Protein Purification Micro to Macro, Burgess, R (ed) Alan R, Liss, Inc. p. 75.
54. Hill RD, Zadow JG (1978) N Z J Diary Sci Technol 113:61
55. Sternberg M (1976) Process Biochemistry 11:11
56. Morawetz H, Sage H (1955) Arch Biochem Biophys 56:103
57. Kokufuta E, Takahashi K (1990) Polymer 31:1177
58. Larionova NI, Unksova LY, Mironov VA, Sakharov IYu, Kazanskaya NF, Berezin IV (1981) Vysokomol Soyed 23:1823

16 Precipitation of Proteins with Polyelectrolytes: Role of Polymer Molecular Weight

Jang-you Shieh and Charles E. Glatz

When used as precipitating agents for proteins, polyelectrolytes serve both to complex the protein into an insoluble form and to flocculate the resulting solid particles. The molecular weight has long been recognized as an important property of a polymer used for flocculation; however, the influence of molecular weight on the dual functionality of polyelectrolytes as precipitants for proteins is less clear. For a series of poly(acrylic acids) of nominal molecular weights from 5 to 4000 kDa, we have determined particle size distributions of protein-polyelectrolyte precipitates, protein recovered in the precipitate, precipitate stoichiometry, and the pH required for the onset of precipitation. Lysozyme, purified and mixed with ovalbumin or in egg white, served as the targeted protein. Molecular weight was found inconsequential to the charge density of the polymer and of minor importance to the stoichiometry of protein-polyelectrolyte complexation. On the other hand, higher molecular weights were important for obtaining high recoveries of proteins in the presence of other solutes and were more effective in the initial stages of complexation.

16.1 Introduction

Protein precipitation is an important operation for the laboratory and industrial scale recovery and purification of proteins. The use of polyelectrolytes as precipitating agents offers several advantages. Very low concentrations of polyelectrolytes are required and the fractionation potential appears to be promising. Although the cost of the polyelectrolyte may be high, required amounts are low and recycling of the polyelectrolyte may be possible. Fractional precipitation may be accomplished through careful selection of the polyelectrolyte and precipitation conditions [1]. Where such changes in conditions do not provide sufficient selectivity, fusions of sequences of charged amino acids to the protein may bestow sufficient charge character to allow selective removal [2, 3].

In the polyelectrolyte precipitation of protein, an important but less-studied polyelectrolyte parameter is molecular weight. In non-protein systems, the effect of polymer molecular weight on flocculation has been studied, but these results do not address the precipitation mechanism because, for flocculation, insoluble particles are already present. Observations directly from protein precipitation are that precipitation yields are higher with larger molecular weight polymers [4] and that the binding of poly(acrylic acid) with bovine serum albumin [5] and chymotrypsin [6] increases with PAA molecular weight.

Macromolecular Complexes
in Chemistry and Biology
Eds Dubin/Bock/Davis/Schulz/Thies

Formation of the solid phase begins with binding of protein and polyelectrolyte. Using turbidimetric titrations Dubin et al. [7] found no binding of proteins to polycations below the isoelectric point of the protein. Near the isoelectric point of the protein, primary intrapolyion complexes are formed with a number of protein molecules binding to a single polymer chain without substantial collapse or expansion of the polyion. The critical pH for bulk phase separation seems to be significantly larger than the pH at which intrapolymer complexes are first formed. Near the isoelectric point of the protein, each polyion chain binds protein molecules in a number which depends on polymer contour length, chain flexibility, and protein dimensions. The net charge of this complex, however, remains positive until the pH substantially exceeds the isoelectric point [8].

The purpose of this study was to investigate the effect of polyelectrolyte molecular weight on protein precipitation by using a series of poly(acrylic acids) (PAA) of *MW* 5–4000 kDa to precipitate egg white proteins (EWP) and lysozyme (one component of EWP). In the course of this work, we have also been able to make observations on the stoichiometry of complex formation and factors influencing the rate of precipitation. Determined were the solubilities of egg white and lysozyme in the presence of PAA, PAA left in solution, complex composition, and particle size distributions of the lysozyme-PAA precipitates. The early stage of precipitation was studied by turbidimetric titrations to examine the binding force between protein and different molecular weight PAA. The ionic strength, pH, protein/polyelectrolyte ratio, and charge density are also factors in the complexation.

16.2 Materials and Methods

16.2.1 Materials

PAA (Polysciences, Inc., Warrington, PA) of nominal weight average molecular weights of 5, 50, 500, and 4000 kDa were prepared as 1 mg/ml solutions in deionized water. Dried egg white solids (type P-110, Henningsen Foods, Inc., White Plains, NY) were dissolved by suspending 10 g in 100 ml of pH 5.4, 0.07 M ionic strength sodium acetate buffer, followed by centrifugation at 15 000 rpm (Sorvall R2C, Dupont Co., Wilmington, DE) for 20 minutes to remove insolubles. Lysozyme (from chicken egg white, Sigma Chemical Co., St. Louis, MO) was dissolved in pH 7.5, 0.07 M ionic strength phosphate buffer. For experiments at higher ionic strength, NaCl was added to the required level.

16.2.2 PAA Molecular Weight

The average molecular weights of the PAA were calculated from measurements of the intrinsic viscosity. A 10 ml solution of PAA dissolved in 2N NaOH was run through an Ubbelohde viscometer (Fisher Scientific, Itasca, IL) at 25 °C,

Table 16.1 Poly(acrylic acid) molecular weight (*MW*) characteristics. Included are the manufacturer's catalog values, reported polydispersity (ratio of weight to number-averaged *MW*) and those determined from the measured intrinsic viscosities

Nominal *MW*	Polydispersity	*MW* by viscosity
4 000 000	2–4	—
500 000	2–4	230 000
50 000	2–4	54 000
5000	2–4	3300

obtaining four replicates of the efflux time. Repeating the procedure for a series of polymer concentrations, we were able to determine the intrinsic viscosity from which the viscosity-averaged molecular weight was obtained using literature values of the Mark–Houwink parameters [9]. Characteristics of the polymers as supplied by manufacturer and as determined here are contrasted in Table 16.1. Clear solutions of PAA *MW* 4 000 000 were not obtained in 2 N NaOH and therefore no viscosity average molecular weight of PAA *MW* 4 000 000 was obtained. PAA concentrations were standardized by titration of PAA solution with 0.01 N NaOH in the presence of 0.1 M NaCl [10].

16.2.3 Protein Assays

Soluble protein content was determined by the biuret method [11] after centrifugation of the sample at 15 000 rpm for 20 minutes. Lysozyme activity was determined by following the rate of lysis of a suspension of *Micrococcus lysodeikticus* cells [12]. PAA remaining in solution was assayed for turbidimetricity by a method described below.

16.2.4 Precipitation

Precipitations were carried out at room temperature in a 150 ml beaker. Agitation was provided by a magnetic stirrer for 3 minutes and then by a shaker at 100 rpm for 1 hour. PAA was prepared as a 1 mg/ml solution in deionized water and pipetted into the beaker over a period of 1–3 seconds. The proteins were in the buffered solutions described above.

16.2.5 Particle Size Distribution

The precipitations for particle size distribution analyses were done by adding 20 ml PAA solution into a 400 ml baffled vessel containing 286 ml of lysozyme solution (1 $g\,l^{-1}$, pH 7.5, 0.07 M ionic strength phosphate buffer). Agitation was provided by a 45° pitched, 6-bladed turbine at 500 rpm for one hour. After a 1500-fold dilution of the precipitated slurry with the same prefiltered (0.2 μm capsule filter, Sartrobran, Hayward, CA) phosphate buffer, the particle size distributions were obtained using a Model TAII Coulter Counter (Coulter Electronics, Hialeah, FL) equipped with 50, 70, or 100 µm aperture tubes. The counter was calibrated with 4.06 µm latex particles (Polysciences, Inc.).

16.2.6 Turbidimetric Titrations

Turbidimetric titration curves were obtained by successively adding HCl (1 N at first and then 0.1 N as the critical pH was approached) dropwise to a 250 ml solution of protein (1 $g\,l^{-1}$ for the individual proteins; 3 $g\,l^{-1}$ for the EWP; added NaCl concentrations of 0, 0.07, and 0.2 M were used to examine ionic strength effects) and PAA (2.5 mg dissolved in 2.5 ml), initially at pH 11.5, until the critical pH, below which association began to take place, was reached. The transmittance at 436 nm (Bausch and Lomb Spectronic 710 spectrophotometer, Rochester, NY) and pH were recorded after each addition. The critical pH was taken as that pH where the turbidity began to rise. Near the critical pH readings did not stabilize, so readings were recorded each minute without further acid addition.

16.2.7 PAA Assay

To determine PAA remaining in solution after precipitation, an assay was developed based on the complete precipitation of PAA by an excess of lysozyme under proper conditions. PAA-containing sample (from 2–15 ml containing about 0.1 mg of PAA) was added to 5 ml of lysozyme solution (2 $g\,l^{-1}$) in an appropriate amount of phosphate buffer (pH 7.5, 0.07 M) to make the final volume 20 ml. The mixture was agitated by magnetic stirrer and after 3 minutes the OD_{436} was measured. PAA concentration was calculated from a standard curve prepared by addition of known amounts of PAA. Though lysozyme is nearly the only component of egg white to bind PAA, the minor influence of residual EWP was corrected for by including the same background level of such proteins in preparing the standard curve. Hence a different standard curve was prepared for each molecular weight and dosage of PAA.

16.3 Results and Discussion

16.3.1 PAA Assay

Figures 16.1a and 16.1b show the stability of the OD reading and a set of standard curves, respectively, for the turbidimetric assay for residual PAA.

16.3.2 Protein Removal

The results of the EWP precipitation are shown in Fig. 16.2. For *MW* 500 000 and *MW* 50 000 the maximum precipitation level occurred at PAA/total protein

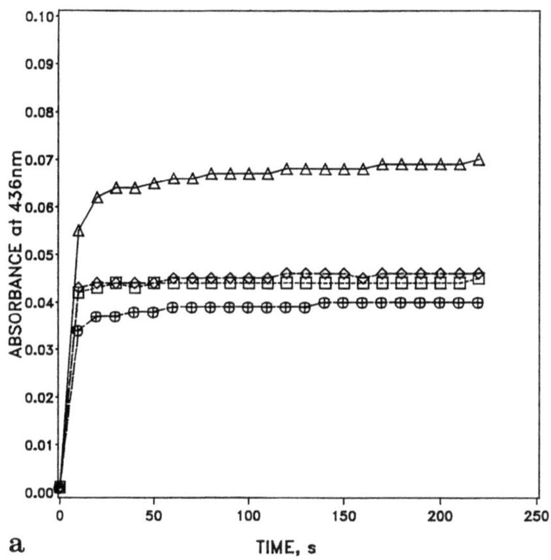

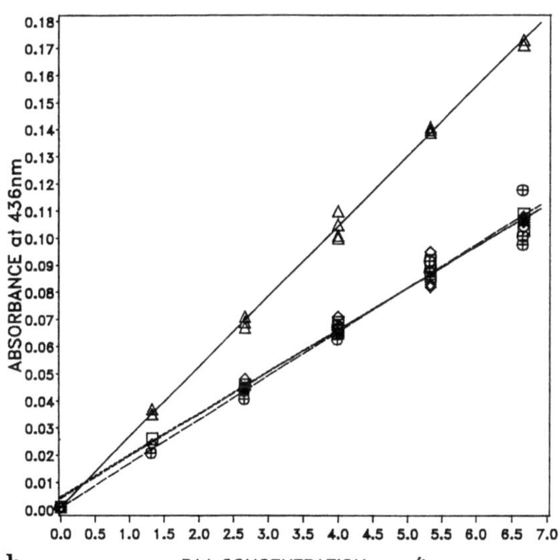

Fig. 16.1a, b. Poly(acrylic acid) assay based on turbidity as measured by OD_{436} resulting from PAA-lysozyme precipitation for PAA of various molecular weights for assay solution with $0.5\,g\,l^{-1}$ lysozyme in pH 7.5 phosphate buffer. △ PAA, 4000 kDa; □ PAA, 500 kDa; ◇ PAA, 50 kDa; ⊕ PAA, 5 kDa. **a** Stability of turbidity at a mass ratio of PAA to lysozyme of 0.00533. **b** Standard curves

mass ratio near 0.0125, and further addition of polymer resulted in lower protein removal. However, the decreases of protein removal with further addition of polymer were not obvious for *MW* 5000 and *MW* 4 000 000. In the latter case, this may be a result of centrifugation bringing down otherwise soluble complexes.

For EWP precipitation, larger molecular weight PAA had higher efficiency in protein removal which is consistent with the results of Sternberg and

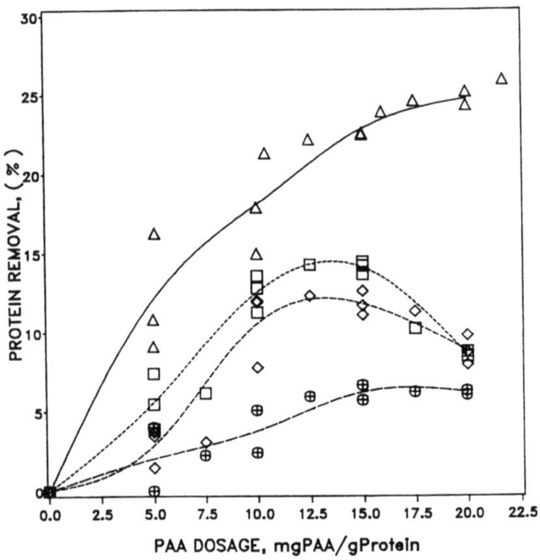

Fig. 16.2. Effect of polymer dosage and molecular weight on egg white precipitation at pH 5.4, 0.07 M NaCl. △ PAA, 4000 kDa; □ PAA, 500 kDa; ◇ PAA, 50 kDa; ⊕ PAA, 5 kDa

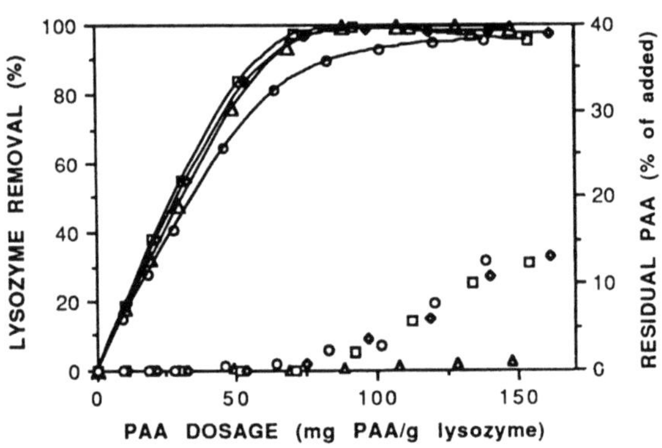

Fig. 16.3. Effect of polymer dosage and molecular weight on lysozyme precipitation (curves) and PAA remaining in solution (points only) at pH 7.5, 0.07 M NaCl. △ PAA, 4000 kDa; □ PAA, 500 kDa; ◇ PAA, 50 kDa; ○ PAA, 5 kDa

Hershberger [4] showing high molecular weights to be more effective precipitants. However, lysozyme removal was largely independent of PAA molecular weight (see Fig. 16.3). Maximum precipitation occurred at a PAA/lysozyme mass ratio near 0.09.

The favorable binding of lysozyme by all sizes of PAA is an indication that the high positive charge density of this protein is sufficient to ensure strong electrostatic binding. Polyelectrolyte size becomes a minor factor under such conditions. In the precipitation of the EWP at pH 5.4, most of the proteins are still negatively charged or have only small positive charge. Under these conditions, larger polyelectrolyte size brings about measurable improvement in precipitation of protein. The charge density of the polymer is not a factor in this

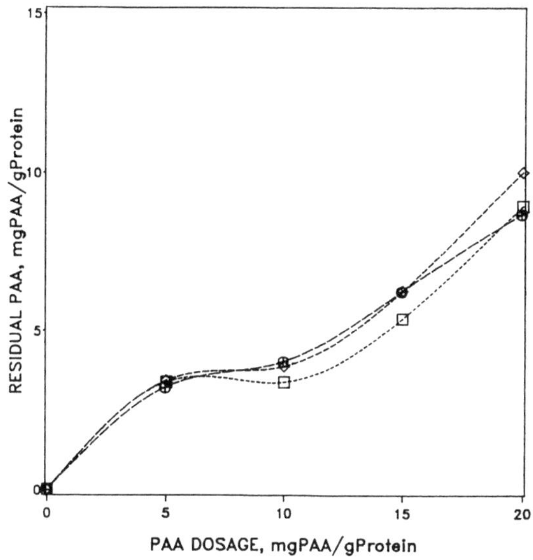

Fig. 16.4. Poly(acrylic acid) left in solution after precipitation of egg white proteins by PAA at pH 5.4, 0.07 M NaCl. △ PAA, 4000 kDa; □ PAA, 500 kDa; ◇ PAA, 50 kDa; ⊕ PAA, 5 kDa

comparison, since the titration curves of PAA were found to be independent of molecular weight, as has also been noted by Nagasawa et al. [13].

An additional indication of the difference in strength of association of polymer and protein for the two different protein solutions is seen in the comparison of Figs. 16.3 and 16.4. Figure 16.4 shows the appreciable levels of PAA left in the supernatant when precipitating from egg white. In contrast, PAA does not appear in measurable amounts (with the exception of *MW* 5000 which does begin to appear earlier) in the supernatant from lysozyme precipitations until lysozyme removal is essentially complete (Fig. 16.3). As will be seen later, weakening of electrostatic associations by increases in ionic strength also retards precipitation.

16.3.3 Stoichiometry

For lysozyme precipitation, the protein content of the precipitate calculated from a mass balance is shown in Fig. 16.5. Protein concentration decreases with increasing PAA dosage, indicating the incorporation of more polyelectrolyte into the complex as polymer dosage is increased.

With the following assumptions, the ratio of protein to PAA charge can be estimated.

1. The charge on lysozyme at pH 7.5 is 5.8 per lysozyme molecule [14].
2. The degree of dissociation of PAA is 0.8 at $I = 0.07$ M, pH 7.5 (from titration curve, not shown).
3. At a degree of dissociation of 0.8, counterion binding occupies 60% of the charged carboxyls [15].

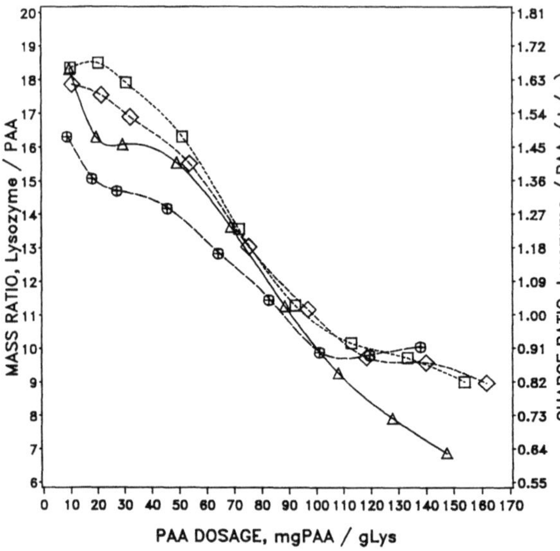

Fig. 16.5. Composition of lysozyme-PAA precipitates and calculated charge ratio as a function of polymer dosage. △ PAA, 4000 kDa; □ PAA, 500 kDa; ◇ PAA, 50 kDa; ⊕ PAA, 5 kDa

The results of the precipitate charge balances (Fig. 16.5) show that the net charge of precipitate is about zero (i.e. charge ratio of 1.0) at the polymer dosage where the maximum lysozyme removal is reached. The highest lysozyme/PAA charge ratio under these assumptions is 1.67. However, the precipitation of protein by polyelectrolyte should have some ion exchange behavior, i.e. some bound counterions would be displaced by lysozyme and lower the charge ratio. By assuming the precipitate to be electrostatically neutral, the extent of displacement of bound counterions was calculated. At the ratio of 1.67, 46% of counterions would be displaced to form a neutral complex. All the precipitates are negatively charged beyond the optimum dose even if no bound counterion is displaced.

The highest lysozyme/PAA mass ratio, ca. 18, corresponds to one molecule of lysozyme associating with about eleven carboxyl groups. A PAA segment of 11 repeat units has a length (28 μm) a little shorter than the diameter of lysozyme. These calculations suggest that, for the strong complexation between lysozyme and PAA, steric hindrance rather than charge considerations limited the binding of lysozyme to PAA when lysozyme was present in excess. Steric hindrance and counterion binding may also explain why the fraction of lysozyme precipitated was the same (the rates, however, were slower as the critical pH was approached) at pH 8.85 as at 7.5. In contrast, protein recovery from solutions of egg white solids showed a sharp maximum at pH 4.5.

16.3.4 Turbidimetric Titrations

Turbidimetric titration curves were obtained for lysozyme in the absence (Fig. 16.6) and presence (Fig. 16.7) of ovalbumin; the critical pH values are

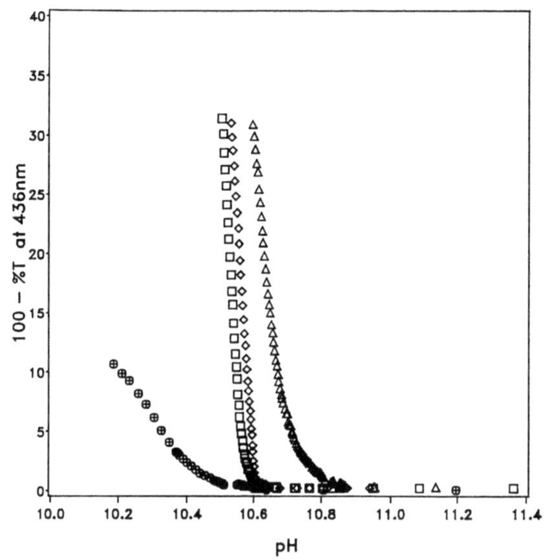

Fig. 16.6. Turbidimetric titrations for lysozyme ($1\ g\,l^{-1}$) with PAA/lysozyme mass ratio = 0.01 and 0.07 M added NaCl. △ PAA, 4000 kDa; □ PAA, 500 kDa; ◇ PAA, 50 kDa; ⊕ PAA, 5 kDa

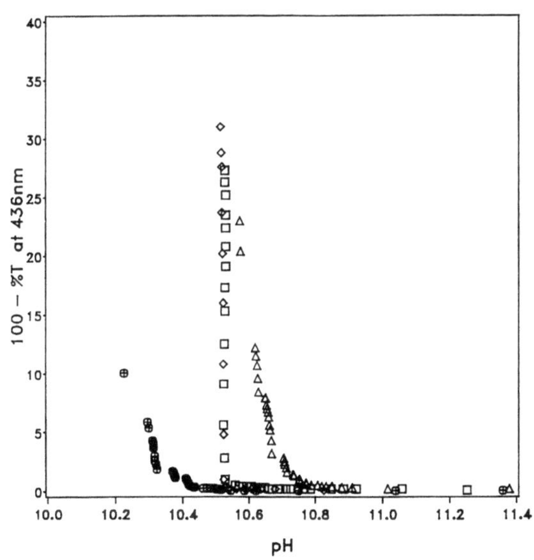

Fig. 16.7. Turbidimetric titrations for lysozyme ($1\ gl^{-1}$) in the presence of ovalbumin ($1\ gl^{-1}$) with PAA/lysozyme mass ratio = 0.01 and 0.07 M added NaCl. △ PAA, 4000 kDa; □ PAA, 500 kDa; ◇ PAA, 50 kDa; ⊕ PAA, 5 kDa

summarized in Table 16.2. Higher critical pH values again were obtained with the higher molecular weight PAA. This trend may be the result of stronger binding or it may stem from the higher local concentration of protein molecules bound to a single chain rather than the higher entropy situation of binding to the equivalent weight of smaller chains. In effect the entropic barrier to precipitation is overcome at the binding step.

Table 16.2 Critical pH values for lysozyme precipitation

PAA *MW*	Without ovalbumin	With ovalbumin
4 000 000	10.81	10.76
500 000	10.65	10.53
50 000	10.61	10.53
5000	10.51	10.42

The presence of ovalbumin only shifted the critical pH to a slightly lower value, less than 0.1 pH unit. Similar experiments performed by Dubin et al. [7] showed a critical pH for a binary mixture of proteins to fall between that of the individual components (BSA and ribonuclease (RNAse), whose individual critical pHs were separated by about one pH unit. In this work, the critical pH value of ovalbumin was 6 pH units less than that of lysozyme reflecting the similar difference in isoelectric points. Clearly the two proteins did not coprecipitate at an intermediate pH; such coprecipitation might be expected if the two proteins were interacting electrostatically. The same absence of coprecipitation from mixtures of ovalbumin and lysozyme was reported (1) with CMC as the precipitant. At pH 5.8 and 7.5, where only lysozyme is positively charged, only lysozyme was recovered in the precipitate. When at pH 4.2, where lysozyme has a net charge of 11 and ovalbumin of 6, only lysozyme is precipitated at low polymer dosage. Only at polymer dosages where virtually all the lysozyme is removed from the solution does ovalbumin begin to precipitate. In contrast to Dubin's system, then, the behavior of lysozyme is not greatly disturbed by the presence of a second protein.

For EWP, the turbidity did not rise sharply; hence the critical pH was not so obvious. The results of turbidimetric titrations for egg white at 0.07 M added salt are shown in Fig. 16.8. *MW* 4 000 000 PAA still gave the highest critical pH value and *MW* 5000 gave the lowest (*MW* 5000 was more clearly to the left at 0.2 M, not shown). Critical pH values for *MW* 500 000 and *MW* 50 000 were very close as was the case with lysozyme alone. Critical pH values should somehow reflect the efficiency of PAA in egg white precipitation. In fact, parallel behavior was seen in Fig. 16.2, where *MW* 4 000 000 PAA removed the most protein while the *MW* 5000 PAA removed the least with *MW* 5 000 000 and *MW* 50 000 PAA in between those two.

In the titration curves of lysozyme in the presence of ovalbumin, ovalbumin only weakly interfered with the complexation of lysozyme and PAA. However, in the egg white system, only 3.4% of the protein is lysozyme. Most of the proteins have net charges opposite to that of lysozyme below the isoelectric point of lysozyme, pH 10.7. The likelihood of interactions among the proteins shifting the critical pH is much greater; Fig. 16.8 shows this to occur for all but the highest *MW*. Why the critical pH of *MW* 4 000 000 PAA was not much affected (compare Figs. 16.6 and 16.8) is not understood. However, the absence of a shift is an indication of selective removal of the lysozyme; such behavior is desirable in trying to fractionate proteins by precipitation.

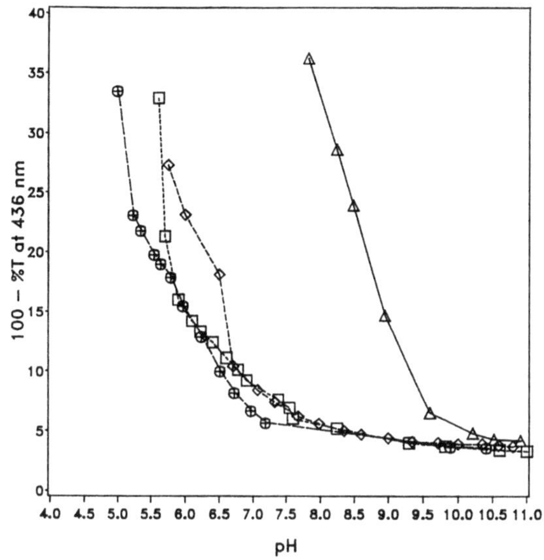

Fig. 16.8. Turbidimetric titrations for egg white proteins with 0.07 M NaCl, protein concentration $= 3\ gl^{-1}$, PAA/protein mass ratio $= 0.0133$. △ PAA, 4000 kDa; □ PAA, 500 kDa; ◇ PAA, 50 kDa; ⊕ PAA, 5 kDa

Turbidimetric titrations of EWP at 0.2 M added salt resulted in lower critical pH values. The decrease in critical pH with increased ionic strength was also seen with lysozyme alone where critical pHs of 11.4, 10.8, and 7.5 were obtained for *MW* 4 000 000 PAA for 0, 0.07, and 0.2 M added NaCl, respectively. The addition of simple electrolytes opposes the interaction between proteins and PAA by electrostatic screening. Consequently, a higher positive surface charge on the protein (corresponding to a lower pH) is required for complex formation at higher ionic strength.

16.3.5 Precipitate Particle Size

Relative to the modest molecular weight effects observed in forming the solid phase, the effects on solid aggregation were quite large. Mean particle sizes were 7.5, 3.5, 2.6, and 1.9 μm for precipitation of lysozyme with PAA of 4000, 500, 50, and 5 kDa nominal molecular weight, respectively. The much larger particles obtained with the largest PAA probably indicate an element of bridging flocculation, which requires high molecular weight, in contrast to the patch or charge neutralization models which simply depend on an overall charge balance [16]. Formation of polymer bridges might result in stronger flocs which are able to resist breakage.

16.4 Conclusions

The titration curves of polyacrylic acids are independent of molecular weight, that is, the degree of dissociation and hence charge density is independent of molecular weight.

For EWP precipitation, larger molecular weight has higher efficiency in protein removal. However, in lysozyme precipitation, protein removal is independent of molecular weight except for *MW* 5000 PAA. It is likely that highly charged protein, which has strong electrostatic interactions with PAA, is less affected by the molecular weight of polyelectrolyte.

Particle size was affected by polyelectrolyte molecular weight. Larger molecular weight PAA gave larger precipitates probably as the result of a bridging effect.

For turbidimetric titrations in the binary system, ovalbumin lowers the critical pH of lysozyme less than 0.1 pH unit perhaps as the result of weak interference with PAA/lysozyme complexation. For turbidimetric titrations of single proteins and mixtures, larger molecular weight PAA gave higher critical pH values. In contrast, increased ionic strength lowers the critical pH value as a result of electrostatic screening.

Acknowledgement. This work was supported by National Science Foundation Grant No. ECE-8514865.

16.5 References

1. Clark KC, Glatz CE (1990) In: Hamel J-FP, Hunter JB, Sikdar SK (eds) Downstream processing and bioseparation. American Chemical Society, Washington, DC, p 170 (ACS Symposium Series, vol 419)
2. Parker DE, Glatz CE, Ford CF, Gendel SM, Rougvie MA, Suominen I (1990) Biotechnol Bioengr 36:467
3. Zhao J, Ford CF, Glatz CE, Rougvie MA, Gendel SM (1990) J Biotechnol 14:273
4. Sternberg M, Hershberger D (1974) Biochim Biophys Acta 342:195
5. Sakamoto M, Kuramoto N, Komiyama J, Iijima T (1982) Int J Biol Macromol 4:207
6. Kuramoto N, Sakamoto M, Komiyama J, Iijima T (1984) Int J Biol Macromol 6:69
7. Dubin PL, Ross TD, Sharma I, Yegerlehner BE (1987) In: Hinze WL, Armstrong DW (eds) Ordered media in chemical separations. American Chemical Society, Washington, DC, p 162 (ACS Symposium Series, vol 342)
8. Dubin PL, Murrell JM (1988) Macromolecules 21:2291
9. Brandrup J, Immergut EH (1975) Polymer Handbook. John Wiley & Sons, Inc., New York
10. Newman S, Krigbaum WR, Laugier C, Flory PJ (1954) J Polymer Sci 14:451
11. Doumas BT (1975) Clin Chem 21:1159
12. Parry RM, Chandan RC, Shahani KM (1965) Proc Soc Exp Biol Med 119:384
13. Nagasawa M, Murase T, Kondo K (1965) J Phys Chem 69:4005
14. Beychok S, Warner RC (1959) J Am Chem Soc 81:1892
15. Huizenga JR, Grieger PF, Wall FT (1950) J Am Chem Soc 72:2636
16. Gregory J (1987) In: Tadros TF (ed) Solid/liquid dispersions. Academic Press, NY, p 163

17 Complex Coacervation: Microcapsule Formation

Diane J. Burgess

Simple and complex coacervation and the various theories of coacervation are reviewed. The Tainaka theory, which is the most broadly based of all the theories, was considered the most generally applicable. However this theory alone does not give a full explanation of the coacervation process.

The application of coacervation for the production of pharmaceutical microcapsules is discussed. There are a number of potential advantages in the microencapsulation of pharmaceutical and other products. Encapsulation by coacervation is an uncomplicated process which does not involve any elaborate manufacturing equipment. However this process usually involves the use of chemical crosslinking agents such as glutaraldehyde which have potential toxicity problems. In addition the application of these agents may be harmful to the encapsulated material. A complex coacervation process which is stable without the use of crosslinking agents or the application of heat is described. Complex coacervate microgels are formed between albumin and acacia under specific pH and ionic strength conditions. These microgel systems maintain stability under the following conditions: aging (six week study period); dilution with distilled water; change in pH (3.0 to 9.0); and change in ionic strength (1 to 80 mM).

17.1 Introduction and Terminology

Coacervation is the separation of a macromolecular solution into two immiscible liquid phases: a dense coacervate phase, which is relatively concentrated in the macromolecules, and a dilute equilibrium phase [1]. The word coacervate is derived from the Latin: *acervus*, meaning aggregate or heap, with the prefix, *co*, to signify the preceding union of the macromolecules. Coacervates may be described as liquid crystals and mesaphases. When only one macromolecule is present, this process is referred to as simple coacervation and when two or more macromolecules of opposite charge are present it is referred to as complex coacervation [1]. Simple coacervation is induced by a change in conditions which results in molecular dehydration of the macromolecules. This may be achieved by the addition of a non-solvent, the addition of microions or a temperature change. Complex coacervation is driven by electrostatic interaction between two or more macromolecules.

Coacervate systems are in dynamic equilibrium and alteration of the conditions may result in either the reformation of a one phase system or the formation of a flocculate or precipitate [1]. For example, the addition of excess non-solvent to an aqueous gelatin simple coacervate will result in precipitation. The phenomenon of coacervation is closely related to precipitation and flocculation.

Macromolecular Complexes
in Chemistry and Biology
Eds. Dubin/Bock/Davis/Schulz/Thies

In both simple and complex coacervation solvent is entrapped between the loops of the macromolecules as they associate. This is known as occlusion of solvent. Coacervates are more fluid compared to other systems with higher structural order, such as micelles. It is assumed that site specific interactions do not occur within the coacervate, therefore allowing freedom of movement of the individual molecules. Site specific interactions may lead to precipitation rather than coacervation if the charge densities of the molecules are sufficiently high.

17.2 Simple Coacervation

Simple coacervation may result on the addition of a dehydrating agent (coacervating agent) which promotes polymer-polymer interactions over polymer-solvent interactions. Bungenberg de Jong [1] has reported the induction of coacervation of aqueous gelation solutions on addition of sodium sulfate or ethanol. According to the Huggins theory of phase separation [2, 3], the interaction parameter (X) is increased on coacervation. X is a dimensionless quantity proportional to the difference in interaction energy between a macromolecular segment-solvent contact and an equivalent number of solvent-solvent and macromolecule-macromolecule contacts. This parameter must be above a certain critical minimum value in order for coacervation to take place. X may be increased by temperature reduction and therefore phase separation may result upon cooling [4]. It has been shown that the critical value of X (X_{cr}) which is necessary for coacervation to take place is dependent on the molecular weight of the polymer [4]:

$$X_{cr} = 1/2(1 + r^{-1/2})^2$$

where r is the ratio of molar volumes of solute to solvent, and therefore r is proportional to the molecular weight of the solute. Higher molecular weight species will give smaller values of X_{cr} and therefore coacervate more readily.

As X is increased intramolecular associations result between the loops of the macromolecules causing a reduction in the volume over which the macromolecules extend. This can be measured by a decrease in the viscosity of the system. As more dehydrating agent is added intermolecular associations also increase, resulting in the production of a coacervate. The coacervating agent tends to be excluded from the coacervate phase, thus allowing more freedom of movement of the polymers within the coacervate. This is possible as the associations between the molecules are dynamic. The two phase coacervate system is in equilibrium and is readily reversible.

Simple coacervation is dependent on the structure of the macromolecules. Khalil et al. [5] and Nixon et al. [6] have shown the importance of the gelatin molecules being in a compact coil structure for simple coacervation to take place. Gelatin molecules attain this structure at pH values close to their isoelectric pH (pI) values and at low ionic strength. The compact coil configuration results from the intermolecular attractive forces which occur at the pI value due to the charges on the molecules being balanced at this pH. When the pH is altered away from the pI value the charges on the molecules are no longer balanced and the resulting repulsive forces cause unfolding of the molecules and this prevents coacervation from taking place. Similarly the addition of salt causes a decrease in the intramolecular attractive forces due to screening of the charged functional groups and results in molecular unfolding.

Therefore simple coacervation will not take place at salt concentrations above a critical maximum salt concentration [6]. Simple coacervation is suppressed at high macromolecule concentrations. Under conditions where the macromolecule concentration approaches that normally found in the coacervate phase there is little energetic advantage in the formation of a coacervate phase.

17.3 Complex Coacervation

Complex coacervate formation is dependent on a number of factors, including: pH, ionic strength, macromolecular weight, concentration and mixing ratio [1, 7]. Charge is a most significant factor for complex coacervation, and unless the two polyions carry opposite charges coacervation will not take place. The amount of charge carried by each polyion is also critical. At very low charge densities coacervation is suppressed [8–10] and at very high charge densities precipitation and/or gelation may occur [1, 9, 11–14]. The pH and ionic strength of the medium affects the charges carried by the polyions and therefore affects coacervation. Coacervate yield is optimized for an equal ratio mixture of macromolecules, by weight, at the pH value where they carry equal and opposite charges [7]. This pH is known as the electrical equivalence pH (EEP). Figure 17.1 shows a typical effect of change in pH on coacervate yield for a gelatin/acacia complex coacervate system (all other parameters are held constant at optimum values) [1, 7]. The mixing ratio of the macromolecules affects the extent of coacervation, since the electrostatic interaction will be altered depending on the amount of each macromolecule available to interact. Figure 17.2 shows a typical effect of macromolecule mixing ratio on coacervate yield for two macromolecules which carry equal charges (all other parameters are held constant at optimum values) [1, 7]. The optimum concentration ratio for maximum coacervate yield will vary depending on the ratio of the charges [7].

The molecular weight of the macromolecules must fall within a critical range for complex coacervation to take place [15–21]. High molecular weight materials will form gels or precipitates and low molecular weight materials will interact by ion pairing rather than coacervation [11, 12]. Coacervation is enhanced at low temperatures due to increase in solvent-solvent, solvent-solute and solute-solute interactions [22–25]. Coacervation is suppressed at high macromolecule concentrations [1, 7]. This effect is also observed for simple coacervation [1]. In highly concentrated systems, the concentration of the mixture will approach that normally found in the coacervate and as a result the

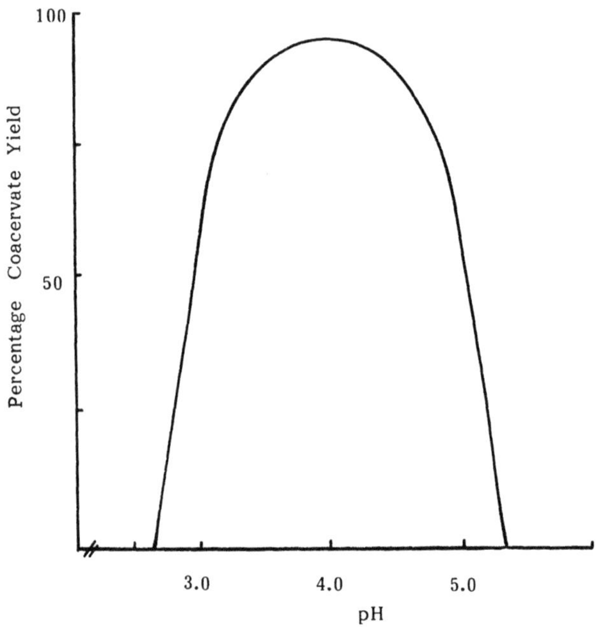

Fig. 17.1. The effect pH on coacervate yield for a typical gelatin/acacia coacervate system. (All other parameters held constant)

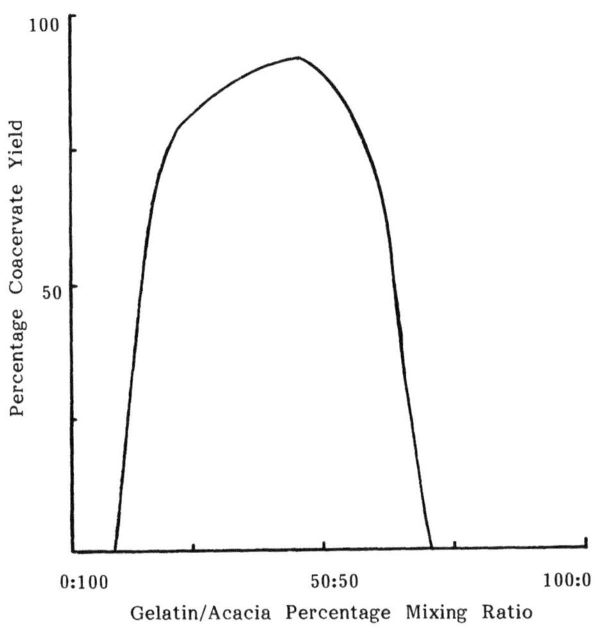

Fig. 17.2. The effect of macromolecule mixing ratio on coacervate yield for a typical gelatin/acacia complex coacervate system. (All other parameters held constant)

energy gain on formation of a coacervate is greatly reduced. In these concentrated mixtures the oppositely charged macromolecules will be in close proximity with one another and the molecular skeins may overlap. Thus there will be no advantage in forming a separate coacervate phase. Therefore, once a critical macromolecule concentration is reached coacervation will be completely suppressed. This phenomenon is termed 'self' suppression [1].

Suppression of coacervation at high salt concentration, known as 'salt' suppression [1], occurs as a result of the microions forming a dense atmosphere around the macromolecules and thus preventing electrostatic interaction. Coacervation may also be suppressed at low salt concentrations [8–10]. This effect is considered to be a result of the higher charge carried by the macromolecules under these conditions. Highly charged molecules will be in the extended molecular conformation which does not favor coacervation.

17.4 Theory of Complex Coacervation

There are four different theories of complex coacervation. The oldest is the Voorn–Overbeek theory [15–21], which was developed using the data of Bungenberg de Jong for gelatin/acacia complex coacervation [1]. The Voorn and Overbeek theory is based on the following assumptions: that there is a random chain distribution of macromolecules in both phases; that solvent-solute interactions are negligible; and that the interactive forces are distributive in nature with the system behaving as though the charges are free to move. If site specific interactions were to take place then precipitation rather than coacervation would be expected to occur. The first assumption implies that coacervation is limited to systems of relatively low charge density unless the ionic strength is high, as otherwise the macromolecules would be in the unfolded state rather than the random coil configuration. If the macromolecules are present in the extended state it would be difficult to entrap water within the molecular skeins and coacervation would be unlikely to occur. Voorn and Overbeek [15–21] interpreted coacervation as a competition between the electrostatic forces which tend to accumulate the charged macromolecules and entropy effects which tend to disperse them. They also regarded the solvent content of the coacervate in terms of its entropy contribution, as this allows a number of possible arrangements of the macromolecules. The electrostatic forces are assumed to be distributive in nature to account for the fluidity of the coacervate. According to the Voorn and Overbeek theory [15–21] the factors which affect complex coacervation are the charge densities and molecular weights of the macromolecules, the temperature, the dielectric constant and the Huggins interaction parameter (which was assumed to be negligible). Coacervation is considered to be a spontaneous process driven by electrostatic interaction.

Voorn and Overbeek [15–21] showed quantitatively that the critical conditions for coacervation were met for a two component system consisting of a polyion salt and water when $\sigma^3 r \geq 0.53$. That is when either the molecular charge density (σ) or the ratio of the molar volumes of solute to solvent (r), [r, is proportional to the polyion molecular weight] or both of these factors are sufficiently large. However complex coacervates have been formed when the above conditions are not met. It is possible to manipulate other factors, such as temperature to achieve coacervation [8, 22]. The assumption that the Huggins interaction parameter is negligible is therefore incorrect. Veis and Aranyi [22–25] developed a theory to explain a practical case of coacervation between two oppositely charged gelatins, where the Voorn–Overbeek theory did not apply and coacervates formed when the critical condition $\sigma^3 r \geq 0.53$ was not met. Coacervation occurred in this gelatin system on temperature reduction. Veis [23] modified the Voorn–Overbeek theory to include the Huggins interaction parameter, which increases significantly on temperature reduction. The Veis–Aranyi theory is limited to systems of low charge density [22]. Veis [23] considered that the nature of the electrostatic interaction was different than that calculated by Voorn and Overbeek from the Debye–Hückel theory [26, 27] since light scattering data indicated that aggregates were present in the equilibrium phase. These aggregates were thought to be ion-paired [22, 25]. Veis replaced the Debye–Hückel derived electrostatic term with a term which is a function of concentration and charge density and represents the electrostatic free energy change on the transfer of a polymer segment from the dilute to the concentrated phase [23].

The three basic assumptions of the Voorn–Overbeek theory do not hold for the Veis–Aranyi model. The molecules are not randomly distributed in both phases nor are the charges, since ion-paired aggregates are present in the dilute phase. The Huggins interaction parameter is not negligible, in fact it is the driving force for coacervation, since coacervation is induced on temperature reduction. Veis and Aranyi [22] proposed that complex coacervation of oppositely charged gelatins occurred in two steps. First the gelatins spontaneously aggregate by electrostatic interaction to form neutral aggregates of low configurational entropy, and then these aggregates coacervate to give phase separation. Coacervation is driven by the gain in configurational entropy resulting from the formation of a randomly mixed coacervate phase. In contrast, the Voorn–Overbeek theory [15–21] states that coacervation is a spontaneous process driven by electrostatic interaction. The differences between the Voorn–Overbeek and the Veis–Aranyi theories arise since each is attempting to explain very specific coacervation conditions. The Voorn–Overbeek theory was developed based on data obtained for the spontaneous formation of coacervation in the gelatin/acacia coacervate system, and the Veis–Aranyi theory was developed to account for coacervation between oppositely charged gelatins.

Two other complex coacervation theories have been developed: by Nakajima and Sato [11, 12], which is an adaptation of the Voorn–Overbeek theory, and by Tainaka [13, 14], which is an adaptation of the Veis–Aranyi theory. Nakajima and Sato agreed with the Voorn–Overbeek theory in that the charges should be treated as distributed uniformly in both phases, but included the Huggins interaction parameter and they altered the electrostatic term. The

Tainaka theory differs from that of Veis and Aranyi in that aggregates are considered to be present in both phases, which possibly neutral, are not formed by specific ion pairing. According to Tainaka, phase separation is driven by attractive forces among aggregates, which are stronger the higher the molecular weights and the greater the charge densities of the macromolecules. Unlike the Veis–Aranyi theory the Tainaka theory is not restricted to low charge density, however both the charge density and polyion molecular weight should fall within a critical range. Where the charge density or molecular weight values are higher than this range, a concentrated gel or precipitate will form since strong long range attractive forces will exist among the aggregates. At macromolecule charge densities and molecular weights below this range the dilute solution will be stabilized by short range repulsive forces and coacervation will not occur. The Tainaka theory is more general than the other theories and is applicable to both high and low charge density systems. However, it does not provide a full explanation of the coacervation process. Notably it does not explain suppression of coacervation at low ionic strength, which has been reported by Burgess and coworkers [8–10]. This theory does provide an adequate explanation of the complex coacervation process for a large number of systems.

17.5 Coacervation as a Method of Microencapsulation

Coacervation is a common method of microencapsulation [1, 8, 28–30]. The potential of encapsulation by coacervates was first noted by Bungenberg de Jong and Kaas [31] who observed that organic liquids suspended in the equilibrium phase of a coacervate system were taken up by the coacervate droplets. Bungenberg de Jong et al. [32] showed that solid particles could also be entrapped in coacervate systems. On phase separation by simple or complex methods tiny coacervate droplets are formed which coalesce and sediment to form a separate coacervate phase. If a core material, such as drug particles, is present in a polyion system prior to coacervation then the coacervate will deposit on and coalesce around these core particles (Fig. 17.3). Agitation of the coacervate system by stirring or other means can prevent coalescence and sedimentation of the coacervate droplets. The coacervate droplets can be crosslinked to form stable microcapsules by addition of a crosslinking agent, such as glutaraldehyde, or the use of heat [8, 28–30]. The first commercial application of this encapsulation technique was by the National Cash Register Company for the manufacture of carbonless copying paper [33]. The potential use of microencapsulation in the pharmaceutical industry has been considered since the 1960s [5, 34, 35]. Drug microencapsulation by coacervation has been reviewed by Madan [29] and by Nixon [30].

In the pharmaceutical industry, microencapsulation has many advantages, such as: to mask odor and taste; to convert oils into solid entities; to protect drugs against moisture, light, heat or oxidation; to delay volatilization; to separate incompatible materials during compounding of a medicine; to improve

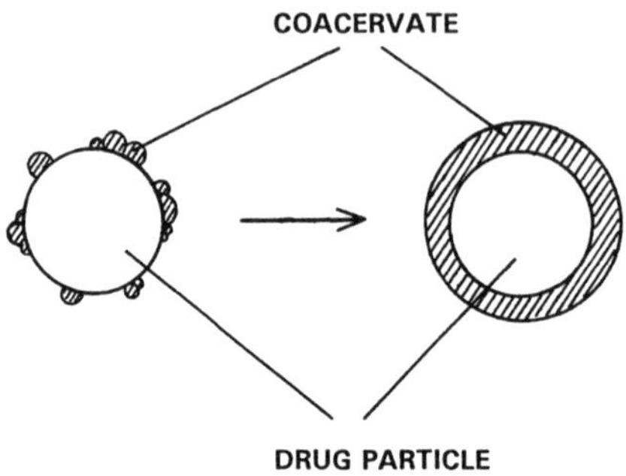

Fig. 17.3. Coacervate formation around a drug particle

the flow ability of powders; to handle toxic materials; to aid in the dispersion of water insoluble substances in aqueous media; and to produce sustained or controlled release medications [30, 36–38]. Microencapsulation has also been used medically for the encapsulation of live cells and vaccines [37, 39, 40]. The immobilization of mammalian cells within microcapsules has use in organ transplantation therapy [41]. The immobilized cells should be protected from the hosts' immune system and therefore should survive to deliver their product (such as insulin) in a bioresponsive manner for extended periods of time. In addition, the entrapped cells may be derived from a species other than the host. For example, bovine pancreatic islets could be used in the treatment of human diabetes.

Complex coacervate microencapsulation does not require any elaborate manufacturing equipment. The large number of variables which affect complex coacervation (pH, ionic strength, macromolecule concentration, macromolecule ratio and macromolecular weight [1, 7] affect microcapsule production, with the result that there are a large number of controllable parameters in this process. These parameters can be manipulated to produce microcapsules with specific properties. Complex coacervate systems which have been investigated as potential microencapsulation methods include: gelatin/acacia [42]; carbopol-gelatin [43]; pectin-gelatin [44]; gelatin/gelatin [8]; and sodium carboxymethylcellulose-gelatin [45]. Complex coacervate microcapsules have been formulated as suspensions or gels [46], and have been compounded within suppositories [47, 48] and tablets [49, 50]. Tablets containing microcapsules have been utilized to obtain a sustained release action. For example, sustained release sodium phenobarbitone tablets have been prepared containing simple coacervate, ethylcellulose, sodium phenobarbitone microcapsules [50]. Sustained release suppositories containing microencapsulated indomethacin have been successfully formulated [48]. These indomethacin suppositories showed a zero order release profile which did not decrease for 9 hours and approximately 100% of the indomethacine was released. The microcapsules were prepared by simple coacervation using ethylcellulose. Sustained release indomethacin suppositories have the following advantages: reduction of gastrointestinal irritation, avoidance of both disagreeable taste and first-pass effects; and reduction in the frequency of drug administration.

Hemoglobin-containing coacervate vesicles have been prepared by complex coacervation of gelatin with: acacia, pectin and dextran sulfate [51]. The hemoglobin is anchored into the vesicle walls by interaction of its polyanion binding site with the carbohydrates. Oxygen binding of the immobilized HbA is reversible and cooperative. Kinetic studies on CO_2 binding have shown that CO_2 uptake is similar to that of human erythrocytes. These microcapsules slowly dissolve in isotonic salt solutions unless they are stabilized by crosslinking with glutaraldehyde [51]. Hemoglobin incorporated into glutaraldehyde stabilized coacervates has functional properties similar to those of intra-erythrocytic hemoglobin [51].

Although many successful coacervate microencapsulation systems have been prepared coacervate microcapsules have a number of limitations: they can be produced only at specific pH values; they require stabilization by use of crosslinking agents or heat; and the extent of crosslinking determines the retention of the encapsulant. The limitation of pH can be overcome on addition of water soluble nonionic polymers, such as polyethylene glycol [52, 53]. The addition of a small amount of these polymers, allows microencapsulation to occur over an expanded pH range (for example, the pH range for coacervation of gelatin and acacia can be extended from pH 2.6 to 5.5 [7] to pH 2 to 9 [52]. In addition these polymers induce simple coacervation [52], and this has been shown for macromolecules such as: gelatin, caboxymethylcellulose and ethylene-maleic anhydride copolymer. The pH range for simple coacervation is also expanded when these water soluble nonionic polymers are used. For example, the pH range for simple coacervation of gelatin was increased from the vicinity of the isoelectric point to the pH range 5.5 to 9.5 [52].

Glutaraldehyde and formaldehyde are commonly used as crosslinking agents for protein/polysaccharide complex coacervate systems such as gelatin and acacia. A condensation reaction occurs between the amino groups of the protein and the aldehydes. However there are potential toxicity problems when these chemicals are used in phramaceutical products. In addition, both the use of chemical crosslinking agents and the application of heat may be harmful to the encapsulant materials, such as thermo- and chemically labile drugs and live cells. Examples of drugs which may not be able to withstand these processing conditions are the many genetically engineered protein and polypeptide drugs.

Crosslinking of coacervates is necessary to stabilize the coacervate emulsion droplets and hence form microcapsules. A stable coacervate system formed without the use of chemical crosslinking agents or the application of heat would be potentially useful as a delivery system for protein and polypeptide drugs and other materials unable to withstand these crosslinking procedures. The relative stability of coacervate emulsion systems, as other emulsion systems, is dependent on: dispersed phase droplet size and viscosity, interfacial tension, interfacial charge and interfacial rigidity. Small droplet size, high viscosity, low interfacial tension, high interfacial charge and rigid interfacial films all contribute to more stable emulsion systems. The interfacial tension of aqueous coacervate systems is usually low and therefore this parameter cannot be effectively manipulated to enhance stability. Values of the order of 1 mNm^{-1} have been reported for gelatin/acacia complex coacervate systems [54]. Complex coacervates have a net zero charge since the charges carried by the macromolecules neutralize one another. Alteration of pH or ionic strength to change the charge carried by the coacervates will only affect the extent of coacervation and will not affect the net charge carried by the coacervates. The parameters which can be manipulated to affect coacervate stability are particle size, viscosity and interfacial rigidity.

Viscous coacervate systems include albumin and alginic acid [9] and albumin and acacia [55]. The albumin/alginic acid system tends to form complex precipitates in addition to complex coacervates except under very restricted conditions of pH, ionic strength and concentration and this limits the usefulness of this system. The albumin/acacia coacervate system also varies significantly with the conditions, particularly pH and ionic strength. pH and ionic strength control the charge carried by the macromolecules and therefore the electrostatic interactive forces. Charge also determines macromolecular shape, which affects the type of interaction and the amount of solvent entrapped within the associated macromolecules. The appearance of the albumin/acacia coacervate system changes markedly as the pH and ionic strength conditions are altered [55]. Interaction was observed over the pH range 2.0 to 6.0 and ionic strength range 0 to 100 mM [55]. The optimum conditions for maximum coacervate yield were pH 3.9 and ionic strength 10 mM. Under these conditions the system rapidly separates into a viscous coacervate phase and a dilute equilibrium phase. Close to the optimum conditions for maximum coacervate yield the system has a cloudy appearance and as the conditions are altered away from the optimum conditions the system becomes less cloudy and eventually clear.

The viscosity of the albumin/acacia system under the optimum conditions for maximum coacervate yield was too high to prepare microcapsules, the coacervate phase could not be emulsified into the equilibrium phase [55]. However it is considered that the highly dispersed coacervate systems which form at pH values close to the optimum conditions may be useful for microencapsulation. The coacervate yields obtained for these systems are high (between 80 and 89% w/v, [55]) and these highly dispersed systems appear to be relatively stable in the dispersed state compared to other coacervate systems. It was therefore decided to investigate this coacervate system as a potential microencapsulation system.

17.6 Materials and Methods

Bovine serum albumin (mol. wt. 6.7×10^4 [osmotic pressure], isoelectric pH 5.6), acacia (mol. wt. 2.4×10^5 [56]), Amberlite IR-120P (cation-exchanger), Amberlite IR-400 (anion-exchanger) and glutaraldehyde were obtained from Sigma Chemicals, USA. Gelatin B (Bloom No. 250, mol. wt. 4.6×10^4 [osmotic pressure], isoelectric pH 4.8), was obtained from Gelatin Products Ltd, U.K.

The isoelectric pH values of albumin and gelatin were measured by microelectrophoresis and by ion exchange [10]. Hydrochloric acid, sodium hydroxide, sodium chloride and other chemicals used were of analytical grade and obtained from Fisher Scientific, USA. The polyion solutions were prepared by dispersion in distilled water at $40 \pm 0.1\,°C$. The polyions were allowed to hydrate completely; this took 30 min to 1 h. Following hydration the solutions were deionized by mixing for 30 min at $40 \pm 0.1\,°C$ with Amberlite resins IR-120P and IR-400 before use. This method is an adaptation of that of Janus et al. [57].

17.6.1 Albumin/Acacia Complex Coacervation and Microcapsule Production

Albumin/acacia coacervates were prepared by mixing equal volumes of deionized polyion solutions with constant stirring (300 rpm) at $40 \pm 0.1\,°C$ for 1 h, at the appropriate pH and ionic strength conditions. The pH and ionic strength of the mixtures were adjusted using HCl, NaCl and NaOH solutions. The concentration of the polyion solutions was adjusted to give a final total polyion concentration of 2% w/v. The appearance of the caocervates was then examined visually.

Microcapsules were prepared by mixing 100 ml of each polyion solution prepared as above. The pH was adjusted to either pH 3.8 or 4.2, maintaining constant ionic strength at 10 mM using HCl, NaCl and NaOH solutions. The mixtures were treated in one of the following ways: (i) left to stand, no further treatment; (ii) stirred at 450, 600 or 900 rpm for 1 h using a Dyna mixer (Fisher Scientific); (iii) stirred at 450 rpm for 10 min using a Dyna mixer, followed by sonication for 10 min using a Sonifier, Branson Sonic Co., at a power setting of 5. Each batch was separated into two portions; one portion was left to stand, no further treatment; and 5 ml of crosslinking agent (16% glutaraldehyde solution) was added to the other portion, and mixing was continued for a further 5 min. When necessary, commassie blue dye was added to the acacia solution prior to mixing. The percentage of commassie brilliant blue present in the coacervate and equilibrium phases was determined by visible spectrophotometry at 550 nm.

17.6.2 Gelatin/Acacia Complex Coacervate Production

Gelatin/acacia complex coacervates were produced as the albumin/acacia complex coacervates described above, with the exception that the pH was adjusted to pH 3.6 (the optimum pH for maximum coacervate yield [7]) maintaining constant ionic strength at 10 mM using NaCl, NaOH and HCl solutions.

17.6.3 Particle Size Stability Study

The microcapsule samples prepared were stored for periods up to 42 days in constant temperature water baths at 5, 25 and $40 \pm 0.1\,°C$. The particle size was measured on days 1, 14 and 42. The mean diameter (dsn) and the size distribution of the microcapsules were determined using a HIAC/Royco particle sizer (Model 4100). During the HIAC/Royco particle size analysis the microcapsules were suspended in 0.9% w/v sodium chloride solution. Microcapsule samples were examined by optical and scanning electron microscopy. The stability of the microcapsules to dilution and change in pH and ionic strength was studied by altering the pH and ionic strength of the final microcapsule suspension to values within the pH range 3.0 to 9.0 and the ionic strength range 1 to 80 mM. Particle size analysis of the microcapsule suspensions was conducted as above using the HIAC/Royco particle sizer and the samples were examined microscopically.

17.7 Results

As reported previously albumin/acacia coacervate microcapsules could not be prepared at the optimum conditions for maximum coacervate yield, pH 3.9 and ionic strength 10 mM, due to the high viscosity of the coacervate phase [55]. Emulsification was attempted using high speed stirring and ultrasonic energy via an ultrasonic probe, however none of these methods was successful. At pH and ionic strength conditions close to the optimum coacervate yield conditions highly dispersed systems formed and these were investigated as potential microencapsulation systems. Two systems were selected for the preparation of coacervates: pH 3.8, ionic strength 10 mM; and pH 4.2, ionic strength 10 mM. Coacervates were prepared at different manufacturing conditions and particle size analysis was conducted to investigate the stability of these coacervates over a six week period. Each batch was divided in two and one portion was crosslinked with glutaraldehyde and the other portion was left untreated. The crosslinked sample was used as a control to determine any change in the particle size of the non-crosslinked coacervates with time. The batches were prepared under a range of different shear conditions, to determine the effect of shear rate on the particle diameter. Ultrasonication and different stirring speeds were utilized and some batches were left unstirred. The various batches were stored at 25 °C.

The coacervates were analyzed using a HIAC/Royco particle sizer and an optical microscope. All coacervate samples examined by microscopy were spherical and the mean particle diameters did not change over a six hour observation period.

Coalescence of coacervate droplets was not observed. For comparative purposes an identical study was conducted using gelatin/acacia coacervates which are known to coalesce readily. Accurate determination of the mean coacervate diameter of this system could not be made as coalescence occurred very rapidly. Over a period of 10 minutes all the gelatin/acacia coacervate droplets in the field of view had coalesced into one.

The mean particle diameters of different microcapsule batches are given in Tables 17.1 and 17.2 for coacervates prepared at pH 3.8 and 4.2, respectively. The mean particle diameters did not vary significantly. There was no apparent correlation between stirring rate and mean particle diameter. An increase in applied shearing force usually reduces the particle size of emulsion systems, thus smaller coacervate droplets would be expected with higher stirring speeds. The particle size data for the non-crosslinked and crosslinked coacervates were not significantly different, indicating that the non-crosslinked coacervate droplets are relatively stable.

Table 17.1. The effects of stirring rate and aging on the mean particle diameter of albumin/acacia non-crosslinked and crosslinked complex coacervates prepared at pH 3.8

Stirring speed (rpm)	Mean particle diameter ($dsn \pm \delta g$ in µm)		
	Day 1	Day 14	Day 42
NS	5.6 ± 1.4	5.6 ± 1.4	5.9 ± 1.3
450	5.7 ± 1.4	5.6 ± 1.4	5.5 ± 1.4
600	5.3 ± 1.5	5.5 ± 1.4	5.5 ± 1.4
900	5.4 ± 1.5	5.5 ± 1.3	5.6 ± 1.4
sonicated	5.5 ± 1.3	5.9 ± 1.2	5.9 ± 1.2
NS*	5.3 ± 1.5	5.0 ± 1.3	5.0 ± 1.4
450*	5.7 ± 1.4	5.6 ± 1.4	5.5 ± 1.4
600*	5.1 ± 1.4	5.3 ± 1.5	5.3 ± 1.4
900*	5.4 ± 1.5	5.4 ± 1.5	5.6 ± 1.4
sonicated*	5.5 ± 1.3	5.7 ± 1.2	5.7 ± 1.3

* – crosslinked coacervates.
NS – Not stirred

Table 17.2. The effects of stirring rate and aging on the mean particle diameter of albumin/acacia non-crosslinked and crosslinked complex coacervates prepared at pH 4.2

Stirring speed (rpm)	Mean particle diameter ($dsn \pm \delta g$ in µm)		
	Day 1	Day 14	Day 42
NS	5.7 ± 1.2	6.0 ± 1.7	6.2 ± 1.5
450	5.7 ± 1.4	5.9 ± 1.6	5.8 ± 1.4
600	5.9 ± 1.5	6.1 ± 1.4	6.1 ± 1.4
900	5.1 ± 1.3	5.1 ± 1.3	5.2 ± 1.4
sonicated	5.5 ± 1.4	5.7 ± 1.4	5.5 ± 1.4
NS*	5.9 ± 1.6	6.1 ± 1.4	6.2 ± 1.5
450*	5.8 ± 1.4	6.9 ± 1.5	6.4 ± 1.6
600*	5.6 ± 1.5	5.9 ± 1.3	5.8 ± 1.4
900*	5.2 ± 1.5	5.5 ± 1.4	5.6 ± 1.6
sonicated*	5.7 ± 1.4	5.8 ± 1.5	6.2 ± 1.6

* – crosslinked coacervates.
NS – Not stirred

The mean particle diameters did not change significantly on aging over a six week period (Tables 17.1 and 17.2). This data suggests that the non-crosslinked systems are extremely stable to coalescence. The slight variation in the mean particle diameters with time are probably a consequence of flocculation (as observed by optical microscopy). Non-crosslinked liquid coacervate systems generally coalesce rapidly and form a separate phase [1, 8, 58]. Coalescence was observed microscopically for the gelatin/acacia coacervate system studied, but not for the albumin/acacia system. The relative stability of the albumin/acacia system may be due to high viscosity, which results from strong interactions between albumin and acacia and between albumin and water. Albumin has a high percentage of polar side groups including: aspartic acid, glutamic acid and threonine; all of which contribute to strong interactions with water. It appears that on mixing albumin and acacia solutions, under specific conditions, complex coacervates form and these aggregate up to a critical size. This aggregation process is almost instantaneous. Once formed the aggregates are not susceptible to further aggregation, coalescence and/or Oswald ripening. Judging from the spherical shape of the coacervates they must be in the liquid state when they form. However once the critical size is reached the consistency of the droplets change and they gel. The semisolid nature of these coacervates is consistent with their unusual behavior, such as: no evidence of particle size growth with time; and particle size does not vary with applied shear. The gel consistency of the albumin/acacia coacervates was confirmed by observation of the coacervates using an optical microscope. The coacervates deform elastically on application of finger pressure via a coverslip. When this pressure is removed the coacervates regained their original shape. Figure 17.4 is a scanning electron micrograph of albumin/acacia coacervate microgels prepared at pH 4.2.

The albumin/acacia coacervates were stable to dilution with distilled water and were unaffected by change in pH and ionic strength over the ranges studied: pH 3.0 to 9.0 and ionic strength 1 to 80 mM. The mean particle diameters of the diluted coacervate systems were measured initially and after two weeks to

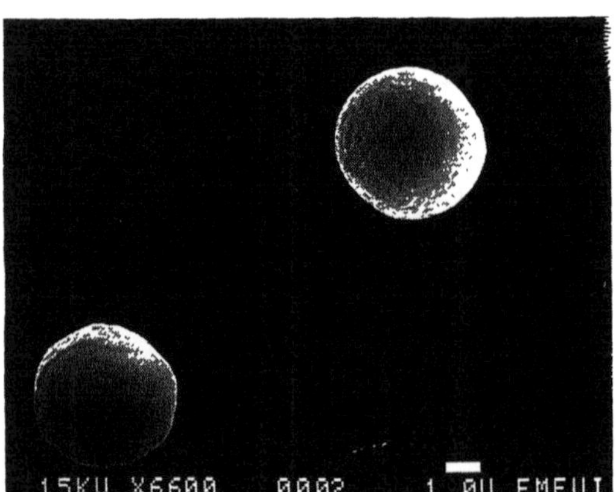

Fig. 17.4. Scanning electron micrograph of albumin/acacia coacervate microgels prepared at pH 4.2. The bar represents 1 μm

determine the stability of these systems. There was no significant difference in the mean particle diameters of the initial and the diluted samples. Liquid coacervate systems such as the gelatin/acacia system are unstable to change in pH and ionic strength and when these parameters are changed to values outside the coacervation range a one phase system forms.

Commassie brilliant blue, a water soluble dye was successfully encapsulated within the albumin/acacia coacervate system. Greater than 95% of the added dye was present within the microcapsules, as determined by spectrophotometric analysis. This indicates the usefulness of the albumin/acacia complex coacervate system as a method of microencapsulation.

The albumin/acacia system described appears to be in agreement with the Tainaka theory of complex coacervation. According to the Tainaka theory aggregates are present in both phases and coacervation is driven by attractive forces among the aggregates.

17.8 Conclusions

Albumin/acacia complex coacervates formed at pH and ionic strength values close to but not at the optimum conditions for maximum coacervate yield are stable with respect to: aging, temperature, pH, ionic strength and dilution (with and without glutaraldehyde crosslinking); as determined by particle size analysis. The relative stability of the coacervate microgels is apparently a consequence of the high viscosity of this system. This coacervate system appears to fit the Tainaka theory of complex coacervation. Albumin/acacia coacervate microgels, with average particle sizes of approximately 6 μm, have been prepared at pH and ionic strength values close to but not at the optimum conditions for maximum coacervate yield. Commassie brilliant blue dye was successfully encapsulated within these coacervates.

The albumin/acacia coacervate system is stable without the use of any crosslinking agents or the application of heat and therefore this system may be suitable for the encapsulation of drugs, biological agents and cells, which are unable to withstand the harsh processing conditions involved in other microencapsulation methods.

Acknowledgments. The author thanks Ms. H.K. Bhowra, Ms. T. Marek and Mr. O.N. Singh, University of Illinois at Chicago for technical assistance.

17.9 References

1. Bungenberg de Jong MG (1949) In: Kruyt GR (ed) Colloid Scinece, vol II Reversible systems. Elsevier, New York, p 335
2. Huggins ML (1942) J Chem Phys 46:151
3. Huggins ML (1942) J Chem Phys 64:1712

4. Flory PJ (1953) Principles of polymer chemistry. Cornell University Press, Ithaca, New York, p 317
5. Khalil SAH, Nixon JR, Carless JE (1968) J Pharm Pharmac 20:215
6. Nixon JR, Khalil SAH, Carless JE (1968) J Pharm Pharmac 20:215
7. Burgess DJ, Carless JE (1984) J Colloid and Interface Sci 98:1
8. Burgess DJ, Carless JE (1985) Int J Pharm 27:61
9. Singh ON, Burgess DJ (1989) J Pharm Pharmacol 41:670
10. Burgess DJ (1990) J Colloid and Interface Sci 140:227
11. Nakajima A, Sato H (1972) Biopolymers 10:1345
12. Sato H, Nakajima A (1974) Colloid Polym Sci 252:944
13. Tainaka K (1979) J Phys Soc Japan 46:1899
14. Tainaka K (1980) Biopolymers 19:1289
15. Overbeek JThG, Voorn MJ (1957) J Cell Comp Physiol 49:7
16. Voorn MJ (1956) Rec Trav Chim 75:317
17. Voorn MJ (1956) Rec Trav Chim 75:405
18. Voorn MJ (1956) Rec Trav Chim 75:427
19. Voorn MJ (1956) Rec Trav Chim 75:925
20. Voorn MJ (1956) Rec. Trav Chim 75:1021
21. Voorn MJ (1959) Fortschr Hochpolym Forsch Bd 1:192
22. Veis A, Aranyi CJ (1960) J Phys Chem 64:1203
23. Veis A (1961) J Phys Chem 65:1798
24. Veis A (1963) J Phys Chem 67:1960
25. Veis A, Bodor E, Mussell S (1967) Biopolymers 5:37
26. Debye P, Huckel E (1923) Phys Z 24:185
27. Debye P, Huckel E (1923) Phys Z 25:49
28. Nixon JR, Nouh A (1978) J Pharm Pharmacol 30:533
29. Madan PL (1978) Drug Dev & Ind Pharm 4:95
30. Nixon JR (1976) Microencapsulation. Marcel Dekker, New York
31. Bungenberg de Jong HG, Kaas AIW (1931) Bioch Z 232:338
32. Bungenberg de Jong HG, Kruyt HR, Lens J (1932) Kolloidchem Beih 36:429
33. Green BK, Schleicher L (1957) U.S. Patent 2,800,458
34. Luzzi LA, Gerraughty RJ (1964) J Pharm Sci 53:429
35. Phares RE, Sperandio GJ (1964) J Pharm Sci 53:515
36. Gutcho MH (1979) Microcapsules and other capsules, advanced since 1975. Noyes Data Corp, New Jersey
37. Lim F (1983) Biomedical applications of microencapsulation. CRC Press, Boco Raton, FL
38. Kondo T (1979) Microencapsulation: new techniques and applications. Techno Books, Tokyo, Japan
39. Lim F, Sun AM (1980) Science 210:908
40. Kowk KK, Groves MJ, Burgess DJ (1991) Pharm Res 8:341
41. Lafferty KL, Prowse SE, Simeonovic CJ, Warren H (1983) Ann Rev Immunol 1:143
42. Rowe JS (1980) PhD Thesis, University of London, United Kingdom
43. Elgindy NA, Elegakey MA (1981) Drug Dev Ind Pharm 7:587
44. McMullen JN, Newton DW, Becker CH (1982) J Pharm Sci 71:628
45. Koh G, Tucker IG (1989) J Pharm Pharmac 233
46. Calanchi M (1976) In: Nixon JR (ed) Microencapsulation. Marcel Dekker, New York, p 93
47. Umeda T, Matsuzawa A, Yokoyama T, Kuroda K, Kuroda T (1983) Chem Pharm Bull 31:2793
48. Nakajima T, Takashima Y, Iida K, Mitsuta H, Koishi M (1987) Chem Pharm Bull 35:1201
49. de Sabata V (1976) In: Nixon JR (ed) Microencapsulation. Marcel Dekker, New York, p 155
50. Nixon JR, Jalsenjak I, Nicolaidou CF, Harris M (1978) Drug Dev Ind Pharm 4:117
51. Brouwer M, Cashon R, Bonaventura J (1990) Biotechnology and Bioengineering 35:831
52. Jizomoto H (1984) J Pharm Sci 73:879
53. Jizomoto H (1985) J Pharm Sci 74:469
54. Burgess DJ, Megremis PT, Patel I, Yoon JK (1989) Pharm Res 6:S-150
55. Burgess DJ, Kwok KK, Megremis PT (1991) J Pharm Pharmacol 43:232
56. Rees DA, Welsh EJ (1977) Angew Chem Int Ed Engl 16:214
57. Janus JE, Kenchington AW, Ward AG (1951) Research (London) 4:247
58. Luzzi LA (1976) In: Nixon JR (ed) Microencapsulation. Marcel Dekker, New York, p 193

18 Complexation of Proteins with Polyelectrolytes in a Salt-Free System and Biochemical Characteristics of the Resulting Complexes

E. Kokufuta

The present chapter deals with the complexation of proteins with polyelectrolytes in a salt-free system with regard to the following two aspects. The first is how an irregular distribution of ionizable groups attached along the length of inflexible protein molecules affects the stoichiometry of salt-linkage formation with oppositely charged flexible polyelectrolytes. And the second is to what extent biochemical ability is maintained in the resulting complexes. Three proteins, human serum albumin (HSA), human or bovine hemoglobin (Hb), and bovine trypsin (BT) were chosen as samples in view of the information available on their amino acid sequences, three-dimensional conformations, structures of biochemically active sites, etc. Poly(diallyldimethylammonium chloride) (PDDA) and potassium poly(vinyl alcohol sulfate) (KPVS), the polyion charges of which are independent of pH within the range 2 to 13, were employed as the polyelectrolytes. The stoichiometry of complexation was studied as a function of pH using a colloid titration method. The biochemical properties of the resultant stoichiometric complexes were examined in terms of the binding affinity of cyanide ions with the heme (ferri-protoporphyrin IX) in the ternary KPVS/Hb/PDDA complex, and also of the protease activity of the BT/KPVS complex. A model for stoichiometric complexes of proteins with polyelectrolytes is proposed on the basis of the results obtained. Also presented are the functional capabilities of the ternary KPVS/Hb/PDDA complex as a cyanide ion exchanger, and of the BT/KPVS complex as an immobilized enzyme.

List of Symbols and Abbreviations

ATEE	$N\alpha$-acetyl-tyrosine ethyl ester
BAEE	$N\alpha$-benzoyl-arginine ethyl ester
BANA	$N\alpha$-benzoyl-arginine-*p*-nitroanilide
BT	bovine trypsin
BTNA	$N\alpha$-benzoyl-tyrosine-*p*-nitroanilide
C	molar concentration
D	optical density
Hb	hemoglobin
HbCN	cyanide methemoglobin
HbOH	alkaline methemoglobin
$Hb^{+}OH_2$	acidic methemoglobin
HSA	human serum albumin
K	equilibrium constant
KPVS	potassium poly(vinyl alcohol sulfate)
LEE	leucine ethyl ester

Macromolecular Complexes
in Chemistry and Biology
Eds. Dubin/Bock/Davis/Schulz/Thies

M_s	mole number of polyion charges bound by salt linkages to 1 g of oppositely charged polyelectrolyte titrant
M_s^0	M_s under the conditions where all the ionizable groups of polyion are completely charged and bound to polyelectrolyte titrant (M_s at $\alpha_c = 1$)
PDDA	poly(diallyldimethylammonium chloride)
Y	degree of saturation of the heme in Hb with cyanide ions
α	degree of dissociation of polyelectrolyte estimated by colloid titration or potentiometric titration

18.1 Introduction

Proteins may interact with polyelectrolytes in aqueous media to result in either soluble or insoluble complexes, the latter in the form of a complex coacervate or an amorphous precipitate. Such phenomena will not only be important in technological fields such as protein separation using polyelectrolytes, but will also provide a basis for a better understanding of the macromolecular interactions available only in living systems. Many researchers and engineers are thus interested in studying the complexation of proteins with polyelectrolytes. This chapter summarizes our recent studies [1–5] dealing with the stoichiometry of the complexation of proteins with strongly acidic and basic polyelectrolytes in salt-free aqueous solutions of varying pH levels, together with biochemical functions of the resulting water-insoluble protein complexes.

In the course of our studies on polymer-polymer complexation, stoichiometry is of primary importance since it helps to clarify the composition, structure and mechanism of the formation of complexes. In the case of polyelectrolyte complex formation, the stoichiometry is generally based on the moles of ion pairs or salt linkages that are formed between oppositely charged polyions through Coulombic interactions. The moles of the bound ionic groups are determinable from the mixing ratio of anionic and cationic polymer solutions under the conditions in which a complex is formed, e.g., the phase separation point. This approach is also available for the investigation of protein-polyelectrolyte complexation, even if the proteins depart further from the usual polyelectrolyte norms. That is, they exhibit nonrandom conformations and are polyampholytes with nonuniform charge distributions.

In order to investigate the stoichiometry of the complexation of proteins with polyelectrolytes, it is essential to have detailed information concerning the number of ionizable groups in the proteins. In the initial stage of our studies, we thus employed human serum albumin (HSA) as the protein sample in view of the information available on its amino acid sequence. The number of moles of acidic groups (carboxyl, mercapto and phenolic OH) and basic groups (amino, imidazolyl and guanidyl) in HSA can be identified from its amino acid sequence, and it has also been found that they bind stoichiometrically to oppositely charged strong polyelectrolytes, poly(diallyldimethylammonium chloride), PDDA, and potassium poly(vinyl alcohol sulfate), KPVS, under conditions where the acidic or basic protein groups are completely ionized in a salt-free system [4].

Another important aspect in protein/polyelectrolyte complexation is to what extent the biochemical capacity of the protein is maintained in the resulting complex. This is because biochemical activity is very sensitive to slight variations in protein ternary structure, which may occur as a result of complexation with polyelectrolytes. Hence, knowledge gained from investigations of the biochemical activity of protein/polyelectrolyte complexes should help clarify the configuration of the protein within the complex. The results of these investigations are also applicable to the preparation of biochemically active immobilized protein complexes. For these reasons, we used human (or bovine) hemoglobin, Hb, and bovine trypsin, BT, both of which exhibited definitive biochemical properties in the next stage of our study of protein complexation with polyelectrolytes [1–[3, 5]. The protein biochemical activity was examined in terms of the binding affinity of the ferri-protoporphyrin IX with cyanide ions for the ternary KPVS/Hb/PDDA complex, and of protease activity for the BT/KPVS complex.

18.2 Experimental Section

18.2.1 Polyelectrolyte Samples

Poly(diallyldimethylammonium chloride), PDDA, and potassium poly(vinyl alcohol sulfate), KPVS, were used as the strongly basic and acidic polyelectrolytes within our complexes, respectively. The properties of both polyelectrolytes have been studied in detail in previous investigations [6, 7]. The structural formulas are shown in Fig. 18.1 along with various important physical data.

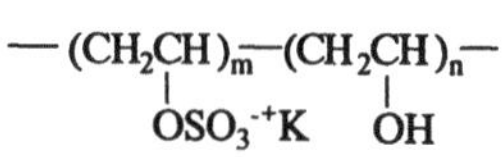

KPVS

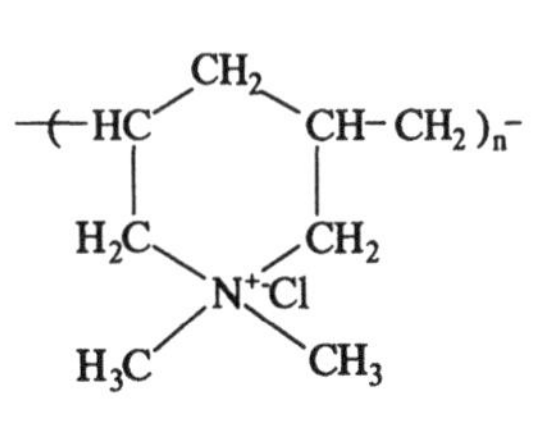

PDDA

Fig. 18.1. Structures of polyelectrolytes studied. The physical data for KPVS and PDDA are as follows: DP_n for KPVS, 1500; $[\eta]$ (in 1 N NaCl at 25 °C) for PDDA, 1.67 $dl \cdot g^{-1}$; equivalent weight referring to the molecular weight of polymer assigned to one mole of ionizable groups, 166 for KPVS and 158 for PDDA; degree of esterification for KPVS which is expressed as $m/(m+n)$, 0.925. Both KPVS and PDDA maintain a completely dissociated state in the pH range 2 to 13

18.2.2 Protein Samples

All protein samples were obtained from commercial sources and used after their purity was verified by amino acid analysis and electrophoresis. In the case of BT, the active site titration method using *p*-nitrophenol-*p'*-guanidinobenzoate was also employed to confirm enzymatic purity. Some of the details of the protein properties [8–14] that are particularly invaluable in our studies are summarized in Table 18.1, together with the commercial sources from which we obtained the samples.

18.2.3 Preparation of polyelectrolyte Complexes with Hemoglobin and Trypsin

The complex of Hb/KPVS, used as the sample in colloid titration with PDDA, was prepared at pH 2.8 where the complexation follows a stoichiometric relationship. The polymer solution (0.6 l) containing the calculated amount (0.252 g) of KPVS was slowly added to the protein solution (5 l) including 1 g of human Hb while stirring. The resulting complex was separated by decantation and centrifugation, washed with a 7:3 mixture of methanol and 0.1 N HCl, and dried at 50 °C for 1–2 days under reduced pressure. The sulfur (1.38 $mmol \cdot g^{-1}$) and iron (52.6 $\mu mol \cdot g^{-1}$) contents are in agreement with the values (S,

Table 18.1. Protein samples studied

Property	Human serum albumin (HSA)[b]	Hemoglobin (Hb)		Bovine trypsin (BT)[c]
		human	bovine	
Number of ionizable groups:				
ε-amino (lysyl)	59	44	46	14
α-amino (N-terminal)	1	4	4	1
imidazolyl (histidyl)	16	38	33	3
guanidyl (arginyl)	24	12	14	2
γ-carboxyl (glutamyl)	61	24	26	4(2)
β-carboxyl (aspartyl)	36	30	38	7(4)
α-carboxyl (C-terminal)	1	4	4	1
phenolic OH (tyrosyl)	18	12	10	10
mercapto (cysteinyl)[a]	1	6	2	0
Absolute molecular weight	66,436	64,503	64,421	23,268[d]
Isoelectric point	4.7–5.2	6.8–7.0	—	10.1–10.8
Subunit structure	monomer	tetramer		monomer
Commercial origin	Sigma Chemical Co.	ICN Pharmaceuticals, Inc.		Sigma Chemical Co.
References	8, 9	10–12		13, 14

[a] Mercapto groups bound to each other by disulfide linkages are omitted.
[b] Number of ionizable groups and absolute molecular weight were established by Meloun et al. (Ref. 8).
[c] Data in parentheses were established by Mikeš et al. (Ref. 13).
[d] Obtained from Walsh's data (Ref. 14)

1.42 mmol·g^{-1}; Fe, 51.9 μmol·g^{-1}) calculated by assuming stoichiometric complexation between Hb and KPVS.

The ternary complex consisting of KPVS/Hb/PDDA was also prepared to investigate its coordination reaction with cyanide ions. In this preparation, bovine Hb was initially complexed with KPVS at pH 2.2, then with PDDA at pH 12. The calculated amounts of KPVS (2.50 g for 10 g Hb) and PDDA (1.65 g for 10 g Hb/KPVS complex) were used. Preparation procedures similar to those mentioned above yielded the ternary complex as a water-insoluble amorphous precipitate; Fe content 43.1 μmol·g^{-1}.

The BT/KPVS complex was prepared under the conditions at which the complexation follows a stoichiometric relationship. The preparation was carried out by the addition of 8.21 ml of 0.00249 N KPVS solution to 167 ml of the enzyme solution including 25 mg BT, using the apparatus for colloid titration. The resulting water-insoluble complex was separated by centrifugation and, if necessary, dried under the same conditions as used for the Hb complex preparation. The sulfur content (1.15 mmol·g^{-1}) was in fair agreement with that calculated by assuming a stoichiometric complexation (1.16 mmol·g^{-1}).

In order to study the functional property of the BT/KPVS complex as an immobilized enzyme, the complex was entrapped within calcium alginate gel beads of approximately 1 mm diameter. The entrapment was carried out by dropping an aqueous suspension including sodium alginate (2%) and the complex (0.05%) into a mildly stirred 10 mM $CaCl_2$ solution adjusted to pH 2.4, followed by curing the resultant gel beads in the same $CaCl_2$ solution for 4 h.

18.2.4 *Colloid Titration*

Initially used by Terayama for measuring the concentration of polyions in aqueous media [15], the method of colloid titration is based on complexation between oppositely charged polyions. For example, the titration of a polyanion with a polycation is:

$$\underset{\text{polyanion}}{\int\!\!\text{-COO}^{-}\,{}^{+}\text{Na}} + \underset{\text{polycation}}{\text{Cl}^{-}\,{}^{+}\text{N(CH}_3)_3\text{-}\!\!\int} \rightarrow \underset{\text{polyion complex}}{\text{-COO}^{-}\,{}^{+}\text{N(CH}_3)_3\text{-}\!\!\int} + \text{NaCl}\,. \tag{18.1}$$

This is similar to the pH titration of a polymeric acid with a monomeric base:

$$\underset{\text{polyacid}}{\int\!\!\text{-COOH}} + \underset{\text{base}}{\text{NaOH}} \rightarrow \underset{\text{salt (polyanion)}}{\int\!\!\text{-COO}^{-}\,{}^{+}\text{Na}} + \text{H}_2\text{O}. \tag{18.2}$$

The number of moles (M_s in mol·g^{-1}) of charges fixed to the polyelectrolyte sample that are bound by salt linkages to 1 g of titrant with the opposite charges can be determined volumetrically. When the pH dependence of M_s obtained with a weak polyelectrolyte as the sample, is studied using a strong polyelectrolyte titrant, such as KPVS or PDDA, for which the dissociation is independent of pH, one can estimate the degree of dissociation (α_c) of the sample as a function of pH;

$$\alpha_c = \frac{(1/M_s)}{(1/M_s^0)} \tag{18.3}$$

where M_s^0 represents the value of M_s under the conditions in which all the ionizable groups of the sample are completely charged and bound to the titrant, i.e. M_s at $\alpha_c = 1$. This is true in cases where the polymer titrant forms salt linkages with only the ionized groups (e.g. COO^-) but not with the uncharged groups (e.g. COOH), both of which coexist within the same polyion sample. On the other hand, the degree of dissociation (α_p) from potentiometric titration can be estimated by

$$\alpha_p = \alpha_n + \frac{(C_{H^+} - C_{OH^-})}{C_M} \tag{18.4}$$

where α_n represents the degree of neutralization and C_{H^+} and C_{OH^-} are the molarity of proton and hydroxyl ions, respectively. C_M is the molar concentration of acidic or basic groups fixed to the polymer. By comparing α_c with α_p

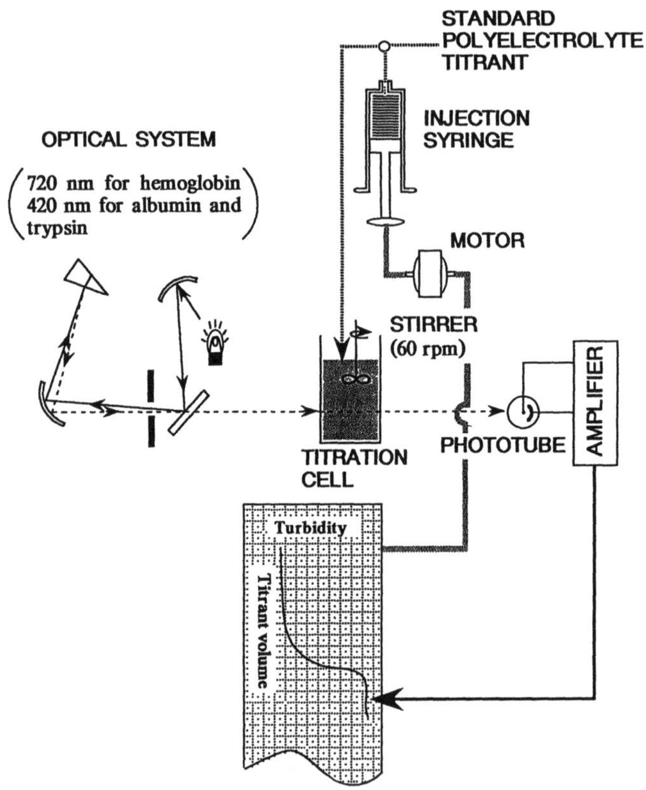

Fig. 18.2. Schematic illustration of turbidimetric titration apparatus used for colloid titration

values, the stoichiometry of the complexation between the sample and titrant polyions can be evaluated. By taking these factors into account, we have determined the stoichiometry of complexation as a function of pH for several polyelectrolytes [16, 17]. Thus the same approach is applicable for the complexation of proteins with polyelectrolytes.

In the original study, the end point of the titration was indicated by a metachromic color change in toluidine blue [15]. In order to avoid possible interactions of toluidine blue with protein molecules, however, we employed turbidimetric measurements to determine the endpoint of titrations. Figure 18.2 shows a schematic illustration of the apparatus for the measurement of turbidity during colloid titration. In addition to turbidity, measurements of the change in conductivity due to the increase in the concentration of small ions as a result of complexation [see Eq. (18.1)] also served as an independent means for indicating the end point. The conductometric method, however, could only be used at the neutral pH range where concentrations of H^+ and OH^- ions were relatively low. As is detailed below, there is good agreement between the end points determined by both turbidimetric and conductometric titration methods throughout all our experiments.

18.2.5 Physicochemical Characterization of Complexes

Infra-red (IR) spectroscopy was carried out to investigate the salt linkages in the complexes, *Protein* $-COO^- \ {}^+N(CH_3)_2(CH_2=)_2$–*PDDA* and *Protein* $-NH_3^+ \ {}^-OSO_3$–KPVS. The complexed proteins were washed with 0.01 N NaOH (for protein/KPVS complexes) or 0.01 N HCl (for protein/PDDA complexes), lyophilized and then dried at 50–60 °C for 1–2 days under reduced pressure. Control samples were prepared by lyophilizing protein from solutions containing 0.01 N NaOH or HCl.

The sulfur content of the complexed proteins was measured by elemental analysis. The iron content of the ternary complex of KPVS/Hb/PDDA was determined by observing the color development between O-phenanthroline and iron after decomposition of the complex by heating in a 1:3 mixture of HNO_3 and HCl.

The dissociation of the acidic groups remaining in a stoichiometric KPVS/BT complex was examined by potentiometric titration. The complex (27.75 mg containing 25.00 mg of trypsin) was dispersed into 80 ml of distilled water and titrated with 0.0109 N NaOH in a nitrogen atmosphere at 25 ± 0.1 °C.

18.2.6 Biochemical Characterization of Complexes

The coordination of cyanide ions to heme (ferri-protoporphyrin IX) in a ternary complex of KPVS/Hb/PDDA was examined as a function of pH and then compared with the results for uncomplexed native Hb. The degree of saturation (Y) of the heme with cyanide ions was measured by cyanide adsorption experiments for aqueous suspensions including the complex and potassium cyanide, as well as by spectrophotometric analysis at 540 nm of native Hb solutions containing potassium cyanide. In the latter case, the Y value was calculated using the relation:

$$Y = (D - D_0)/(D_\infty - D_0) \tag{18.5}$$

where D_0 is the optical density of the sample free of potassium cyanide (KCN), D_∞ is the optical density of the sample including an excess of KCN and D is the optical density of the sample with KCN not in excess. Various buffer solutions of ionic strength 0.15 were used for both adsorption and spectrophotometric measurements: CH_3COOH/CH_3COONa (pH 3.2 to 5.5); $Na_2B_4O_7/KH_2PO_4$ (pH 5.5 to 9.2); $Na_2B_4O_7/Na_2CO_3$ (pH 9.2 to 11.0); NH_4Cl/NH_4OH (pH 8.0 to 11.0); $Na_2HPO_4/NaOH$ (pH 11.0 to 12.0); and NaOH/NaCl (pH 12.0 to 13.7). Throughout these experiments, we studied the effects of complexation of Hb with the polyelectrolytes on the three-dimensional structure of the globin subunits and also the configuration in the vicinity of the heme. In addition, the functional capability of the KPVS/Hb/PDDA complex as a cyanide ion exchanger was also evaluated.

The protease activities of BT before and after the complexation with KPVS were assayed at different pH levels using *N*α-benzoyl-DL-arginine-*p*-nitroanilide hydrochloride (BANA):

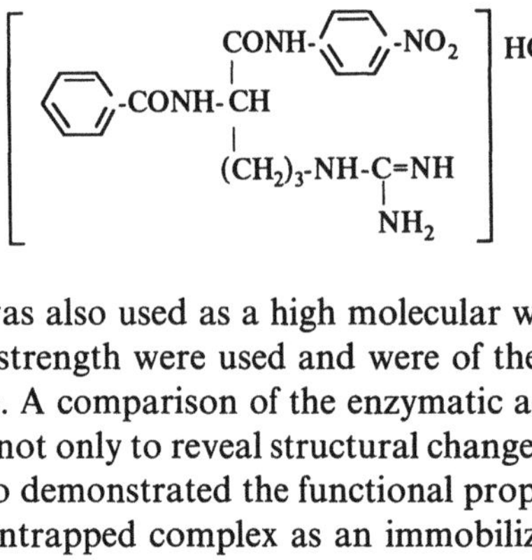

Hammarsten casein was also used as a high molecular weight substrate. Buffer solutions of 0.1 ionic strength were used and were of the same composition as those described above. A comparison of the enzymatic activities for native and complexed BT served not only to reveal structural changes in the active site after complexation, but also demonstrated the functional property of the complex or calcium alginate gel-entrapped complex as an immobilized enzyme.

18.3 Results and Discussion

18.3.1 Stoichiometry of the Complexation

Upon the addition of polyelectrolyte to a protein solution, an increase in turbidity is observed due to the formation of an insoluble polyelectrolyte/protein complex, followed by precipitation at a specific polyelectrolyte/protein mixing ratio. Such a change in turbidity is often accompanied by an increase in the measured solution conductivity (Fig. 18.3), which is particularly pronounced

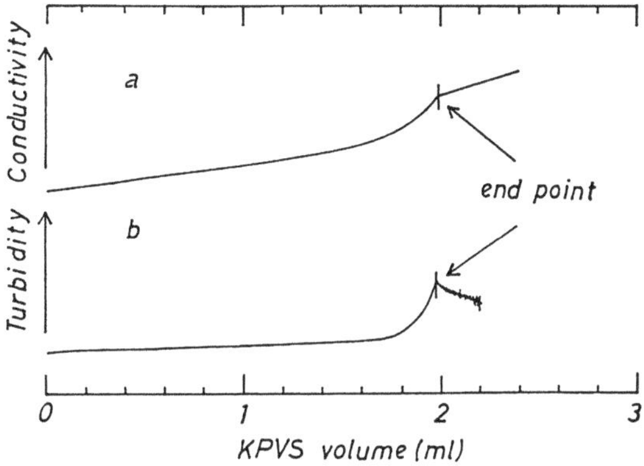

Fig. 18.3a, b. Titration curves of Hb with KPVS titrant. Fifty milliliters of the sample solution containing 3.95 mg of Hb were titrated with 0.002495 N KPVS. The pH of the titration system was adjusted to 3.48 with 0.1 N HCl. The end point was determined by the methods of conductivity **a**) and turbidity **b**) (quoted from Ref. 1)

for solutions containing low concentrations of small ions (mainly H^+ and OH^-) in the range of pH 3.5 to 9.5. Both these measured trends are indicative, yet not conclusive, of the formation of salt linkages or ion pairs between oppositely charged ionic groups, as described in Eq. (1), but culminate at an inflection point which probably determines the end point of complexation between polyelectrolyte and protein. Hence, we further measured the end point of colloid titration as a function of protein concentration at various pH values to determine the existence of a stoichiometric correlation between the titrant volume and protein concentration. In all measurements performed for each complexation system making use of different proteins, the titrant volume was directly proportional to the protein concentration at various pH values (see Fig. 18.4), thereby indicating quantitatively the completion of the formation of complexes between protein and polyelectrolyte which may well be due to salt linkage formation.

In order to confirm the above hypotheses of a stoichiometric correlation between salt linkage formation and measured inflection points which determine the completion of complexation in both conductometric and turbidimetric measurements, we further examined salt linkage formation between the protein amino groups and KPVS, as well as between carboxyl groups and PDDA, by infra-red (IR) spectroscopy. The protein/KPVS complexes obtained at the end point of colloid titration were washed thoroughly with 0.01 N NaOH, dried in vacuo and subjected to spectroscopy. The washing with alkaline solution was performed to remove protons from the protein NH_3^+ groups, which were free of salt linkages with KPVS. A control sample was prepared by lyophilizing a 0.01 N NaOH solution containing proteins. The absorption band assigned to amine salts at 2400–2700 cm^{-1} (a broad shoulder) was observed for samples containing the complexed proteins but not for samples prepared from the native protein solutions. On the other hand, the protein/PDDA complexes washed with 0.01 N HCl showed two absorption bands assigned to carboxylate ions at 1640 and 1420 cm^{-1}, whereas the spectra of the lyophilized, uncomplexed proteins from 0.01 N HCl solutions showed an absorption band assigned to the

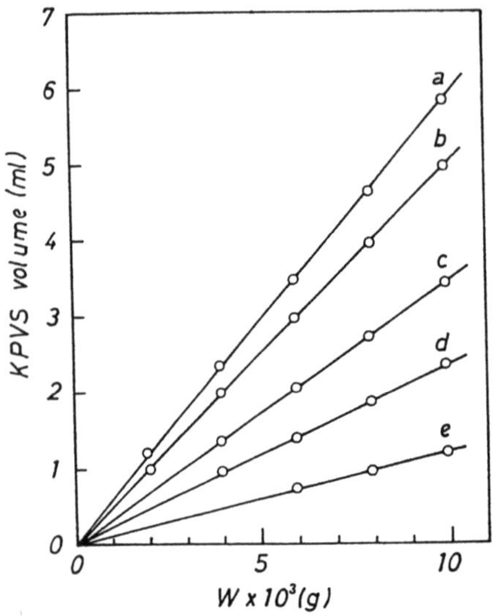

Fig. 18.4. Linear relationship between the volume of KPVS titrant and the weight (W) of Hb at different pH levels (*a* 2.12; *b* 3.48; *c* 4.12; *d* 4.71; and *e* 5.79). The value of W was represented by grams of Hb in 50 ml of sample solution (quoted from Ref. 1)

protein COOH groups at 1720 cm^{-1}. These results are conclusive of the formation of $-NH_3^+ \ ^-OSO_3-$ salt linkages in the protein/KPVS complexes, and also of $-COO^- \ ^+N(CH_3)_2(CH_2=)_2-$ salt linkages in the protein/PDDA complexes. Hence, they are supportive of our hypothesis that both the conductometric and turbidimetric end points are directly indicative of the completion of salt-linkage formation between proteins and polyelectrolytes. And further, the results of the end point of colloid titration as a function of protein concentration at various pH values measured by conductometric and turbidimetric measurements can be used to determine the stoichiometric relation involved in the formation of the protein/polyelectrolyte complex.

The results from colloid titration and IR spectroscopy prompted us to study the complexation of proteins with polyelectrolytes *in terms of the stoichiometry of formation of salt linkages*, even though proteins are polyampholytes with nonuniform charge distributions. For this purpose the M_s values are most suitable which can be determined as the slope of the titrant volume at the end point as a function of protein concentration at various pH levels (see Fig. 18.4 and Sect. 2.4). The colloid titration curves were constructed by plotting M_s versus pH, shown in Fig. 18.5, for both HSA with KPVS as well as with PDDA titrants. The M_s values obtained for HSA with PDDA titrant show a gradual increase with increasing pH, followed by a plateau in the range of pH > 12. For the case of KPVS, however, the M_s value increases with decreasing pH and attains a plateau in the range of pH < 2.5–3. Both KPVS and PDDA maintain a completely dissociated state in the pH range 2 to 13 [6, 7]. The observed changes in M_s with pH can thus be related to the dissociation changes of the acidic and basic groups in HSA (see Eq. (18.3)). In addition, the M_s values in the plateau range at pH > 12 and pH < 3 for PDDA and KPVS, respectively, can be regarded as the M_s^0 value in Eq. (18.3).

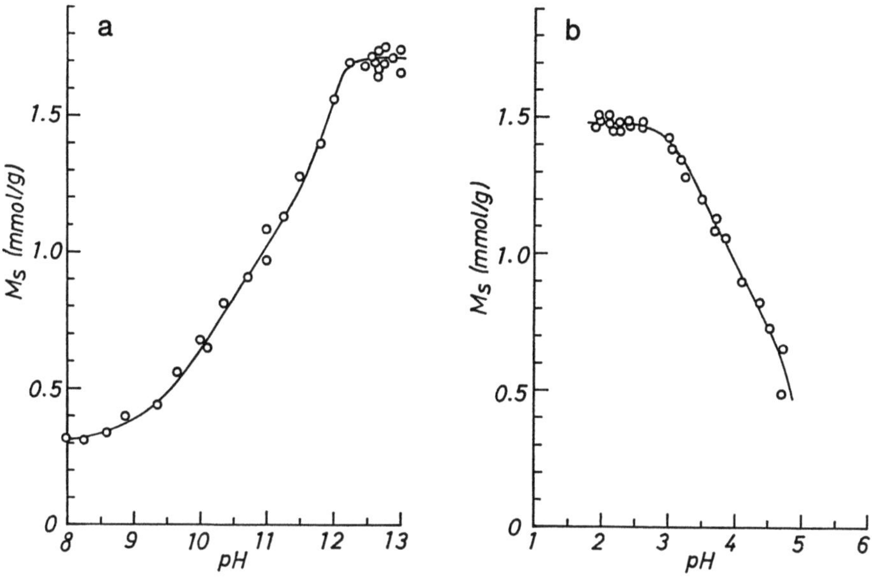

Fig. 18.5a,b. Colloid titration curves of HSA with 0.00501 N PDDA titrant **a**) and with 0.00497 N KPVS titrant **b**) (quoted from Ref. 4)

For the case of weak polyelectrolytes, the stoichiometry of salt linkage formation during complexation with KPVS or PDDA can be studied as a function of pH curves by comparing the pH curves of α_c and α_p [16, 17]. For proteins, however, a pH-induced conformation change usually influences the relation between pH and α_p [18]. We thus examined the M_s^0 values quantitatively to study the stoichiometry of the formation of salt-linkage between acidic groups in HSA and quaternary ammonium ions in PDDA, as well as between basic groups in HSA and sulfate groups in KPVS, under the conditions where the acidic or basic groups in the protein are completely dissociated. The number of acidic and basic groups in HSA was estimated from the M_s^0 value using the relation $M_s^0 \times M_w$, where M_w is the absolute molecular weight. We then compared the result obtained with the values known from the amino acid sequence (Table 18.2). The agreement between both values was found to be within the usual limit of the variation ($< \pm 3\%$) of the analytical method employed here. Therefore, we have shown that salt-linkage formation of the acidic groups in HSA with PDDA, and of the basic groups with KPVS, respectively, follows a 1 : 1 stoichiometric relationship under the conditions reported here [4].

The stoichiometry of complexation for the other proteins was also investigated according to the same analytical procedures [1, 2, 5]. There was good agreement between the number of protein ionizable groups estimated from the M_s^0 values and those from the known amino acid sequence (see Table 18.2), except in the case of the Hb/KPVS complex. The Hb/KPVS complex was investigated in order to further determine the properties of the ternary KPVS/Hb/PDDA

Table 18.2. Number of acidic and basic groups estimated from M_s^0 values and identified from amino acid sequence

	M_s^0, mmol·g^{-1}		Acidic groups		Basic groups	
	with PDDA[a]	with KPVS[b]	from M_s^0 with PDDA	by amino acid sequence	from M_s^0 with KPVS	by amino acid sequence
HAS	1.71	1.48	114	117	98	100
Hb (human)	—	1.47	—	—	95	98
Hb (human)/ KPVS complex	1.51	—	115[c]	76	—	—
BT	—	0.818	—	—	19	20[d]

[a] Obtained from the colloid titration curve at pH > 12.
[b] Obtained from the colloid titration curve at pH < 3.
[c] Calculated as $Mw = 76\,000$, which was determined by the iron content (52.6 μmol·g^{-1}) (see Sect. 2.3).
[d] Obtained from Walsh's data (see Table 18.1)

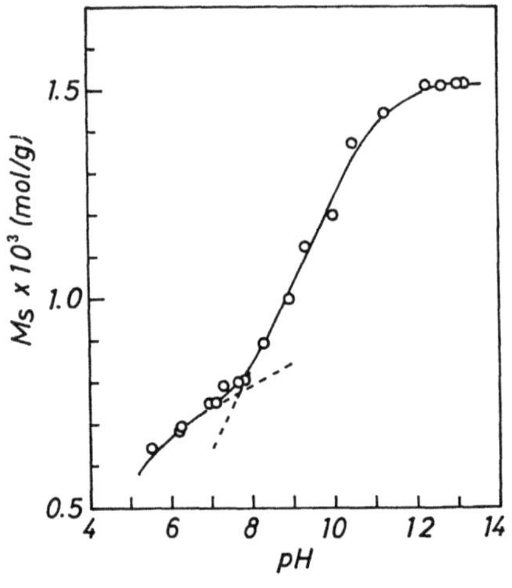

Fig. 18.6. Colloid titration curve of stoichiometric Hb/KPVS complex with 0.00513N PDDA titrant (quoted from Ref. 2)

complex [2]. The colloid titration curve for the Hb/KPVS complex with PDDA is shown in Fig. 18.6. Hb stoichiometrically complexed with KPVS is not soluble below pH 5, but is soluble and titratable with PDDA above pH 6. The titration curve with PDDA exhibits an inflection at pH 7.7, a gradual increase in the pH range 8 to 11, and a plateau beyond pH 12. The value of M_s at the inflection point is approximately 0.78 mmol·g^{-1}. It is estimated that 59 acidic groups take part in the formation of salt linkages with PDDA at the inflection point. The value estimated from M_s is comparable to the number of carboxyl groups remaining in the Hb component of the binary complex. It can therefore be said that in the pH range below the inflection point, the carboxyl groups in the binary complex preferentially form salt linkages with the quaternary ammonium ions in PDDA. In addition, the formation reaction follows a 1 : 1 stoichiometric relationship specifically at the inflection point. The other acidic groups, SH in the cysteinyl residues and phenolic OH in the tyrosyl residues, further participate in the complexation with PDDA in the pH region above the inflection point. However, as shown in Table 18.2, there is little agreement between the total number of acidic groups estimated from the M_s value in the plateau (i.e. M_s^0) and those of the amino acid sequence. This result can be explained by assuming that in the basic region, the salt linkages between Hb and KPVS are severed during the course of the complexation with PDDA. Thus the isolated $-OSO_3-$ groups in the KPVS component form new salt linkages with the quaternary ammonium ions in PDDA. If this assumption is valid, we can then estimate that 56 acidic groups, including 12 phenolic OH and 6 mercapto groups, form salt linkages with PDDA in the pH region between the inflection point and the plateau. Thus, 38 sulfate groups isolated from the salt linkages with the basic groups in Hb/KPVS complex take part in salt linkage formation with PDDA, which is in agreement with the number of imidazolyl groups in the Hb amino acid sequence (see Table 18.1).

In conclusion, we have indicated that KPVS or PDDA is complexed through 1 : 1 stoichiometric salt linkages with the non-flexible protein molecules having a non-random distribution of ionizable groups, under conditions in which the ionizable groups are fully dissociated in aqueous salt-free solutions. In addition, PDDA was found to form a ternary KPVS/Hb/PDDA complex stoichiometrically with the carboxyl groups remaining in the stoichiometric Hb/KPVS complex when the pH was adjusted to 7.7. At alkaline pH, however, PDDA linked not only to the other acidic groups (phenolic OH and mercapto), but also to the sulfate groups of KPVS resulting from partial cleavage of salt linkages in the Hb/KPVS complex. Therefore, the salt linkages in the protein/polyelectrolyte complexes are reversible.

18.3.2 Coordination of Cyanide Ions with Ternary Complex of Hemoglobin

It is well known that the affinity of cyanide ions for Hb, in which iron (III) carries an extra positive charge, is very high because the CN^- ion can form a coordination bond with iron to neutralize the extra charge [19]. This led us to investigate the coordination of CN^- to the heme (ferri-protoporphyrin IX) in the complexed and uncomplexed Hb, in order to be able to determine the effects of complexation with polyelectrolytes on the ternary conformation of Hb [3].

The theoretical estimation of the equilibrium reaction of Hb with potassium cyanide over a wide range of pH was first studied. In aqueous solutions with different pH values, Hb coexists in two different forms: Hb^+OH_2 (acidic methemoglobin) in which the water molecule is attached to the 6th coordination site of ferri-protoporphyrin IX, and HbOH (alkaline methemoglobin) in which the H_2O of Hb^+OH_2 is replaced by a hydroxyl ion. The equilibrium reaction between both types is represented by [20]:

$$Hb^+OH_2 + H_2O = HbOH + H_3O^+ \ . \qquad (18.6)$$

The ionization equilibrium of hydrocyanic acid is also represented by:

$$HCN + H_2O = H_3O^+ + CN^- . \tag{18.7}$$

Thus, the formation of cyanide methemoglobin (HbCN) from Hb and KCN over a wide pH range may reasonably be pictured as:

$$Hb^+OH_2 + HCN = HbCN + H_3O^+ \tag{18.8}$$

$$Hb^+OH_2 + CN^+ = HbCN + H_2O \tag{18.9}$$

$$HbOH + HCN = HbCN + H_3O^+ \tag{18.10}$$

$$HbOH + CN^- = HbOH^- + OH^- \tag{18.11}$$

The equilibrium constants for the six reactions described above may be defined as

$$K_1 = \frac{[HbOH][H_3O^+]}{[Hb^+OH_2][H_2O]} \tag{18.12}$$

$$K_2 = \frac{[H_3O^+][CN^-]}{[HCN][H_2O]} \tag{18.13}$$

$$K_3 = \frac{[HbCN][H_3O^+]}{[Hb^+OH_2][HCN]} \tag{18.14}$$

$$K_4 = \frac{[HbCN][H_2O]}{[Hb^+OH_2][CN^-]} = \frac{K_3}{K_2} \tag{18.15}$$

$$K_5 = \frac{[HbCN][H_2O]}{[HbOH][HCN]} = \frac{K_3}{K_1} \tag{18.16}$$

$$K_6 = \frac{[HbCN][OH^-]}{[HbOH][CN^-]} = \frac{K_4 K_w}{K_1} \tag{18.17}$$

where the brackets signify the molar concentrations of the species and K_w is the ionic product of water. We now consider a system containing Hb^+OH_2, HbOH, HbCN, HCN and CN^-. The total concentrations of Hb (C_1) and of cyanide (C_2) can be given by

$$C_1 = [Hb^+OH_2] + [HbOH] + [HbCN] \tag{18.18}$$

$$C_2 = [HCN] + [CN^-] + [HbCN] \tag{18.19}$$

Since the degree of saturation (Y) of the 6th coordination site of ferri-protoporphyrin IX with cyanide ligand can be defined by

$$Y = [HbCN]/C_1 , \tag{18.20}$$

we obtain the following equation which is useful in determining Y as a function of pH:

$$\frac{1}{Y}(1 - Y)(C_2 - C_1 Y) = \frac{[H_3O^+]}{K_3}\left(1 + \frac{K_2}{[H_3O^+]}\right)\left(1 + \frac{K_1}{[H_3O^+]}\right) \tag{18.21}$$

Several studies have previously reported the equilibrium constants of Eqs. (18.12–18.17), particularly for bovine Hb under the condition of ionic strength 0.15 (see Table 1 in ref. 3). When the triplet of the equilibrium constants is introduced into Eq. (18.21), we obtained

$$\frac{1}{Y}(1 - Y)(C_2 - C_1 Y) = \left(1 + \frac{10^{-9.18}}{[H_3O^+]}\right)(10^{1.26}[H_3O^+] + 10^{-6.85}) \tag{18.22}$$

The curves of Y vs. pH at $C_2/C_1 = 1$ calculated from Eq. (18.22) are shown in Figs. 18.7 and 18.8. The reaction of Hb with KCN approximately follows a 1:1 stoichiometric relationship between pH 8 to 9 because $Y = 1$ at $C_2/C_1 = 1$. An increase in acidity or basicity inhibits the reaction due to the coordination of water molecules (acidic range) or of hydroxyl ions (alkaline range).

The reactions of KCN with Hb and also with a ternary KPVS/Hb/PDDA complex were studied as a function of pH. Bovine Hb was used because the equilibrium constants in Eq. (18.22) are known for this substance. The ternary complex obtained from bovine Hb is of the same composition as that obtained from human Hb in the colloid titration at pH > 12 (see Fig. 18.6). This was confirmed by the result that the iron content (43.1 μmol·g^{-1}) of the ternary bovine Hb complex is close to the value (46.1 μmol·g^{-1}) calculated from the data of bovine Hb in Table 18.1 according to the same manner as used in the ternary human Hb complex. When comparing Figs. 18.7 and 18.8, it is found that in the alkaline range (pH > 9), the experimental points for the complexed and uncomplexed Hb agree with the calculated curves. A considerable difference is, however, found within the pH 4 to 8 range; that is, the theory underestimates the experimental points for the uncomplexed Hb, but overestimates for the complexed Hb.

It is known that in the acidic region, two protons associate with or dissociate from the pyrrole nitrogens of protoporphyrin IX [21]. This means the existence of two different types of Hb^+OH_2 in acidic solutions. If both species have

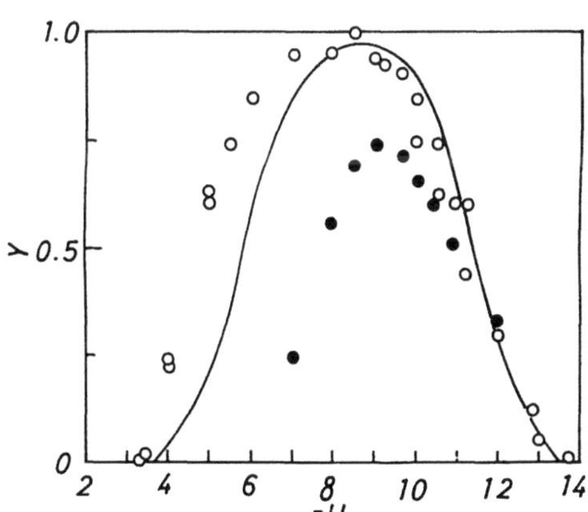

Fig. 18.7. Curves of Y vs. pH for bovine Hb (open circles) at $C_2/C_1 = 1$ ($C_1 = C_2 = 6.52 \times 10^{-5}$ mol·l^{-1}) and hemin (solid circles) at $C_2/C_1 = 100$ ($C_1 = 8.40 \times 10^{-5}$ mol·l^{-1} and $C_2 = 8.40 \times 10^{-3}$ mol·l^{-1}): plots, experimental points; full line, calculation curve using Eq. (18.22) (quoted from Ref. 3)

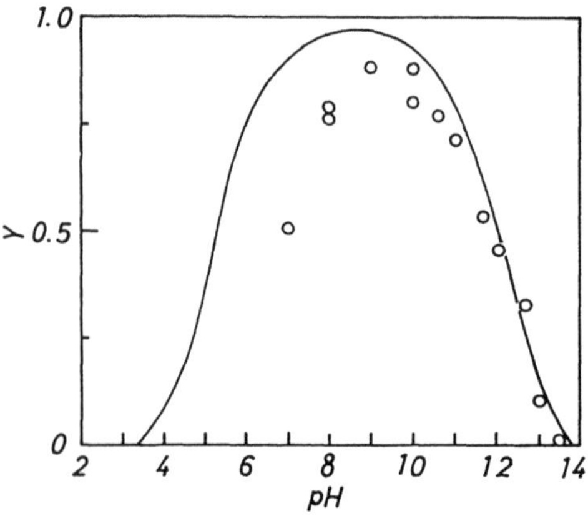

Fig. 18.8. Curves of Y vs. pH for the ternary KPVS/Hb/PDDA complex at $C_2/C_1 = 1$ ($C_1 = C_2 = 2.14 \times 10^{-4}$ mol·l^{-1}): plots, experimental points; full line, calculation curve using Eq. (18.22). The Y value for the complex could not be determined at pH < 7, because volatilization of hydrocyanic acid is non-negligible during adsorption experiments (quoted from Ref. 3)

different equilibrium constants, this could cause the difference between the theoretical and the experimental results for native Hb in the acidic range.

The observed Y values in the acidic range for the complexed Hb differed not only from the theoretical values but also from the experimental results for the uncomplexed native Hb. It is certain that a change in the globin conformation due to any interaction which occurs affects the binding of ligands to Hb in which the iron atom is either divalent or trivalent, because some of the amino acid residues in close proximity to the heme are affected by the conformation change [19]. Thus, we put forth the suggestion that the complexation of Hb with KPVS and PDDA more or less affects a change in the ternary conformation of the globin chain, thereby causing the difference between the coordination reactions of cyanide with the native and the complexed Hb. However, as shown in Fig. 18.7, the binding of the cyanide ligand to hemin (ferri-protoporphyrin IX chloride), which has no globin subunits, is obviously distinguishable from the results for both the complexed and uncomplexed Hb; that is, the Y value did not approach unity even at $C_2/C_1 = 100$. It could therefore be said that the condition of globin in the vicinity of the heme in the complex is approximately analogous to that in the native Hb, although a change in the ternary conformation of the globin chain is, in fact, produced by the complexation.

18.3.3 Protease Activity of Stoichiometric Complex of Trypsin

Enzymes are more appropriate than Hb for studying the effects of complexation with polyelectrolytes on the ternary conformation of proteins, because one can expect to monitor slight variations in protein ternary structure due to the complexation through changes in enzyme activity. This section describes the change in the protease activity of trypsin before and after complexation with KPVS [5]. The complex was prepared under conditions where the formation of salt linkages between BT and KPVS obeyed a 1:1 stoichiometric relation. The activities of the complexed and uncomplexed BT were investigated as a function of pH using low and high molecular weight substrates.

Figure 18.9 shows the pH-activity profiles of the native and complexed enzymes using BANA as the low molecular weight substrate. The complexed BT is found to have an appreciable retention of enzymatic activity. This finding indicates that one imidazolyl group (histidine), which cooperates with both COOH (aspartic acid) and OH (serine) in acylation-deacylation as an intermediate step in the enzyme catalytic action [22], is free of salt linkages with KPVS.

However, the complexation with KPVS results in two characteristic differences in the pH-activity curves of the native and complexed enzymes. First, we found from Fig. 18.9 that in the acidic region the curve of the complexed BT shifts towards a higher pH range than the curve of the native enzyme. Goldstein et al. [23] have observed a similar shifting in the pH-activity curves for two modified trypsin derivatives: acetyltrypsin obtained from acylation of the ε-amino groups of trypsin; and copolymer-bound, water-insoluble trypsin in which the enzyme was covalently bound via its ε-amino groups to the COOH groups of a 1:1 copolymer of maleic acid and ethylene. The substrate used in their study is $N\alpha$-benzoyl-L-arginine ethyl ester (BAEE):

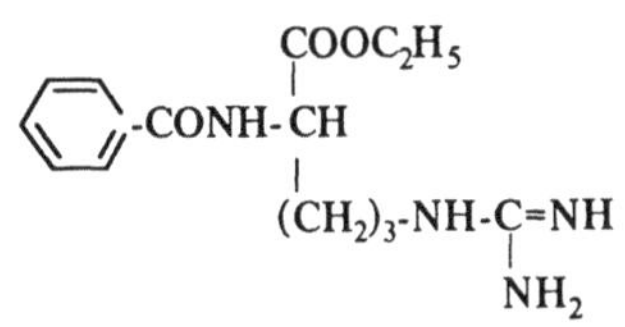

The magnitude of the shifting of the activity curve for copolymer-bound trypsin was reduced with increasing ionic strength, while the ionic strength did not alter the shifted curve for acetyltrypsin, in which the negative charges prevail over the positive charges because of the acylation of the amino groups. Goldstein et al. thus concluded that the electrostatic potential due to the negatively charged maleic acid copolymer affects the local concentrations of both H^+ ions and positively charged BAEE molecules in the microenvironment of the copolymer-bound enzyme molecules. In the case of the complexed BT, as observed in acetyltrypsin, the ionic strength had little influence on the activity curve. It therefore seems most likely that the shifting of the activity curve for the BT/KPVS complex is due not to the negative charges of the KPVS component but to the preponderance of negative charges arising from a 1:1 stoichiometric neutralization of the positive charges of the enzyme molecules through complexation with KPVS.

Another important feature of Fig. 18.9 is that in a narrow pH range from 8 to 9, a rapid increase in complexed enzyme activity was observed, whereas the native enzyme activity gradually decreased. In order to understand such abnormal activity on the part of complexed BT, the potentiometric titration curve with NaOH for the acidic groups (COOH and phenolic OH) remaining in the complex was examined. As shown in Fig. 18.10, the titration curve is characterized by two inflection points. Analysis of the titration data using the same method as for evaluating the number of ionizable groups from the M_s^0 values shows that 12

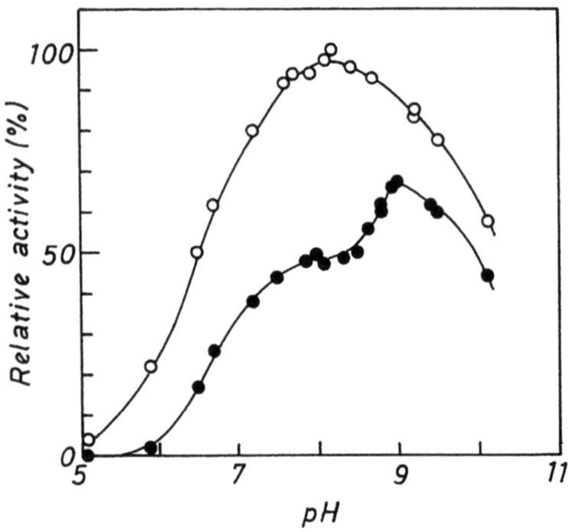

Fig. 18.9. pH-activity curves of native (open circles) and complexed BT (solid circles) determined at 30 °C by using a 2.66 mM BANA solution. Relative activity was calculated from the initial rate of BANA hydrolysis and 100% activity refers to the hydrolyzing rate (4.61 $\times 10^{-2}$ mM·min^{-1}) of the native enzyme at pH 8.18 (quoted from Ref. 5)

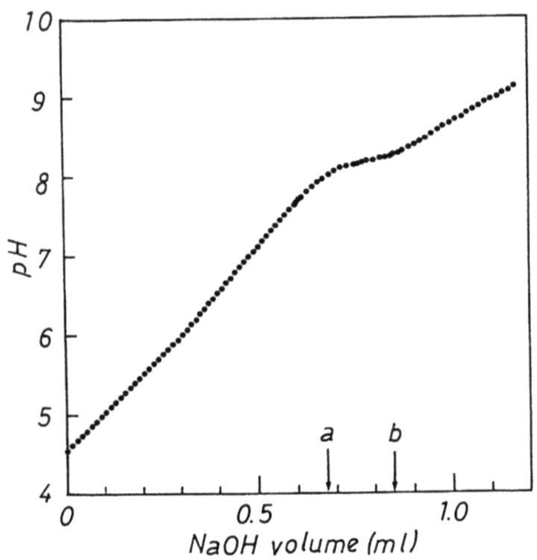

Fig. 18.10. Potentiometric titration curve of the stoichiometric BT/KPVS complex with 0.0190 N NaOH as the titrant. Arrows (*a*) (0.679 ml) and (*b*) (0.848 ml) on the abscissa indicate the neutralization points of 12 carboxyl groups and of 12 carboxyl groups plus 3 imidazolyl groups, respectively, which were calculated from Walsh's data shown in Table 1. The complex containing 25 mg of BT was subjected to titration (quoted from Ref. 5)

acidic groups are titrated up to the first inflection point. These titrated groups are regarded as the carboxyl groups when consulting Table 18.1. Three additional groups are found to be titrated between the first and second inflection points. Since the phenolic OH groups in polypeptides or proteins usually dissociate in the alkaline region (pH > 9.5–10, see ref. 18), these three titrated groups seem to be the imidazolyl groups, i.e. one of them lying in the active center, free from the salt linkages, and the other two bound to KPVS. It can thus be said that the salt linkages between the imidazolyl and sulfate groups in the BT/KPVS complex are severed during the deprotonation of the charged imidazolyl groups caused by an increase in pH. Such a cleavage could cause a restoration of a complexation-induced and altered conformation of BT molecules, thereby resulting in a possible recovery of their enzymatic activity, as can be seen in Fig. 18.9.

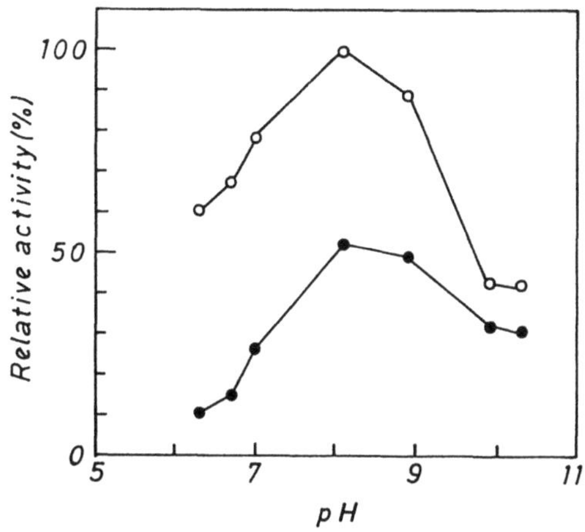

Fig. 18.11. pH-activity curves of native (open circles) and complexed BT (solid circles) determined at 30 °C by using a 4% casein solution. Relative activity is represented as denoting 100% activity of the initial rate of casein digestion [0.14 (absorbance change)·min^{-1}] by the native enzyme at pH 8.10 (quoted from Ref. 5)

A comparison of pH-activity profiles examined with low and high molecular weight substrates should help to clarify the ternary structure of BT in the BT/KPVS complex. Haynes and Walsh [24] have reported a large influence of substrate molecular size on the activity of a water-insoluble trypsin as an immobilized enzyme. The immobilized trypsin was prepared by adsorbing the enzyme molecules as a monolayer onto colloidal silica, followed by crosslinking with glutaraldehyde. This preparation had about an 80% retention of BAEE-hydrolyzing activity compared with native trypsin, whereas the activity towards casein as a high molecular weight substrate was approximately 17% that of the native enzyme. The observed low activity towards casein was interpreted in connection with a lower availability of active sites and a lower efficiency at each site, both of which could arise from a limited diffusion rate of the high molecular weight substrate due to a steric hindrance effect from the supporting matrix. A similar conclusion has also been drawn from enzymological studies of the copolymer-bound trypsin system with casein and hemoglobin [25].

In contrast to these previous results, a large difference in the activities caused by the molecular weight of the substrate is not observed in the BT/KPVS system (see Fig. 18.11 and compare it with Fig. 18.9). Also, in the pH range 8 to 9, where some of the salt linkages has been severed, the fall in the activity of the complexed BT towards casein is much smaller than that of the native enzyme. This appears to be evidence supporting the recovery of enzyme activity via the restoration of the complexation-induced conformation change of BT, although it showed up more clearly when using BANA. Both results reveal that the usual diffusion limitation of polymer substrates does not have a large influence on the complexed BT-catalyzed degradation of casein. Consequently, the salt linkages maintaining the structure of the complex are found to be very "loose" even if the enzyme was stoichiometrically complexed with KPVS.

18.3.4 Proposed Model for Stoichiometric Protein/Polyelectrolyte Complexes

It has become apparent that the complexation of proteins with polyelectrolytes studied here occurs through salt linkages between oppositely charged groups to yield complexes as amorphous precipitates. And further, the formation reaction of salt linkages follows a 1:1 stoichiometric relationship, even when the carboxyl groups remaining in the stoichiometric complex of Hb/KPVS form salt linkages with the quaternary ammonium ions in PDDA. In order to account for these results, a model for a complex of protein with polyelectrolyte is considered and proposed in this section.

In modeling, it must be kept in mind that proteins as polyampholytes depart further from the usual polyelectrolyte norms with respect to conformation and charge distribution. In the complexation systems studied here, for example, the number of the acidic or basic groups per molecule of protein is much smaller than that of the ionizable groups per molecule of polyelectrolyte (see Fig. 18.1 and Table 18.1). Nevertheless, the experimental results have revealed the stoichiometric formation of salt linkages between the protein and polyelectrolyte molecules. In addition, the studies on the coordination of cyanide ions with the ternary KPVS/Hb/PDDA complex and also on the protease activity of the stoichiometric BT/KPVS complex demonstrated that the biochemical capacity of the protein is considerably maintained in the resulting complex. Although complexation via salt linkage formation brings about some alterations in the ternary conformation of protein molecules, both the above mentioned facts imply that this stoichiometric formation is not accompanied by serious protein denaturation, such as an unfolding of protein chains.

It may be deduced from these results that one polyelectrolyte molecule forms a complex with many of the protein molecules until all of the polyion charges are stoichiometrically neutralized with the opposite charges of the proteins. An outline of such a situation is given by the schematic illustration shown in Fig. 18.12, in which the complex consists of a number of inflexible and global protein

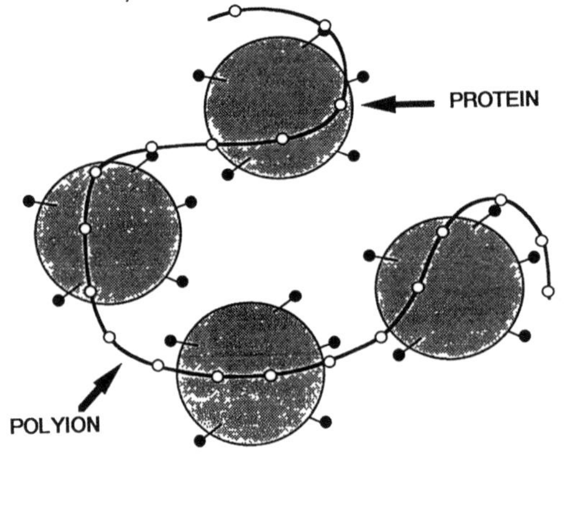

Fig. 18.12. Schematic illustration of stoichiometric complex consisting of protein and polyion. Open and solid circles represent ionizable groups in the polyion and protein, respectively

molecules bridged with loosely extended polyelectrolyte ions. The salt linkages maintaining the structure of the complex as an amorphous precipitate seem to be very "loose", because changes in pH or additions of other polyions sever some of the salt linkages, particularly the linkages between the protein imidazolyl and the KPVS sulfate groups. This looseness may make it possible for the protein and polyion molecules to undergo stoichiometric neutralization or 1 : 1 binding with their oppositely charged groups through thermal motion.

The model proposed here is distinct from the ladder model [26] that was originally proposed by Michaels for interpreting a stoichiometric complexation between oppositely charged flexible polyions. A model [27] similar to that proposed in the present study has been found, however, to satisfactorily account for a 1 : 1 stoichiometric relationship observed in the formation of salt linkages of strong polycations with acidic groups located on the surface of colloid particles [28] or ion exchangers [29] in salt-free systems over a wide pH range. At that time it was reasonable to take into account the possibility that cationic groups bound to the polymer chain form loosely linked ion pairs with anionic groups fixed to the colloid surface.

18.3.5 Functional Capabilities of Hemoglobin and Trypsin Complexes

Several studies have dealt with the application of protein/polyelectrolyte complexation in technological fields: protein separation using polyelectrolytes [30, 31]; enzyme immobilization in polyelectrolyte complexes [32, 33]; the modification of protein/substrate affinity [34]. In this section, two applications of water-insoluble protein/polyelectrolyte complexes are described: the ternary KPVS/Hb/PDDA complex as a cyanide ion exchanger [3], and the BT/KPVS complex as an immobilized enzyme [5].

It is strongly suggested by the results in Fig. 18.8 that the KPVS/Hb/PDDA complex in the form of an amorphous precipitate has a functional capability as a cyanide ion exchanger. In order to test the separation of cyanide ions with the complex, an aqueous solution (pH 9.0) containing 20 ppm of cyanide ions together with borate and phosphate ions as common ions was effused through a column of the complex. As shown in Fig. 18.13(a), the cyanide concentration in the fractions is kept below 0.1–0.2 ppm until the total fraction volume goes up to over 600 ml. The amount of cyanide ions adsorbed is 4.58×10^{-5} mol·(g-complex)$^{-1}$, as estimated from the results in Fig. 18.13(a). This is in accordance with the value [4.31×10^{-5} mol·(g-complex)$^{-1}$] calculated using the iron content of the complex. Next, the elution pattern with 0.1N NaOH for the complex saturated with cyanide ions was examined [Fig. 18.13(b)]. The cyanide concentration reaches a maximal value (96 ppm) at a fraction volume of 40 ml, and about 80% of the cyanide ions bound to the heme in the complex are recovered when the elution is continued until 500 ml. This adsorption and desorption were reproduced reversibly in different runs of the column experiment. It can thus be said that the KPVS/Hb/PDDA complex functions as a cyanide ion exchanger.

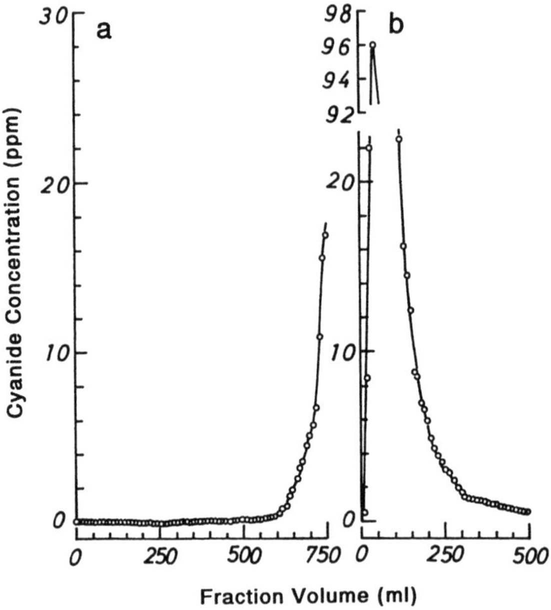

Fig. 18.13a, b. Separation of cyanide ions with the KPVS/Hb/PDDA complex: **a**) obtained by passing $Na_2B_4O_7/KH_2PO_4$ buffer solution (pH 9.0, ionic strength 0.15) containing 20 ppm cyanide ions through a column of the complex (12 g); and **b**) obtained by passing 0.1 N NaOH solution through a column of the complex saturated with cyanide ions (quoted from Ref. 3)

On the basis of this idea, we have synthesized a heme-containing polymer and used it as a cyanide ion exchanger [35, 36].

The BT/KPVS complex is not soluble over a wide range of pH It also exhibits considerable enzymatic activities towards BANA and casein (Figs. 18.9 and 18.11). Thus, the complex can be expected to function as an immobilized enzyme. However, several difficulties arise in handling the complexed enzyme, because it is in the form of an amorphous precipitate. It seems desirable to modify its morphological properties through an appropriate method such as gel entrapment.

Calcium alginate gel is employed both widely and generally in whole microbial immobilization; gel beads of this material can be prepared by simplified techniques under mild conditions. Unfortunately, this gel is not applicable to most of the enzymes due to a large porosity through which enzyme molecules leak out [37]; for example, a protein with a molecular weight of 300 000 easily diffuses from the gel support through the pores. For the complexed BT, nevertheless, this disadvantage seems to be non-problematic because the complexation occurs through the bridging of many BT molecules with a loosely extended KPVS ion. In fact, the extraction of the enzyme from the gel beads into which the complexed BT had been entrapped showed no enzyme dissolution under the conditions where the uncomplexed enzyme was almost extracted from the gel beads [38]. The enzymatic activities of both complexed BT and gel-entrapped complexed BT towards BANA and casein are summarized in Table 18.3, together with those of previously reported immobilized trypsin and α-chymotrypsin which were prepared by binding the enzymes to the

Table 18.3. Activities of trypsin and α-chymotrypsin immobilized by binding to various supports

Support	Enzyme	Substrate	Relative activity[e] (%)	Reference
BT/KPVS complex		casein	56[f]	present study
		BANA	77[f]	
Gel-entrapped BT/KPVS complex		casein	44[f]	present study
		BANA	61[f]	
Sephadex (NCO)[a]	trypsin	casein	2.5–4.0	39
		BAEE	17–24	
CM-Cellulose (OCH_2CON_3)[a]	trypsin	casein	1.5–3.0	40
		BAEE	15–30	
	α-chymotrypsin	casein	1.6–2.8	
		BTNA[b]	19–33	
Cellulose ($OCOCH_2Br$)[a]	trypsin	casein	0.23	41
		LEE[c]	0.6–0.7	
Silica crosslinked by $OCH(CH_2)_3CHO$	trypsin	casein	17	24
		BAEE	80	
Sepharose (NH_2)[a]	α-chymotrypsin	casein	20	42
		ATEE[d]	50	

[a] Immobilized by binding the enzymes to the functional groups shown in parentheses.
[b] *N*α-benzoyl-tyrosine-*p*-nitroanilide; $4\text{-}(HO)C_6H_4CH_2CH(NHCOC_6H_5)CONHC_6H_4NO_2$.
[c] *L*-leucine ethyl ester; $(CH_3)_2CHCH_2CH(NH_2)CO_2C_2H_5$.
[d] *N*α-acetyl-tyrosine ethyl ester; $4\text{-}(OH)C_6H_4CH_2CH(NHCOCH_3)CO_2C_2H_5$.
[e] Denotes percentage of native enzyme activity.
[f] Determined at pH 9

functional groups of supporting materials [24, 39–42]. The complexed or gel-entrapped complexed BT is superior to the other immobilized enzymes with regard to their enzyme activities towards low and high molecular weight substrates, especially towards casein as the high molecular weight substrate. This is because the enzymes bound to the supporting matrix, such as colloidal silica and cellulose particles, are strongly subjected to external and internal diffusion limitations by the polymer substrate (see Sect. 3.3). In contrast, these diffusion limitations show little effect on the enzymatic degradation of casein in the present immobilized enzyme system in which the BT loosely salt-linked with KPVS is entrapped by an alginate gel with a porosity large enough to diffuse casein molecules. As a result, the complexation of enzymes with polyelectrolytes, followed by the entrapment of the resulting complexes into calcium alginate gel would be useful in enzyme immobilization. This hints at the application of polyelectrolyte complexes, as well as of protein/polyelectrolyte complexes, in the preparation of immobilized enzymes [33] or microorganisms [43–46].

18.4 Conclusions and Topics for Future Research

The flexible polyelectrolytes, KPVA and PDDA, are complexed through a 1:1 stoichiometric formation of salt linkages with the inflexible protein molecules having an irregular distribution of ionizable groups, HSA, Hb, and BT, when their ionizable groups are completely ionized in salt-free solutions. There is an appreciable retention of biochemical capacity in the resultant complexes, indicating that changes in the three-dimensional conformations of the protein molecules caused by the complexation are not so large as to cause the loss of their original functions. Therefore, the salt linkages between the polyelectrolyte and protein molecules are very loose, and some of the salt linkages are severed by changes in pH or upon addition of polyions.

In order to understand the nature of salt linkages between protein and polyelectrolyte, more systematic investigations of salt effects will be of great importance. Protein samples of varying charge distributions as well as charge densities will be essential. In addition, detailed knowledge of polymer structural effects, such as charge density, molecular weight, hydrophobic character and hydrogen-bonding ability, will provide key insights into the mechanism of protein/polyelectrolyte complexation through salt linkages. A few authors have already dealt with these subjects [31, 47–49].

Acknowledgements. The author wishes to thank Dr. S. Gorti for valuable discussions and R. Craig for his critical reading of this manuscript. This work was supported by Grants-in-Aid for Scientific Research from the Ministry of Education, Japan.

18.5 References

1. Kokufuta E, Shimizu H, Nakamura I (1980) Polym Bull 2:157
2. Kokufuta E, Shimizu H, Nakamura I (1981) Macromolecules 14:1178
3. Kokufuta E, Watanabe N, Nakamura I (1981) J Appl Polym Sci 26:2601
4. Kokufuta E, Shimizu H, Nakamura I (1982) Macromolecules 15:1618
5. Kokufuta E, Takahashi K (1990) Polymer 31:1177
6. Kokufuta E, Kokubo S, Hirata M, Iwai S (1975) Koubunshi Ronbunshu (Jpn Edn) 32:665; (1975) ibid. (Engl Edn) 4:880
7. Kokufuta E, Kokubo S, Iwai S (1976) Nippon Kagaku Kaishi 1335
8. Meloun B, Moıávek L, Kostka V (1975) FEBS Lett 58:134
9. Behrens PQ, Spiekerman AM, Brown JR (1975) Fed Proc Fed Amer Soc Exp Biol 34:591
10. Dayhoff MO (1972) Atlas of Protein Sequence and Structure. National Biomedical Foundation, Washington DC, Vol 5; (1976) ibid Vol 5, Suppl 2
11. Braunitzer G, Gehring-Müller R, Hilschmann N, Hilse K, Hobon G, Rudloff V, Wittmann-Liebold B (1961) Z Physiol Chem 325:283
12. Hill RJ, Konigsberg W, Guidotti G, Graig LC (1962) J Biol Chem 237:1549
13. Mikeš O, Holevšovskv̌ V, Tomášek KV, Šorm F (1966) Biochem Biophys Res Commun 24:346
14. Walsh KA (1970) Meth Enzymol 19:42
15. Terayama H (1952) J Polym Sci 8:243
16. Kokufuta E, Iwai S (1977) Bull Chem Soc Jpn 50:3043
17. Kokufuta E (1979) Macromolecules 12:350
18. Tanford C (1969) Physical Chemistry of Macromolecules. University of Tokyo Press, Tokyo, pp 548–561
19. Rifkind JM (1973) Inorg Biochem 2:832
20. Beetlestone JG, Irvine DH (1964) Proc Roy Soc (London) A277:401
21. George P, Hanania GIH (1953) Biochem J 55:236
22. Laidler KJ, Bunting PS (1973) The Chemical Kinetics of Enzyme Action. 2nd Edn., Oxford University Press, pp 329–341
23. Goldstein L, Levin Y, Katchalski E (1964) Biochemistry 3:1913

24. Haynes R, Walsh KA (1969) Biochem Biophys Res Commun 36:235
25. Levin Y, Pecht M, Goldstein L, Katchalski E (1964) Biochemistry 3:1905
26. Michaels AS (1965) Ind Eng Chem 57:32
27. Kokufuta E, Takahashi K (1986) Macromolecules 19:351
28. Kokufuta E, Hirai Y, Nakamura I (1981) Makromol Chem 182:1715
29. Kokufuta E, Fujii S, Nakamura I (1982) Makromol Chem 183:1233
30. Clark KC, Glatz CE (1990) In:Hamel J-FP, Hunter JB, Sikdar (eds) Downstream processing and bioseparation. Amer Chem Soc, Washington, DC, p 170
31. Strege MA, Dubin PL, West JS, Daniel-Flinta CD (1990) in:Ladisch M, Willson RC, Painton CC, Builder SE (eds) Protein Purification: from Molecular Mechanisms to Large-Scale Processes. Amer Chem Soc Washington, DC, Chap. 5
32. Zezin AB, Izumrudov VA, Kabanov VA (1989) Makromol Chem Macromol Symmp 26:249
33. Kokufuta E, Shimizu N, Tanaka H, Nakamura I (1988) Biotechnol Bioeng 32:756
34. Ruckpaul K, Rein H, Jänig G-R, Pfeil W, Ristau O, Damaschum B, Damaschum H, Müller J-J, Pürschel H-V, Bleke J, Scheler W (1972) Studia Biophys 34:81
35. Kokufuta E, Hasegawa H, Nakamura I (1984) Polym Bull 11:209
36. Kokufuta E, Saito K (1987) J Appl Polym Sci 34:517
37. Klein J, Stock J, Vorlop KD (1983) Eur J Microbiol Biotechnol 18:86
38. Kokufuta E, Nakamura I, Takahashi K (1987) Abstr Polym Soc Jpn 34:556
39. Axén R, Porath J (1966) Nature 210:367
40. Takami T, Ando E (1968) Seikagaku (in Japanese) 40:749
41. Patchornik A (1965) Israel Patent 18207
42. Axén R, Vretblad P, Porath J (1971) Acta Chem Scand 25:1129
43. Kokufuta E, Matumoto W, Nakamura I (1982) J Appl Polym Sci 27:2503
44. Kokufuta E, Matumoto W, Nakamura I (1982) Biotechnol Bioeng 24:1591
45. Kokufuta E, Shimohashi M, Nakamura I (1988) Biotechnol Bioeng 31:382
46. Kokufuta E, Suzuki S, Nakamura I (1988) J Ferment Technol 66:433
47. Dubin PL et al. (1988) Macromolecules 21:2291; (1990) ibid 23:2500
48. Glantz CE et al., in this book
49. Kokufuta E, Nishimura H (1991) Polym Bull in press

Part IV

Ionomers in Solution

19 Ionomer Solutions = Polyelectrolyte or Ionomer Behavior?

G. Gebel

For almost ten years the properties of ionomer solutions have been studied and lightly sulfonated polystyrene (S-PS) was chosen as the model compound. The properties of S-PS ionomer solutions in polar and nonpolar solvents, which are briefly reviewed, were associated with polyelectrolyte and ionomer behavior respectively. In both cases, structural models have been proposed to explain the rheological and scattering data. A similar study on perfluorinated ionomer solutions is then presented and compared in order to define some general ionomer behavior.

19.1 Introduction

Hollyday [1] defined an ionic polymer as "a polymer, either organic or inorganic, which contains both covalent and ionic bonds in its chain or network structure". But ionic polymers, like a duck-billed platipus, are easier to recognise than to define [2]. All attempts to classify ion-containing polymers [1–4] using the nature of the chain or network, the type of ionic interactions or the charge densities were not really satisfactory. If we focus on organic chains without crosslinks, two types of polymers are usually considered depending on the charge densities [1–4]: the ionomers and the polyelectrolytes.

The polyelectrolytes are usually homopolymers containing almost one charge per monomer. The polymer chain is a macroion (or polyion) accompanied by the counterions. The main property of these polymers centers around their hydrophylic nature, namely they are water-soluble [1–4]. The thermodynamical and rheological properties of polyelectrolyte solutions have been extensively studied [5–8] especially since all the important biological macromolecules are polyelectrolytes [8]. The polyelectrolytes are also of great industrial importance, particularly in four main areas: water treatment, paper, textiles and oil recovery. As a consequence of the concentration of electric charges along the polyelectrolyte chain, there is a strong electrical interaction between these charges and the surrounding small ions, therefore, the solutions show large non-idealities in their osmotic pressure, ion activities and electrical transport [9]. The theory of counterion condensation [10] postulates that if the linear charge density exceeds a critical value, a fraction of the counterions will "condense" on the macroion so as to reduce the effective charge density to this critical value. The counterion condensation depends on the solvent, the quantity

Macromolecular Complexes
in Chemistry and Biology
Eds. Dubin/Bock/Davis/Schulz/Thies

of added salt and the concentration. Substantial changes in the number of bound counterions may lead to a conformation change which modifies the solution properties. The most famous modification is the so-called "polyelectrolyte effect" which correspond to an increase of the "reduced viscosity" with dilution.

The ionomers are polymer chains containing up to 10 mol % ionic groups which are usually neutralized by metals ions. These polymers can either be homopolymers partially substituted with ionic groups or a random copolymer composed of a neutral monomer and a small quantity of an ionic comonomer [1–4]. The low polarity of the polymer matrix leads to an aggregation of non-ionized dipoles to form multiplets and clusters of multiplets [11]. The properties which can be attributed to ionic aggregation (which acts as a physical crosslink) include a high melt viscosity, thermoplasticity, toughness, outstanding abrasion resistance, oil resistance, . . . [12, 13]. Although industrial uses of ionomers are essentially based on these properties (packaging, adhesives, impact modifiers, . . .), the permeability, selectivity and ionic conductivity of some ionomers immersed in polar media lead to unique applications as membranes for separation (for example in chlor-alkali production [14]). As a consequence of industrial uses, ionomers are essentially studied as bulk materials [12]. Although viscoelastic properties of ionomers have primarily been studied [12, 15], structural investigations have been performed [12] either on dry materials or on swollen membranes in order to determine the shape and the distribution of the ionic aggregates but this question is still subject to controversy.

Polyelectrolytes are considered as particular ionic materials characterized by the existence of macroions and are used and studied in solution. Ionomers are, however, considered as modified polymers in which the ionic species act as physical crosslinks and are generally studied in the solid state. These considerations have led to two different fields of research involving different physical concepts and even the symposia devoted to ion-containing polymers are clearly divided in two parts.

More recently, the solution behavior of ionomers has gained interest. The solution behavior was first studied almost ten years ago in nonpolar solvents. It was shown that introducing a small amount of charge along the polymer chain produces interesting changes in the rheological properties. For example, the solution gelifies at very low concentration compared to the unmodified polymer [16] or the solution exhibits shear thickening [17]. The effect of the solvent, the counterion, the temperature and the ionomer on the rheological properties were extensively studied. The most studied ionomers were the so-called "halatotelechelic" which are made of linear chains terminated at each end by a carboxylated group [18–20], the sulfonated EPDM – an ethylene propylene diene terpolymer [21–24] – and the sulfonated polystyrene [25–32]. The ionomer solution behavior was also studied in mixed polar/nonpolar solvents [27] and in polar solvents [26–36]. The solubility of ionomers depends on the polarity of the solvent and on the sulfonate level [24, 37]; for example, high sulfonate level ionomers are soluble in polar solvents whilst low sulfonate level ionomers are more soluble in nonpolar solvents.

Two limiting cases of behavior can be defined: in nonpolar solvents, the system is dominated by more or less solvated dipoles which attract each other and lead to an aggregation phenomenon whilst, in polar solvents, the ion pairs are ionized, the polymer chain is more or less extended depending on the concentration and the macroions repel each other. This behavior is analogous to those of bulk ionomers for the former and to polyelectrolyte solutions for the later. Recently Rinaudo and Eisenberg [38] proposed new definitions for polyelectrolytes and ionomers no longer based on charge densities but rather on their behavior. Ionomers can be defined as "polymer in which the bulk properties are governed by ionic interactions in discrete regions of the material (the ionic aggregates)" and polyelectrolytes as "polymers in which solution properties in solvents of high dielectric constants are governed by electrostatic interactions over distances larger than typical molecular dimensions". According to these definitions, a polymer containing a few percent of ionic group dissolved in a polar solvent should be considered as a polyelectrolyte. In fact, these definitions are already accepted. For example, an ionomer dissolved in polar solvent was already often called a weakly charged polyelectrolytes [39, 40] and the term of associating polymer was commonly used to describe an ionomer dissolved in a nonpolar solvent [17, 25].

Although a lot of work has been devoted to ionomer solutions, most of them are concerned with rheological studies of different types of ionomers in different types of solvents. Only the behavior of ionomer solutions in three typical cases will be presented in order to review the structural models corresponding to the ionomer and polyelectrolyte behavior. For this purpose we will compare the results obtained using different techniques on the sulfonated polystyrene solutions in polar and nonpolar solvents and on perfluorinated ionomer solutions. These three cases were chosen for the following reasons: the polystyrene chain is usually soluble in both polar and nonpolar solvents while there is no solvent for the perfluorinated chain; lightly sulfonated polystyrene (S-PS) among ionomers has been considered as the model compound and polar and nonpolar solvents correspond to limiting cases in the ionomer behavior. Finally in all these cases, the system has been studied using different techniques with fairly similar conditions which allow s comparison of the results and the structural interpretations.

19.2 Sulfonated Polystyrene Ionomer Solutions in Nonpolar Solvents

In nonpolar solvents the ionic groups are not dissociated but exist as more or less solvated ion-pairs which form electric dipoles, the attractive interaction of which lead to their aggregation. Therefore, the structure and properties of S-PS solutions in nonpolar solvents are governed, as in the bulk state, by ion-pair aggregation. This aggregation has been studied using rheological, light scattering, small angle neutron scattering and fluorescence experiments. At high polymer concentration the ion-pair aggregation leads to a shear thickening phenomenon [17, 18], then to gelation and ultimately to the demixing of the solution [16, 20]. In the following we will focus on the study of dilute and semi dilute solutions which correspond to the range of concentration where competition between intra- and intermolecular interactions occurs.

The starting anionically polymerized polystyrene usually employed is a commercial homopolymer purchased either from Dow Chemical (Styron 666, $M_W/M_N = 2.05$), from Pressure Chemical and Co. or from Polymer Laboratory Ltd. ($M_W/M_N = 1.05$). The sulfonation reaction is monitored in order to avoid a complete sulfonation as in the polyelectrolyte preparation. The most commonly employed reaction is performed in 1,2-dichloroethane at 50 °C using acetyl sulfate as sulfonating agent [41]. The sulfonate content is determined after reaction and purification. The polymer molecular weight ranges from 3500 to 1 800 000 and the sulfonate content from 0 to 10%. The nonpolar solvents commonly used are THF, toluene, xylene or low polarity solvents like cyclohexanone. The conditions of preparation and the aging of the samples were shown to be important parameters [28, 42].

19.2.1 Rheological Studies

The rheological behavior of ionomer solutions in nonpolar solvents were extensively studied [16–33]. Typical reduced viscosity-polymer concentration profiles of dilute and semidilute solutions of S-PS ionomers and of unmodified polymer are presented in Fig. 19.1. At low concentration the viscosity of ionomer solutions is lower than that of the reference polymer but increases much more

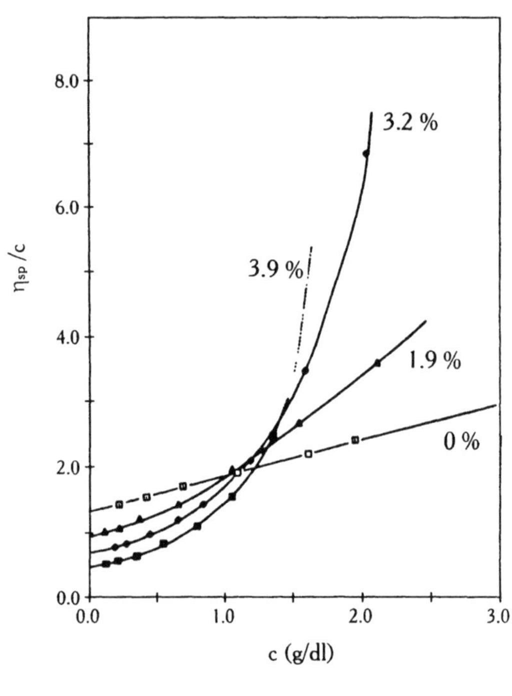

Fig. 19.1. Reduced viscosity as a function of polymer concentration for S-PS (*Mw* 400 000) of various ion contents in THF (from Ref. [30])

rapidly at higher concentrations. Above the crossover point with the nonionic polymer, the viscosity rises rapidly into a viscous gel-like solution. This behavior is attributed to intra- and intermolecular association of ion-pairs. The effect of intramolecular physical crosslinks is to decrease the size of the "gaussian coil" in the solution, and therefore the macroscopic viscosity is lowered at low polymer concentration. At higher concentration, the ionomer coils are close enough to interact, intermolecular associations occur and viscosity increases up to gelation. The crossover point concentration was attributed [28] to the polymer coil overlap concentration, c^*.

This interpretation of the viscosity behavior in the dilute and semi dilute range of concentration was confirmed by the effect of ion content (Fig. 19.1) [25–30]. Increasing the sulfonate level increases both the intra- and intermolecular associations and consequently decreases the intrinsic viscosity which is the extrapolation of the reduced viscosity to zero concentration and increases the slope at high polymer concentrations. Since both intra- and intermolecular aggregations increase with the polymer chain length, the effect of molecular weight [30] which increases the observed phenomenon was also consistent with the previous interpretation. The effect of intramolecular interactions was (demonstrated by the intrinsic viscosity which decreases much more rapidly with ion content when the molecular weight increases. The aggregation is lowered when solvation of the counterion occurs. Therefore, the effect of ion content on the viscosity behavior of polymer solutions was expected to decrease from cesium to lithium and with the addition of simple salt to the solution. These effects were effectively observed with the experimental data [31–33].

The dominance of intra- or intermolecular interactions between ion-pairs depending on the concentration range allows the interpretation of all the results obtained in the viscosity measurements. This rheological behavior is found for all the ionomers studied in the low polarity solvents.

19.2.2 Light Scattering Studies

As this part will be discussed elsewhere in this book, only a general outline of the results obtained with this technique will be presented. Both static and dynamic light scattering have been used to study the aggregation behavior of S-PS ionomers solution in low-polarity solvents [25, 29, 43–46]. From low-angle light scattering experiments [43–46], the plots of Kc/R (Where K is an optical constant, c the concentration and R the reduced scattered intensity) are linear with the concentration (Fig. 19.2). According to the Zimm equation, an apparent molecular weight and the second virial coefficient can be determined. It appears that increasing the ion content leads to an increase in the molecular weight of the aggregates and to a decrease of the virial coefficient which can even become negative [25]. The fact that the apparent molecular weight which is extrapolated to zero concentration is several times greater than that of the starting polystyrene in the same conditions indicates that intermolecular interactions are large even at low concentrations. Consistent results were obtained with quasi elastic light scattering experiments which give hydrodynamic radii [25, 29, 43, 44]. These radii which are presented in Fig. 19.3 are always considerably larger than the corresponding value obtained for polystyrene indicating that intermolecular aggregation occurs even at very low concentration (10^{-6} g/dl) [43].

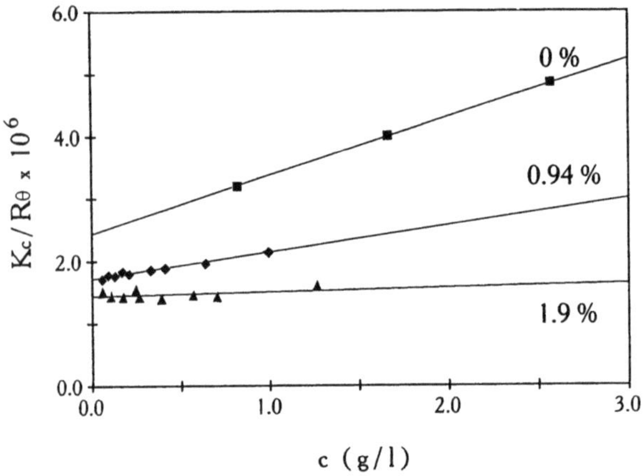

Fig. 19.2. Reciprocal reduced scattered intensity as a function of polymer concentration for S-PS (*Mw* 400 000) of various ion contents in THF (from Ref. [30])

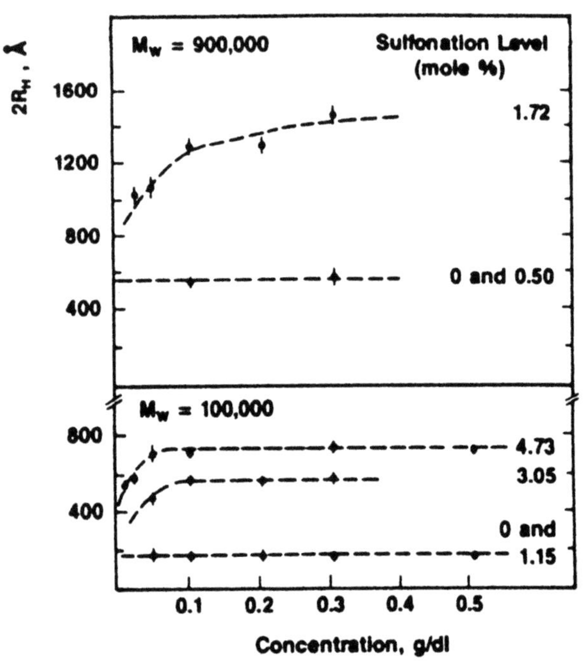

Fig. 19.3. Effective hydrodynamic radii as a function of polymer concentration for a series of S-PS in THF (form Ref. [43])

Even if few ion-pairs interact intermolecularly, and aggregation occurs which increases the apparent molecular weight and produces insolubility of the highest molecular weight S-PS [46]. This aggregation is due to the inability of the sulfonate group to interact in a purely intramolecular manner. This conclusion appears to be inconsistent with the interpretation of the viscosity results. Structural studies were then performed to clarify this point.

19.2.3 Small Angle Scattering Studies

Lantman et al. used Small Angle Neutron Scattering (SANS) to study the aggregation behavior of ionomer solutions [47]. Due to strong association between the chains, conventional SANS experiments only give an average size of the aggregates. The upturn in the small angle scattering data at low angle was considered as a confirmation of the existence of aggregates of large size. The Zimm equation was used to extract for each concentration a radius of gyration and an apparent molecular weight of the aggregates. They were found to increase continuously with the concentration and to be larger than the corresponding values measured for unmodified polystyrene. As the concentration is increased from 0.5 to 4.0 g/dL, the weight average number of chains per aggregate rises from 3 to 12 and the polydispersity of aggregate size was estimated from Zimm plots to be rather large. A mixed labelling technique was then used in order to extract the single chain scattering. The same analysis as in the case of overall scattering data was performed to determine the radius of gyration and the molecular weight of the single chain. These values were found to be roughly constant over the range of concentration. These results led the authors to propose a new model for ionomer aggregation in nonpolar solvents (Fig. 19.4). A primary coil is formed of single chains due to intramolecular association and these primary coils are aggregated to form larger clusters. The previous light scattering data were successfully investigated in the light of this model [44]. The authors proposed a new explanation for the interpretation of the viscosity data: the viscosity of a molecularly dispersed solution can be greater than that of an aggregated solution if the density of material within the

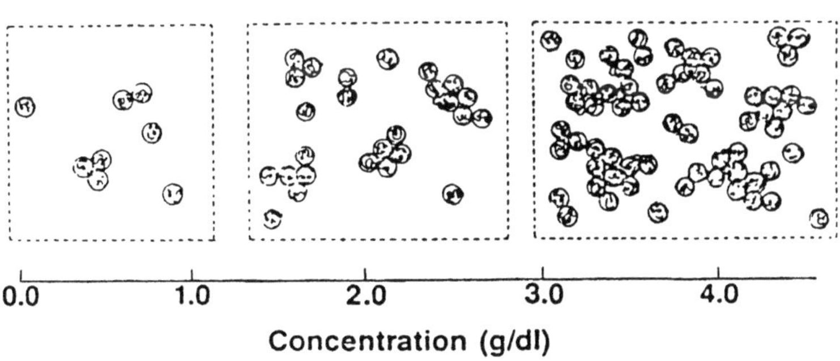

Fig. 19.4. Model for association phenomenon in S-PS/THF solutions (from Ref. [47])

viscometric particles changes. As the molecular weight increases more rapidly than the radius of gyration in S-PS solutions the authors showed, using the Flory equation in which radius of gyration and molecular weight are related, that the reduced viscosity is expected to decrease as was observed in the viscosity measurements. But all the SANS results are based on determination of the radii of gyration which were performed far from the range of validity of approximation. Even if a correction was applied to the data, the results still remain uncertain. Moreover, since the effect of the chain length, the ion content, the solvent and the nature of the ionomer were not studied, it is difficult to generalize these results as the behavior of the ionomer in nonpolar solvents.

The aggregation was supposed then to be controlled by an equilibrium between single collapsed chains and multichain aggregates of all sizes [48]. The equilibrium constants, the apparent molecular weights and the radii of gyration were shown to be independent of the temperature which indicates strong association. The single chain dimensions were also shown [42] to remain constant under shear while the viscosity increases at low shear (shear thickening) and decreases at high shear rates (shear thinning). This result suggests that shear thickening is not due to single chain extension in several intermolecular ion-pair association at the expense of intramolecular associations but to either an elongation of the aggregates under shear or to an interaggregate aggregation. Shear thinning was attributed either to an alignment of the extended aggregates or to an aggregate dissociation.

19.2.4 Fluorescence Studies

Very recently, fluorescence studies were performed [49] using different ionic fluorescence probes associated to different fluorescence quenching molecules. As evidence of microdomain formation, solvated ion-pair are shown to be in close proximity to one another. Indeed a toluene soluble quencher has no effect on the ionic fluorescence probes trapped in the ionic aggregates. The most important results in this work is the study of the fluorescence decay versus the quencher concentration which associated to an appropriate model allows the calculation of the number of ionic species inside an average ionic aggregate (Fig. 19.5). Several features appear: 1) no transition occurs while a transition at 9 g/l between intra- and intermolecular associations was expected for the viscosity data, (2) the ionic species solubility is proportional to the polymer concentration over the range of concentration. So the transition between intra- and intermolecular associations with the crossover point associated to the coil overlap concentration arises probably from a wrong model. These results are in good agreement with the model of Lantman et al. schematized in Fig. 19.4 but several modifications were proposed. The primary particle considered as a single coil could be constituted of more than one polymer chain as suggested by the effect of temperature.

The structure of the ionomer in nonpolar solvents as in the bulk phase is dominated by the ion-pair aggregation. This feature is theoretically and experimentally proved, but the structure of the polymer chain is still not clearly

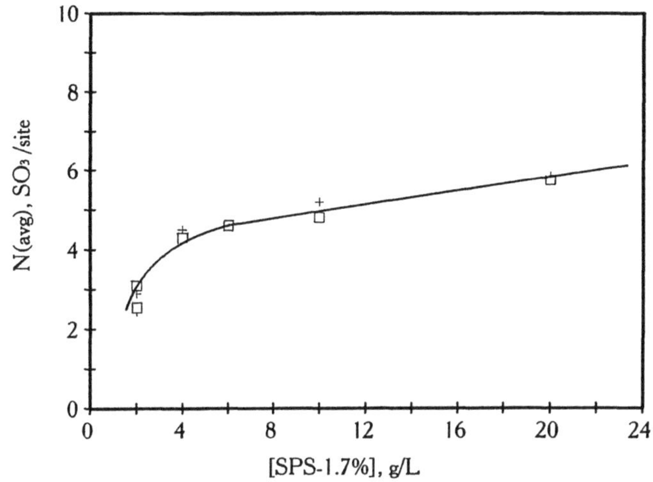

Fig. 19.5. Average number of sulfonate groups per site as a function of ionomer concentration (from Ref. [49])

understood. Even if the model of Lantman et al. [47] is attractive, their general results are far from being proved and the spherical shape for the aggregates is only postulated. Complementary scattering experiments are necessary to clarify the structure of these systems.

19.3 Sulfonated Polystyrene Ionomer Solutions in Polar Solvents

The rheological behavior of ionomer solutions in nonpolar solvent is dramatically modified with the addition of polar solvents [26]. As was shown by Electron Spin Resonance experiments [50], the addition of a few percent of polar solvent is enough to dissociate the ion-pairs and all the phenomena related to the aggregation then disappear. With the addition of more polar solvent a polyelectrolyte effect appears in the viscosity of dilute and semi dilute solutions [27]. Since ionomers are soluble in organic solvents, ionomers offer good model systems for investigating the behavior of polyelectrolytes [30]. Indeed it is possible to analyze the effect of solvent and to avoid the problems associated with aqueous solutions. Moreover, measurements such as light scattering and viscosity are easier to conduct due to the strong scattering and low viscosity of ionomer solutions [30]. The most common polar solvents which are used are dimethylformamide (DMF) and dimethylsulfoxide (DMSO). The preparation of S-PS ionomers is described in the preceding section.

19.3.1 Rheological Studies

The viscosity behavior of S-PS ionomer solutions in polar solvents is presented in Fig. 6. The increase in the reduced viscosity with dilution was always

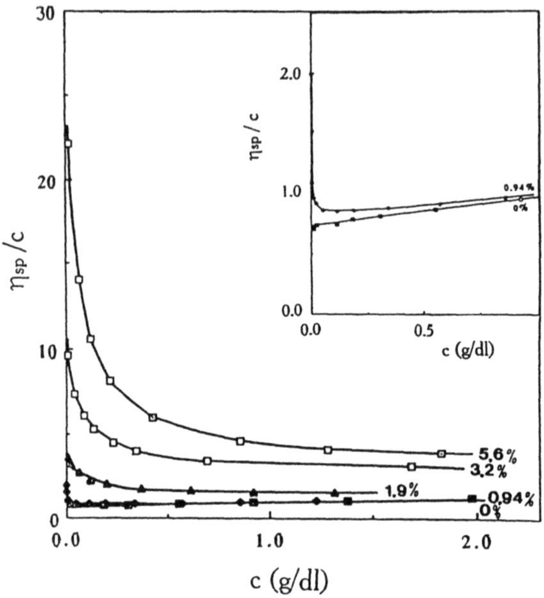

Fig. 19.6. Dependence of reduced viscosity on polymer concentration for S-PS (*Mw* 400000) of various ion contents in DMF (from Ref. [30])

attributed to a polyelectrolyte effect. The interpretation of this effect is that fixed ions on the polymer chain are close to others even at low polymer concentration. In the semidilute and concentrated domains, the repulsions between ions along the polymer chain are partially screened. With dilution the screening effect decreases and the repulsions between fixed ions increase leading to an extension of the polymer chain and consequently to an increase of the reduced viscosity. The existence of the effect in ionomer solutions in polar organic solvent was in itself an interesting result. The effect was first proposed to be related to the dielectric constant of the solvent [27] and then to the solubility parameter [35], but this second hypothesis is far from being convincing. The effect is shown to increase with ion content (Fig. 19.6) and to be rather independent of the molecular weight. The effect of counterions is related to their binding ability [32]. Increasing counterion binding leads to a decrease in the repulsion among the polymer chain and therefore to a reduction of the viscometric particle size and the viscosity.

Studying the effect of molecular weight, Hara et al. [30] have shown that although the number of ions per chain is very small they still show polyelectrolyte behavior. Two examples are given: a 2% sulfonate content with a M_W = 3500 and less than one charge per polymer chain and a 0.2% sulfonate content with a $M_W = 47\,000$ which has only 1.1 ions per chain on average exhibit a polyelectrolyte effect in DMF. It was then concluded that the polyelectrolyte effect is not related to an intramolecular effect but purely to intermolecular effects. An explanation could be that instead of considering the intramolecular repulsive interactions, the intermolecular attractive interactions through the counterions is preferred [30]. This explanation was proposed by Ise et al. [51] to explain the small angle X-ray scattering data in polyelectrolyte systems.

19.3.2 Light Scattering Studies

The light scattering experiments performed on S-PS solutions also exhibit a behavior analogous to polyelectrolytes [45, 46]. It was shown that 0.1% sulfonate content is enough to produce a "polyelectrolyte effect" [46]. The light scattering spectra (Kc/R versus concentration) are not linear as in nonpolar solvents. The reciprocal scattered intensity increases sharply at low polymer concentration (Fig. 19.7) Such curvature suggests that a considerable chain expansion is present and therefore the scattering is no longer described by the conventional quadratic approximation of the virial equation [43]. Hara et al. [45, 46] have determined the effective diameter from the initial slope and shown that this diameter varies linearly with the ion content. The QELS experiment indicates that the diffusion coefficients are constant in the dilute/semidilute concentration range but are decreased by increasing the ion content. Although the dilute/semidilute transition is not known, the fact that the diffusion coefficient ascribed to free diffusion of polymer chain is constant over the concentration range of interest suggests that the intermolecular association between ionomer chains may be very important, even at low concentrations [43]. This result is supported by the analysis of rheological data from Hara et al [30].

19.3.3 Small Angle Scattering Studies

Several experiments of small angle scattering were performed on ionomer solutions in polar solvents [39, 52–54]. As for polyelectrolyte solutions, a peak

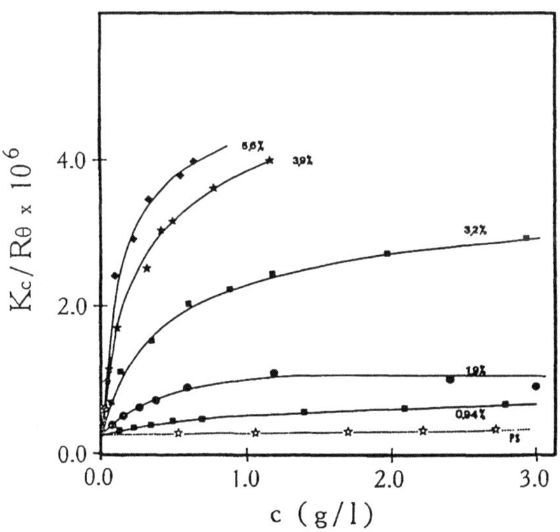

Fig. 19.7. Reciprocal reduced scattered intensity as a function of polymer concentration for S-PS (*Mw* 400 000) of various ion contents in DMF (from Ref. [45])

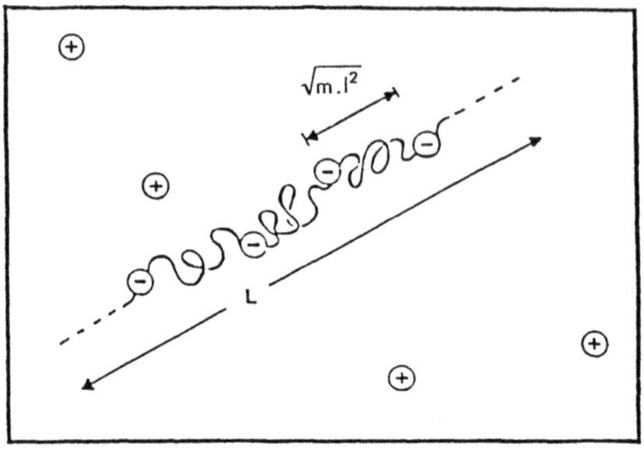

Fig. 19.8. Schematic of S-PS/DMF solutions at infinite dilution (from Ref. [53])

rather broad appears on the scattering spectra. The peak position varies with the concentration and the intensity depends on the ion content. The peak is usually attributed to an interference term [53] and exists even at rather low concentration. Even in polyelectrolyte solutions, the origin of this peak is not clearly understood. The existence of a scattering peak does not allow further analysis without a model for the shape and the distribution of particles. Lantman et al. [53] used the mixed labeling technique in order to extract single chain scattering and therefore to avoid the problems related to the interference term. Single chain scattering data allow the determination of the radii of gyration and molecular weights of polymer particles from Zimm plot analysis. The range of validity of the approximation were not reached in these experiments but a correction was applied to the data. The molecular weight was found to be constant which shows that no aggregation occurs and the radius of gyration continuously increases with dilution. This result was considered to be evidence of the chain extension with dilution and a picture of the chain conformation was proposed (Fig. 19.8).

19.4 Perfluorinated Ionomer Solutions

19.4.1 Introduction

In the early 1960's, Dupont de Nemours produced the first perfluorinated ionomer membrane: the Nafion. The chemical formula of this ionomer is:

$$-[CF_2-CF_2]_n-CF_2-\underset{\substack{|\\ O\\ |\\ CF_2-\underset{\substack{|\\ CF_3}}{CF}-O-CF_2-CF_2-SO_3^-\ X^+}}{CF}-$$

The most interesting properties of this ionomer are acid and alkali chemical resistance, good ionic conductivity and cation versus anion selectivity. These membranes are used in several separation processes including the chlor-alkali production [14] where it acts as a separator between anodic and cathodic compartments. Among the numerous applications, these membranes are also used as solid polymer electrolytes in water electrolysers and fuel cells [55]. Due to their industrial interest, the perfluorinated ionomers have been extensively studied either as dry materials or as swollen membranes [56].

Recently, an efficient method for dissolution has been discovered [57] but very little work has been devoted to the study of solutions. The primary interest of such perfluorinated solutions was first to recast films either to repair existing membranes or to prepare new composite membranes [58–60]. Another interesting use of recast films was the preparation of modified electrodes [61]. Solutions also exhibit very good conductivities and are electrochemically stable [62, 63]. Therefore, solutions can be used as gelatinous polymer electrolytes in lithium batteries or electrochromic displays [62, 63].

The lithium neutralized membranes are dissolved in a water/ethanol mixture at 250 °C for 1 hour in an autoclave. After returning to room temperature, a clear, homogeneous solution (5% w/w) is obtained. These solutions can be dialysed with another counterion and/or another solvent. Another method is to evaporate the solution, crush the remaining blocks and redissolve the powder [64].

19.4.2 Small Angle Scattering

Perfluorinated materials allow the performance of either small angle neutron or X-ray scattering due to the very good contrast between perfluorinated chains and hydrogenated solvents. The scattering spectra of concentrated solutions present a rather well defined maximum whose position varies with the concentration (Fig. 19.9). Aldebert et al. [65] have shown that the peak position varies linearly with the square root of the concentration. A rod-like structure was deduced. Assuming isolated polymer chains with a molecular weight of 2.5×10^5 and that the peak was an interference peak between ordered particles, they suggested a cubic array of rods originally proposed by de Gennes et al. [66]. This cubic array of rods was preferred to the hexagonal array which is usually proposed for polyelectrolytes because the egality of the calculated length and diameter in the case of the hexagonal array was inconsistent with a rod-like structure. The behavior of diluted solutions was studied in a following work [66]. At low polymer concentration, no peak appears, then it was possible to apply a Guinier approximation. The form factor of an infinite perfect cylinder can be approximated by $I(q) = A \cdot \exp(q^2 R^2/4)/q$ where I is the scattered intensity, q the transfer momentum, A a constant and R the radius of the cylinder. Therefore, the plot of $\mathrm{Ln}(qI)$ versus q^2 has to be linear in the Guinier domain and the radius is determined from the slope. A linear behavior was effectively found (Fig. 19.10) confirming a rod-like structure and the values determined for the radii were found to be between 2.0 and 2.5 nm which was in good agreement with those determined from the peak position using a hexagonal array. The

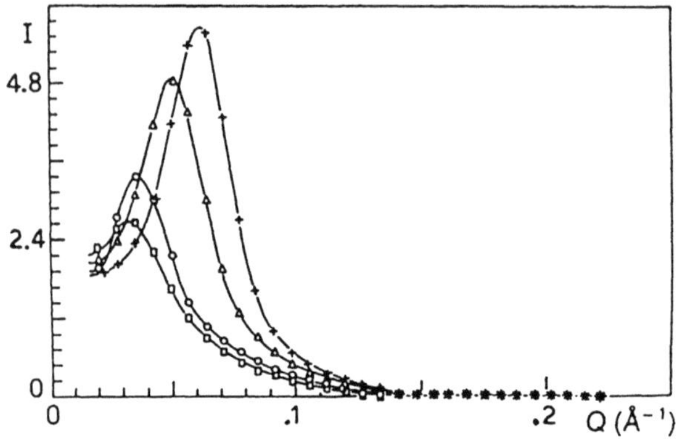

Fig. 19.9. SANS spectra of perfluorinated ionomer solutions in formamide at various volume fractions (4%, 6%, 12%, 20%) (from Ref. [67])

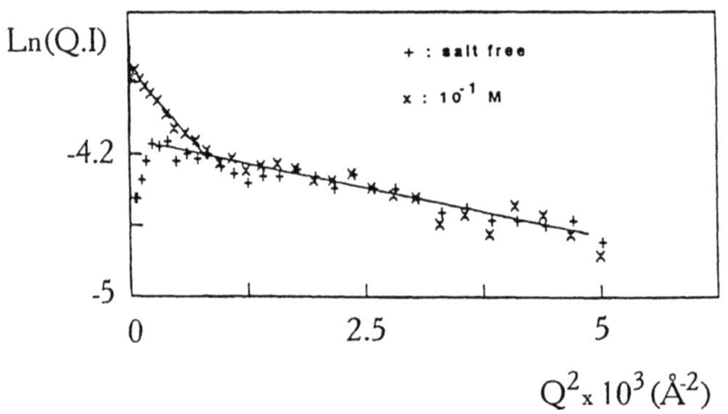

Fig. 19.10. Extended Guinier plot of a dilute perfluorinated ionomer solution in *N*-methylformamide with and without added salt (from Ref. [69])

length on which the cylinder can be considered as perfect was extracted from the deviation to the Guinier law at low angles (100 nm). It was concluded that the rods are made of several polymer chains with the ionic groups at the interface and that these rods are mutually parallel in concentrated solutions (Fig. 19.11).

Changes in the intensity and the position of the peak depending on the solvent were related to the dielectric constant for the former and to the interfacial energy for the latter [60, 68, 69]. This result was confirmed by studying the effect of several mixtures of solvent. The study of short side-chain perfluorocarboxylated ionomer solutions has shown a relation between the side-chain length and the radius of the cylinders [68, 69]. The hypothesis of a perfect two phase system was confirmed by a solvent contrast variation study [68, 69]. Moreover, the study of the Porod limit confirmed the cylindrical structure and allowed the calculation of the radii which were found to be in good agreement with the previous determination [68].

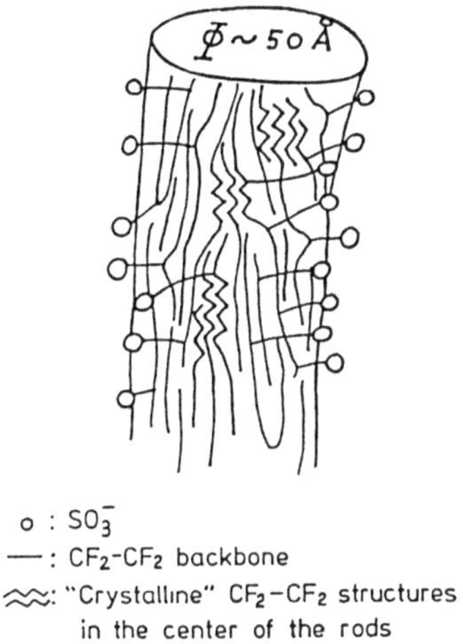

Fig. 19.11. Model of rod-like structure of perfluorinated ionomer in solution (from Ref. [67])

The salt effect on the structure of the solution was also studied [60, 69]. The scattering spectra are dramatically affected – the interference peak disappears and the intensity scattered at low Q increases – when a simple salt is added to a concentrated solution. At low polymer concentration, the Guinier plot remains unchanged except for an increase of the scattered intensity at very low Q values (Fig. 19.10). These effects were attributed to a decrease of the order associated to the beginning of aggregation without any change in the shape and the size of the particles. The quantity of salt necessary to affect the interference peak in concentrated solutions is related to the dielectric constant of the solvent. The same effect occurs respectively with 10^{-3}, 10^{-2} and 10^{-1} mol/l of added salt in ethanol, water and *N*-methylformamide solutions [69].

19.4.3 Rheological Studies

As in S-PS solutions in polar solvent, a polyelectrolyte behavior was evidenced [59, 69, 70]. In alcoholic solutions, the effect was found to decrease with increasing solvent chain length due to the decrease of the dielectric constant [70]. The most important effect was found in dimethylacetamide solutions ($\varepsilon = 38$), but it was surprising to find only a small effect in water solutions and no effect at all in amide solvents of dielectric constants larger than 100. Such a result was also evidenced in S-PS solutions [35]. Two explanations can be proposed: either the effect occurs at lower concentration or the extension in such solvent is already maximum.

The salt effect on the rheological properties is different depending on the concentration [69,70]. At low polymer concentration, the polyelectrolyte effect disappears with the addition of simple salt due to a screening effect (Fig. 19.12), while the solution gelifies in the concentrated solutions due to interparticle aggregation.

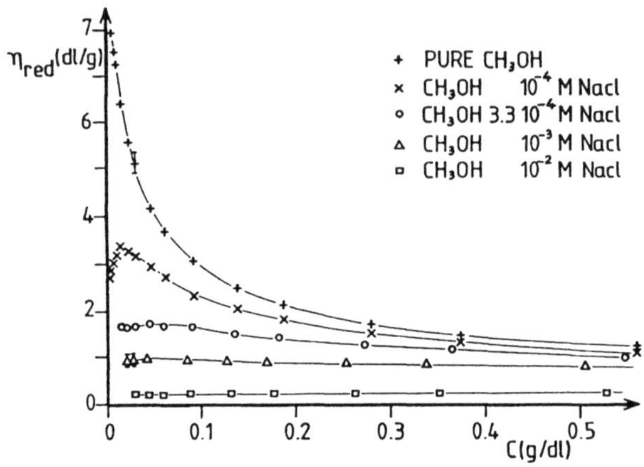

Fig. 19.12. Reduced viscosity as a function of polymer concentration for a perfluorinated ionomer in methanol with added salt (from Ref. [69])

Viscosity experiments, conductivity measurements [62, 63] and ^{19}F NMR data [71] can be interpreted with the rod-like model but these techniques do not actually confirm its validity.

19.5 Conclusion

A recent Small Angle X-ray Study [54] on poly AMPS (acrylamido methylpropanesulfonate) in polar solvent is a good example of the confusion which occurs in the ionomer field of research. The polymer is highly sulfonated (from 16.5 to 66.7 mol % of sulfonate content) and dissolved in *N*-methylformamide, the dielectric constant of which is 182. Therefore, the solution properties should be dominated by the electrostatic repulsions between the ionized ionic groups. Light scattering experiments indicates hydrodynamic radii of several nanometers, the size of which decreases with increasing the temperature. The aggregation was deduced. The polyelectrolyte effect in the viscosity measurements was then interpreted as a consequence of this aggregation phenomenon. The small angle data were also analyzed in terms of aggregation: the interference peak is supposed to be related to an internal structure of the aggregates and several models developed for bulk ionomers were applied because it is well known that the scattering peak in bulk ionomers is due to aggregation. But this aggregation results from the attraction of associated ion-pairs in a low-polarity medium and can certainly not be applied in the present case where dissociated groups repel each others in a very high dielectric medium. Same kind of confusion occurs when models developed for dry ionomers are applied to water swollen membranes (and vice versa).

According to the definitions of Eisenberg and Rinaudo [38], the solution properties of ionomers can be divided in ionomer and polyelectrolyte behaviors depending on the polarity of the solvent. These two behaviors have to be completely dissociated since the former is due to the attractions of associated ion-pairs while the other results from the repulsions between ionized groups. However, ionomers are complex systems which cannot satisfactorily be described by such simple definitions. The S-PS ionomer has always been considered as a model compound to describe the ionomer behavior but the generalization of the results form S-PS to other ionomer solutions is definitively not straightforward. For example, the structural differences between perfluorinated and S-PS solutions in a polar solvent are very significant while the rheological behavior is very similar. Moreover, if a perfluorinated ionomer is exchanged with a hydrophobic alkylammonium cation, the ionomer is not soluble anymore in polar solvents but is easily soluble in THF. In this case the solution behavior is very similar to S-PS solution in nonpolar solvents while the local structure remains unchanged compared to the structure described in Fig. 19.11. Other interesting problems could be the study of: 1) the ionomer structure in mixed polar and nonpolar solvents, 2) the structure of swollen ionomer membranes with polar solvents. In both cases it will probably be difficult to classify their behavior either as ionomer or polyelectrolyte types.

Therefore, ionomer solutions cannot be treated as a whole and cannot even be divided into only two different problems: the ionomer (or associating) behavior in nonpolar solvents and the polyelectrolyte behavior in polar solvents. A new structural problem will probably arise with most of the ionomers even if some of the solution properties appear to be similar. A lot of structural studies including small angle scattering and electron microscopy experiments will be necessary to understand the structure of ionomer solutions and such studies will be very helpful to understand the structure-properties relationships of ionic polymers.

19.6 References

1. Hollyday L (1975) Ionic Polymers. Applied Science, London
2. Wilson AD, Professor HJ (eds) (1986) Developments in ionic polymers-2. Applied Science, London
3. Wilson AD, Prosser HJ (eds) (1983) Developments in ionic polymers-1. Applied Science, London
4. Eisenberg A, King M (1977) Ion-containing polymers. Academic, London
5. Rice SA, Nagasawa M (1961) Polyelectrolyte solutions. Academic, New York
6. Oosawa F (1971) Polyelectrolytes. Marcel Dekker, New York
7. Katchalsky A (1971) Pure and Applied Chem 26:327
8. Selegny E, Mandel M, Strauss UP (eds) (1974) Charged and reactive polymers, vol I..D. Reidel, Dordrecht
9. Overbeek JThG (1976) Pure Appl Chem 46:91
10. Manning GS (1969) J Chem Phys 51:924
11. Eisenberg A (1970) Macromolecules 3:147
12. Pineri M, Eisenberg A (eds) (1986) Structure an properties of ionomers. D. Reidel, Dordrecht
13. Lunderg RD (1987) Encyclopedia of Polymer Science, 2nd Ed 8:393
14. Coulter MO (ed) (1980) Modern chlor-alkali technology. Ellis Horwood, Chichester, England
15. Yeo SC, Eisenberg A (1977) J Appl Polym Sci 21:875
16. Joanny JF (1988) Polymer 21:71
17. Witten TA (1988) J Phys France 49:1055
18. Broze G, Jerome R, Teyssie Ph, Marco C (1983) Macromolecules 16:996

19. Broze G, Jerome R, Teyssie Ph (1982) Macromolecules 15:920
20. Broze G, Jerome R, Teyssie Ph (1982) Macromolecules 15:1300
21. Lundberg RD (1982) J Appl Polym Sci 27:4623
22. Agarwal PK, Makowski, HS, Lundberg RD (1984) Macromolecules 17:1918
23. Agarwal PK, Makowski HS, Lundberg RD (1984) Macromolecules 17:1929
24. Lundberg RD (1986) J Appl Polym Sci 31:1843
25. Peiffer DG, Kim MW, Schulz DN (1987) J Polym Sci: Polym Phys Ed 25:1615
26. Lundberg RD, Makowski HS (1980) J Polym Sci: Polym Phys Ed 18:1821
27. Lundberg RD, Phillips RR (1982) J Polym Sci: Polym Phys Ed 20:1143
28. Lundberg RD, Phillips RR (1984) Polymer Preprints 25 (n°2):303
29. Kim MW, Peiffer DG (1985) J Chem Phys 83:4159
30. Hara M, Wu JL, Lee AH (1988) Macromolecules 21:2214
31. Agarwal PK, Garner RT, Graessley WW (1987) J Polym Sci: Polym Phys Ed 25:2095
32. Hara M, Lee AH, Wu JL (1987) J Polym Sci: Polym Phys Ed 25:1407
33. Hara M, Wu JL, Lee AH (1989) Macromolecules 22:754
34. Rochas C, Domard A, Rinaudo M (1979) Polymer 20:76
35. Peiffer DG, Lundberg RD (1984) J Polym Sci: Polym Phys Ed 22:1757
36. Niezette J, Vanderschueren J, Aras L (1984) J Polym Sci: Polym Phys Ed 22:1757
37. Peiffer DG, Kim MW, Kaladas J (1988) Polymer 29:716
38. Eisenberg A, Rinaudo M (1990) Polym Bull 24:671
39. Martenot JP, Galin JC, Picot C, Weill G (1989) J Phys France 50:493
40. Joanny JF, Leibler L (1990) J Phys France 51:545
41. Makowski HS, Lundberg RD, Shinghal G (1975) US Patent 3,870,841
42. Pedley AM, Higgins JS, Peiffer DG, Rennie AR, Staples E (1989) Polym Commun 30:162
43. Lantman CW, MacKnight WJ, Peiffer DG, Sinha SK, Lundberg RD (1987) Macromolecules 20:1096
44. Pedley AM, Higgins JS, Peiffer DG, Burchard W (1990) Macromolecules 23:1434
45. Hara M, Wu JL (1986) Macromolecules 19:2887
46. Hara M, Wu JL (1988) Macromolecules 21:402
47. Lantman CW, MacKnight WJ, Higgins JS, Peiffer DG, Sinha SK, Lundberg RD (1988) Macromolecules 21:1339
48. Pedley AM, Higgins JS, Peiffer DG, Rennie AR (1990) Macromolecules 23:2494
49. Dowling KC, Thomas JK (1991) Macromolecules 24:4123
50. Fitzgerald JJ, Kim D, Weiss RA (1986) J Polym Sci: Polym Letters 24:263
51. Ise N, Okubo T, Yamamoto K, Kawai H, Haschimoto T, Fujimura M, Hiragi Y (1980) J Am Chem Soc 102:7901
52. Lipatov YS, Tsukruk VV, Dmitruk NV, Polyatskova NV, Shilov VV, Shevchenko VV (1983) Polymer Comm. 24:197
53. Lantman CW, MacKnight WJ, Higgins JS, Peiffer DG, Sinha SK, Lundberg RD (1988) Macromolecules 21:1344
54. Wang J, Wang Z, Peiffer DG, Shuely WJ, Chu B (1991) Macromolecules 24:790
55. Aldebert P, Novel-Cattin F, Pineri M, Millet P, Doumain C, Durand R (1989) Solid State Ionics 35:3
56. Eisenberg A, Yeager HL (eds) (1982) Perfluorinated ionomer membranes. ACS Symp Ser 180, American Chemical Society, Washington
57. Grot WG, Chadds C (1982) European Patent 0 066 379
58. Gebel G, Aldebert P, Pineri M (1987) 20:1425
59. Moore RB, Martin CR (1988) Macromolecules 21:1334
60. Gebel G (1989) Thèse Grenoble
61. Martin CR, Rubinstein J, Bard AJ (1982) J Am Chem Soc 104:4817
62. Guglielmi M, Aldebert P, Pineri M (1989) J Appl Electrochem 19:167
63. Aldebert P, Guglielmi M, Pineri M (1991) Polymer Journal 23:399
64. Aldebert P, Pineri M (1986) French Patent 8 605 792
65. Aldebert P, Dreyfus B, Pineri M (1986) Macromolecules 19:2651
66. de Gennes PG, Pincus P, Velasco R, Brochard F (1976) J Phys France 37:1461
67. Aldebert P, Dreyfus B, Gebel G, Nakamura N, Pineri M, Volino F (1986) J Phys France 49:2101
68. Loppinet B, Gebel G, Aldebert P, Pineri M, Williams CE, to be published
69. Gebel G, Loppinet B, Aldebert P, Pineri M (1991) Polymer Preprints 32(1):612
70. Aldebert P, Gebel G, Loppinet B, to be published
71. Schlick S, Gebel G, Pineri M, Volino F (1991) Macromolecules 24:3517

20 Scattering Studies of Ionomer Aggregates in Nonpolar Solvents

J. Bodycomb and M. Hara

Following a rather extended introduction explaining the aggregation behavior of ionomers studied by scattering methods in low-polarity or nonpolar solvents, some of the recent results on the aggregation behavior of ionomer solutions in this laboratory are discussed. The aggregation behavior of sulfonated polystyrene ionomers dissolved in toulene is analyzed by both static and dynamic light scattering techniques as a function of ion content and ionomer concentration. It is shown that the association between ionomer molecules is enhanced by increasing ion content. Furthermore, the size distribution of the ionomer aggregates obtained to date by using CONTIN analysis of dynamic scattering data suggest that small amounts of very large aggregates coexist with normal size coils, and possibly with collapsed coils, in ionomer solution in the concentration range studied.

20.1 Introduction

20.1.1 Ionomer Solutions

Electrostatic interactions in polymers can cause dramatic changes in the properties of polymeric materials. One approach to creating electrostatic interactions is the introduction of ionic groups onto a neutral polymer chain. Ionomers are a new class of such ion-containing polymers having a few ionic groups up to 10–15 mole% along nonionic backbone chains [1, 2]. Extensive research has been concentrated on the elucidation of the structure-property relationship of ionomers in the solid state during the past 25 years. In addition, because of the realization of unique behavior, the study of ionomer solutions has recently become increasingly active. It has been established that ionomer solutions show two types of behavior, primarily depending on the solvent [3–6]; aggregation behavior in low polarity solvents such as tetrahydrofuran (THF) and "polyelectrolyte" behavior in polar solvents such as dimethylformamide (DMF).

In nonpolar solvents the counterions are not dissociated from the ionomer, instead forming ion pairs, due to the strong ionic interactions in the medium with low dielectric constant. These ion pairs attract each other, leading to aggregation. On the other hand, in polar solvents many of the counterions are dissociated from the ionomer chain, due to weaker ionic interactions in the medium with high electric constant. The electrostatic interactions between ionic

Macromolecular Complexes
in Chemistry and Biology
Eds Dubin/Bock/Davis/Schulz/Thies

groups and counterions cause polyelectrolyte behavior. An extensive review of the latter is available [7]. This paper focuses on the former; a rather extensive introduction covering recent advances in the scattering study of ionomer aggregates in solution will be offered, followed by a discussion of recent experimental results of this laboratory.

20.1.2 Aggregation Behavior of Ionomers in Solution

Initial experimental studies on the aggregation behavior of ionomer solutions was conducted mainly by viscosity measurements [3, 4]. Typical reduced viscosity behavior of ionomers in nonpolar solvent is such that at low concentrations, the reduced viscosity of ionomer solutions is smaller than that of the starting material, suggesting that the average size of the coil is smaller than that of the starting material. This was interpreted as a concentration regime where intramolecular interactions dominate and cause chain contraction. As concentration is increased, the reduced viscosity sharply increases, and surpasses that of the starting polymer, eventually forming gels at higher concentrations. This regime was interpreted as a concentration regime where intermolecular interactions dominate, so the polymer chains predominately interact with each other to form large aggregates. Although the interpretation based on viscosity data is basically correct, new experimental results obtained by scattering techniques, such as light scattering make some modification of the previous view necessary [8, 9].

20.1.3 Scattering of Aggregated Ionomer Solutions

Small angle light scattering results obtained at very low concentrations where the reduced viscosity of ionomer solutions is smaller than that of the parent polymer showed higher molecular weights in ionomer solutions, suggesting that intermolecular aggregation persists down to very low concentrations [8, 9]. A typical Kc/R_0 vs. c curve is a straight line with lower intercept and slope than those of the starting polymer and even with a negative slope for low molecular weight ionomers. Therefore, measured apparent molecular weights of ionomer samples are higher and A_2 values are lower than those of parent polymers. These effects become more pronounced with increasing ion content [9]. The reduction in second virial coefficients may be attributed to a reduced quality of the solvent for the ionomer compared to the neutral polymer system and attraction between ion pairs in the ionomer system. Since ionic groups are attached to these chains, it is expected that the solvent quality in the ionomer systems would decrease, leading to a decrease in A_2. It was pointed out, however, that the higher apparent molecular weight obtained for the ionomer solution whose overall A_2 value is still positive indicate the importance of the latter [9]. Theoretically, it has been pointed out that the overall second virial coefficient of ionomer solution, A_2, consists of two terms; i.e. $A_2 = A_2' + f^2 A_2''$ [10, 11] where A_2' is the second virial coefficient due to the excluded volume of the backbone chain, f is the ion content, and A_2'' is the second virial coefficient due to attractions between ionic dipoles and is therefore always negative. Hara and Wu [9] have suggested that A_2' also decreases with increasing ion content, since the addition of ionic groups (i.e. ionic dipoles) to the polymer chain should lead to a decrease in the solvent quality. This equation provides a useful framework for considering the effect of ion content on the second virial coefficient, specifically by drawing attention to the idea that it is affected by both the excluded volume effect of backbone chains and the association between ion pairs.

Since ionomers are shown to be associating macromolecules, it may be of interest to compare their behavior with that of other associating systems, such as associating proteins in water or neutral polymers in poor solvent below the Θ temperature [12]. While these systems have shown that Kc/R_0 vs. c curves are concave up, reaching to an intercept whose value is the inverse of the molecular weight of single molecules [12], no such curves are reported for ionomer solutions. This may be due to the fact that interactions between ion pairs are much stronger than van der Waals interactions or hydrogen bonding seen in other aggregation systems.

Small angle X-ray scattering (SAXS) has so far been used only for ionomer solutions in polar solvents and no data is available for ionomers in nonpolar solvents [13, 14]

Small angle neutron scattering (SANS) has also been used to study ionomer solutions [15, 16]. Studies on sulfonated PS ionomers in deuterated THF showed that both the ionomer radius of gyration and apparent molecular weight increased with increasing concentration [15]. Furthermore, it was noted that these aggregates were quite polydisperse. SANS is an especially useful tool, as is demonstrated by many examples for polymers both in solid and solution, to study the structure of individual chains in the presence of other chains by using the isotope effect. By using a mixed labeling technique to obtain scattering functions from single chains in molecular aggregates, it was shown that the radius of gyration of the individual chains within an aggregate remain constant with changes of concentration (0.5–4.0 g/dl) [15]. Also SANS studies on a similar system were conducted at various temperatures, and an open association model, which assumes an equilibrium between aggregates of all sizes and single chains, was used to analyze the data [16]. This study concluded that the extent of aggregation of ionomer chains was driven by the increased configurational entropy of the chains in the aggregates compared to that of the single collapsed chain. Finally, it was noted that the sizes of single collapsed chains were very small compared with PS.

Dynamic light scattering has also been used to study ionomer aggregates in solution. It was shown that hydrodynamic radius increases with increasing ion content and ionomer concentration (up to a certain point, above which the hydrodynamic radius becomes almost constant in the case of the THF solution [8]). While Lantman et al. only obtained hydrodynamic radii from scattering data at an angle of 90°, Pedley et al. used the cumulant method to obtain information on the distribution of aggregates and found that the quality factor, a measure of polydispersity, increased with increasing ionomer concentration [17]. Also their data suggested the existence of single collapsed coils, but mostly at concentrations below those where initial viscosity measurements suggested the existence of collapsed coils due to intramolecular interactions [4]. It should be stressed, however, that their analysis is based on the observation that there exists only "single chains and aggregates consisting of two chains" over the concentration range studied (0.1–1.0 g/dl).

So far most studies of aggregation behavior have been conducted with low polarity solvents such as THF [4, 9, 15]. Recently there have been a few reports with nonpolar solvents such as xylene [16–18]. While these systems require low ion contents to form a solution, the ionomer electrical environment is more similar to that found in the solid state (i.e. low dielectric constant). In this work we use static and dynamic light scattering techniques to examine the aggregation behavior of sulfonated polystyrene ionomers in a nonpolar solvent, toluene.

20.2 Experimental

Lightly sulfonated polystyrene samples with ion contents of 0.8 and 1.2 mol% were prepared by solution sulfonation of standard polystyrene ($M_w = 1.06 \times 10^5$, $M_w/M_n < 1.06$). Details about the preparation of lightly sulfonated polystyrene and their characterization were described elsewhere [19].

Polymer solutions were prepared by dissolving the freeze dried ionomer samples in toluene under stirring for two days at room temperature. From this stock solution, dilutions were made and each solution was allowed to stand until two consecutive measurements two days apart by dynamic light scattering gave the same results. Small angle light scattering experiments were conducted with a KMX-6 low-angle light scattering photometer (Chromatix) at a wavelength of 633 nm at 25 ± 0.5 °C. The effect of large dust particles was easily removed by using a flow cell. Dynamic light scattering measurements were conducted with a BI-200SM photogoniometer (Brookhaven) and a BI-2030 digital correlator (Brookhaven) at a wavelength of 633 nm. Measurements were conducted at five angles from 30 to 150 degrees at 25 ± 0.1 °C for cumulant analysis and at an angle of 60 degrees for analysis by CONTIN. Each light scattering cell was cleaned for two hours with condensing acetone vapors using a Thurmond type apparatus [20]. The optical clarification of the solution for both static and dynamic scattering was carried out by passing the solution through two membranes whose pore sizes were 0.5 and 0.2 microns in succession. The specific refractive index increment *dn/dc* was measured at 25 ± 0.1 °C by using a KMX-16 differential refractometer (Chromatix). Details concerning the small angle light scattering experiments were described elsewhere [9].

20.3 Light Scattering Analysis

Since interactions between ion pairs are rather short range, the experimental results may be analyzed as if each aggregate was a separate species. The analyses of light scattering data were described elsewhere [9, 21]; here only essential equations are shown.

In the case of static light scattering, the results in poor solvents can be expressed in terms of a virial expansion [22]

$$Kc/R_\theta = 1/MP(\theta) + 2A_2c + \ldots \tag{20.1}$$

where R_θ is the excess reduced scattered intensity (at an angle θ) from the solution, c represents the polymer concentration (g/cm^3), M the (weight average) molecular weight, and $P(\theta)$ the particle scattering factor. K is the optical constant i.e. $K = 4\pi n_0^2(dn/dc)^2/N_0\lambda_0^4$ where n_0 is the refractive index of the solvent, λ_0 is the wavelength in vacuo, and N_0 is Avagadro's number. For low angle data, $P(\theta)$ approaches 1 and Eq. (20.1) reduces to

$$Kc/R_0 = 1/M + 2A_2c + \ldots \tag{20.2}$$

Dynamic light scattering data can be analyzed by the method of cumulants [23]:

$$\ln[C(\tau)/B - 1]^{1/2} = \ln b^{1/2} - \Gamma\tau + \mu_2\tau^2/2 + \ldots \tag{20.3}$$

where $C(\tau)$ is the measured correlation function, B is the baseline, and b is an optical constant. Γ represents the first cumulant and μ_2 is the second cumulant. An effective diffusion coefficient was obtained from the following equation:

$$\Gamma = D_{\text{eff}}q^2 \tag{20.4}$$

where $q = 4\pi/\lambda \sin(\theta/2)$ is the scattering vector.

By using the well-known Stokes–Einstein relationship [24], effective diffusion coefficient can be converted to an effective hydrodynamic diameter D_{h}:

$$D_{\text{eff}} = \frac{kT}{3\pi\eta D_{\text{h}}} \tag{20.5}$$

where k is Boltzman's constant, T is the thermodynamic temperature, and η is the viscosity of the solvent.

In addition a "quality" factor, defined by μ_2/Γ^2 of Eq. (20.3) offers a measure of the width of the distribution of particle sizes measured by dynamic light scattering [23, 25].

Dynamic light scattering data can also be interpreted by the more exact relation

$$C(\tau)/B - 1 = b_1e^{\Gamma_1\tau} + b_2e^{\Gamma_2\tau} + \ldots \tag{20.6}$$

where b_i is a factor due to the scattered intensity associated with each decay constant, Γ_i (or effective size by Eqs. (20.4) and (20.5)) and can be related to the fraction of species at this size through the use of proper form factors. Equation (20.6) can be rewritten in integral form as

$$[C(\tau)/B - 1]^{1/2} = \int_0^\infty b(\Gamma)e^{-\Gamma\tau}d\Gamma \tag{20.7}$$

In the continuous form it is assumed that b_i is a continuous function of Γ. Solving for $b(\Gamma)$ is a problem of Laplace inversion, which is a well known ill conditioned problem; i.e. there are an infinite number of solutions even for a small experimental error [26]. The most popular approach to this problem is constrained regularization as implemented in Provencher's program CONTIN [26–29]. This technique relies on the idea of parsimony to obtain "Smoothness with the minimum number of peaks." More specifically, a penalty term for the second derivative of $b(\Gamma)$ is included in the least squares calculations of the solution of Eq. (20.7). Thus, the resolution of the technique is rather poor and studies show that it tends to oversmooth the data [30]. Since data was taken at only one angle we will discuss only the nature of $b(\Gamma)$ while noting that at this point a choice of form factors may be merely arbitrary.

20.4 Results and Discussion

Figure 20.1 shows the reciprocal reduced scattered intensity at zero angle, Kc/R_0, against polymer concentration for the ionomer samples as well as for the starting polystyrene in toluene. The data are analyzed by using Eq. (20.2). It is seen that the weight average molecular weight increases and the second virial coefficient decreases with increasing ion content. Similar results were obtained for sulfonated polystyrene ionomers dissolved in a low-polarity solvent, THF

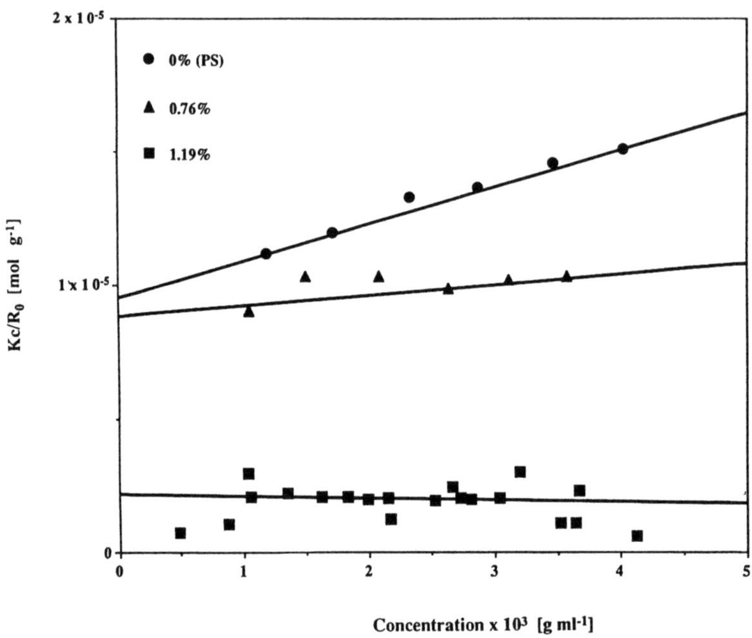

Fig. 20.1. Reciprocal reduced scattered intensity at zero angle, Kc/R_0, versus polymer concentration for ionomers (0.76 and 1.19 mol% ion content) as well as PS in toluene

[9]. A plot of the interaction parameter, A_2, versus ion content is given in Fig. 20.2. As seen in previous studies on ionomers in THF, A_2 decreases with increasing ion content and eventually becomes negative [9]. This decrease can be attributed to the increasing contribution of attractive interactions between ion pairs, as well as the poor solvent quality of toluene for ionomers. As is the case for THF solutions, the higher molecular weight obtained for the system with positive A_2 suggests the important contribution of the attraction between ion pairs.

Figure 20.3 shows the dynamic light scattering data for the system at 25 °C. Here, the effective diameter of the aggregates obtained by using Eq. (20.5) is plotted against ionomer concentration. The same trends as seen in the static scattering results are seen here; the higher the ion content, the larger the effective size of the aggregates. Also, it is seen that the size of the aggregates increases with polymer concentration. Again, similar results were obtained by dynamic light scattering for THF systems [15].

In addition, the "quality" factor that can be a measure of polydispersity for ionomer aggregates is plotted against polymer concentration in Fig. 20.4. By considering the rather large errors involved in quality factors, the difference between ionomer samples with different ion contents may be small. However it is clear that the width of the distribution of particles in the ionomer solution is

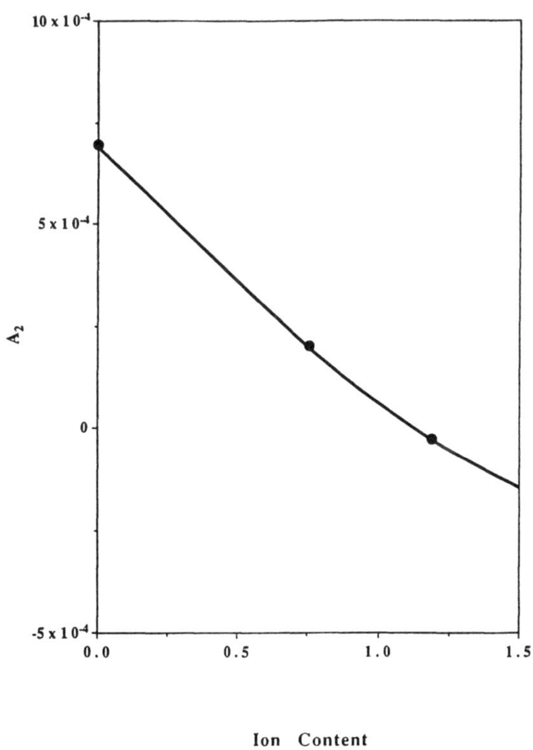

Fig. 20.2. Second virial coefficient, A_2, versus ion content for ionomers in toluene

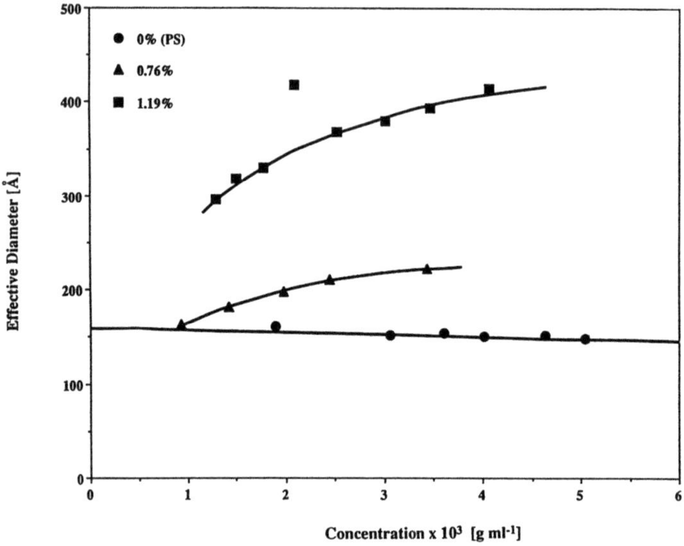

Fig. 20.3. Effective diameter obtained by dynamic light scattering using the method of cumulants versus polymer concentration for ionomers 0.76 and 1.19 mol% ion content) as well as PS in toluene

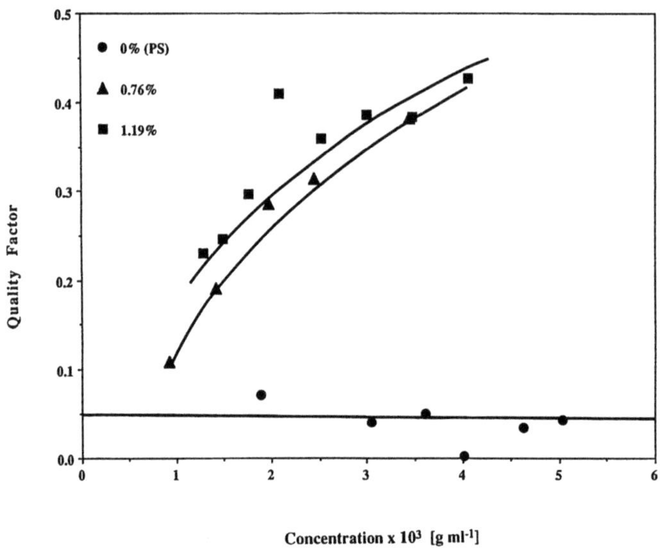

Fig. 20.4. Quality factor obtained by dynamic light scattering using the method of cumulants versus polymer concentration for ionomers (0.76 and 1.19 mol% ion content) as well as PS in toluene

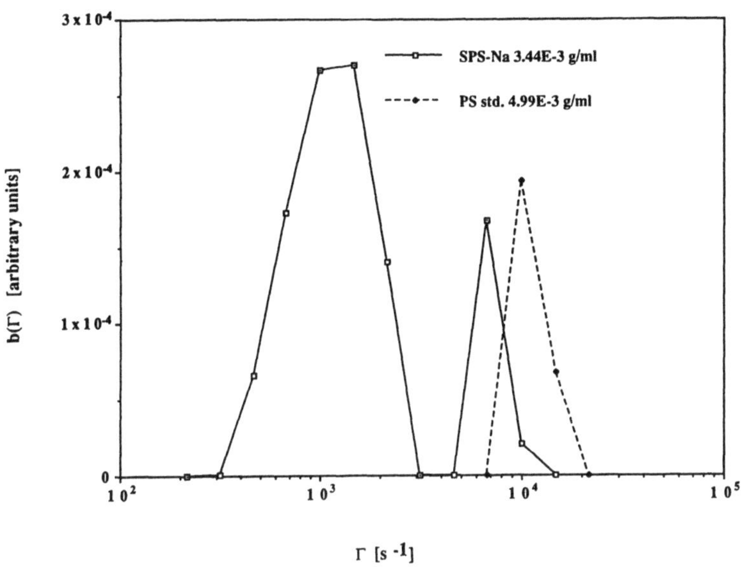

Fig. 20.5. Distribution of decay constant obtained by dynamic light scattering using CONTIN analysis for ionomer (0.76 mol% ion content) as well as PS in Toluene

quite large compared with that of its precursor polystyrene and that this width is increasing with increasing ionomer concentration. These results suggest that aggregation due to intermolecular interactions are indeed occurring at low concentrations and that the widening of the distribution occurs because of an increasing degree of aggregation with increasing concentration.

Finally, the distribution of $b(\Gamma)$ was considered in more detail by using dynamic light scattering data with constrained regularization (CONTIN). Representative results are shown in Fig. 20.5. The ionomer system shows two distinct peaks, one corresponding to a relaxation time close to that of the starting polystyrene solution, and a second, much longer relaxation time, suggesting the existence of some extremely large particles. To test this, a "peak constrained analysis" was performed with CONTIN, during which the ionomer data was forced to take the form of a single peak. The results indicated, however, that the data required two peaks in $b(\Gamma)$. The large particles seem to be caused by ion pair interactions, since the addition of 5% methanol can completely eliminate the peak corresponding to large particles. The effect of alcohol on ionomer aggregates in breaking ion pair interactions is a well known phenomena [3, 16]. Although at this time we cannot obtain more detailed information on very large aggregates, such as sizes and size distribution, partly due to the lack of information on the structure factor of the ionomer aggregates, we may indicate some points. By simply assuming the form factor of rigid spheres for ionomer aggregates, the contribution of each particle to the scattering intensity can be calculated. This analysis shows that the amount of the large aggregates is very small, but the fraction of large aggregates compared with smaller particles (i.e.

the ratio in peak area of the large aggregates to normal size coils in scattered intensity vs. Γ plot) increase with ionomer concentration. This is consistent with general observations of the growth of ionomer aggregates with ionomer concentration. In addition, it should also be noted that there is little indication for the existence of collapsed coils. This may be due to the fact that the contribution of very small particles to scattering is much smaller compared with that of very large aggregates. In fact, some recent results suggest the existence of such small peaks. Because of the preliminary nature of our data, it is difficult to compare our results with various models, including the open association model. Currently, static experiments at higher angles are underway to obtain additional information, including structure factor, as well as more detailed dynamic scattering experiments. These studies will help us clarify these points.

Acknowledgements. We thank Dr. J. Scheinbeim for useful discussions. Acknowledgement is also made to the NSF and Sigma Xi for partial support of our research.

20.5 References

1. Eisenberg A, King M (1977) Ion-Containing Polymers. Academic Press, New York
2. MacKnight WJ, Earnest TR (1981) J Polym Sci: Macromol Rev 16:41
3. Lundberg RD, Makowski HS (1980) J Polym Sci Polym Phys Ed 18:1821
4. Lundberg RD, Phillips RR (1982) J Polym Sci Polym Phys Ed 20:1143
5. Marina MG, Monakov YB, Rafikov SR (1979) Uspekhi Khim 48:722; (1979) Russ Chem Rev 48:389
6. Schade H, Gartner K (1974) Plaste Kautschuk 21:825
7. Hara M (1993) Polyelectrolytes: Science and Technology. Marcel Dekker, New York
8. Lantman CW, MacKnight WJ, Peiffer DG, Sinha SK, Lundberg RD (1987) Macromolecules 20:1096
9. Hara M, Wu J (1988) Macromolecules 21:402
10. Joanny JF (1980) Polymer 21:71
11. Cates ME, Witten TA (1986) Macromolecules 19:732
12. Tanford C (1961) Physical Chemistry of Macomolecules. John Wiley and Sons, New York
13. Wang J, Wang Z, Peiffer D, Shuely W, Chu B (1991) Macromolecules 24:790
14. Aldebert P, Dreyfus B, Pineri H (1986) Macromolecules 19:2651
15. Lantman CW, MacKnight WJ, Higgins JS, Peiffer DG, Sinha SK, Lundberg RD (1988) Macromolecules 21:1339
16. Pedley AM, Higgins JS, Peiffer DG, Rennie AR (1990) Macromolecules 23:2494
17. Pedley AM, Higgins JS, Peiffer DG, Burchard W (1990) Macromolecules 23:1434
18. Peiffer DG, Kaladas J, Duudevon I, Higgins JS (1987) Macromolecules 20:1397
19. Hara M, Lee AH, Wu J (1987) J Polym Sci Polym Phys Ed 25:1407
20. Thurmond CD (1952) J Polym Sci 8:607
21. Hara M, Wu J (1989) In: Weiss R (ed) Multiphase Polymers Blends and Ionomers. American Chemical Society, Washington, DC, p 447
22. Zimm B (1948) J Chem Phys 16:1093; 16:1099
23. Koppel DD (1972) J Chem Phys 57:4814
24. Einstein A (1956) Investigations on the Theory of the Brownian Movement, Dover, New York.
25. Pecora R (ed) (1985) Dynamic Light Scattering. Plenum, New York
26. Provencher SW (1979) Makromol Chem 180:201
27. Provencher SW (1976) Biophys J 16:27
28. Provencher SW (1976) J Chem Phys 16:2772
29. Version 2, March 1984 Courtesy of SW Provencher
30. Stock RS, Ray WH (1985) J Polym Sci: Polym Phys Ed 23:1393

Subject Index

Acrylamide copolymers 137
-/sodium acrylate 124
2-Acrylamido-2-methylpropanesulfonate (AMPS) 36-38, 86, 97, 102
-, poly- 344
Affinity partition 237
- precipitation 241
Affinophoresis 236
Albumin, bovine serum 120, 248, 251, 255, 294
-, human serum 301
Alcoholdehydrogenase 161
Alginate-polylysine, polymannuronate-rich 180
Alginates 175, 178
Alkyl acrylamide 4
Alkyl-4-vinylpyridinium monomers/polymers 6, 75-89
AMPS *see* 2-Acrylamido-2-methylpropanesulfonate
-, poly- 344
8-Anilino-1-naphthalenesulfonic acid (ANS) 23
ANS *see* 8-Anilino-1-naphthalenesulfonic acid
Antibodies 166
Antigens, protein 166
Associative thickness 15, 27, 30
Auramine O 215

Bacillus subtilis 171, 172
Bjerrum distance 145
Block copolymers, hydrophobic 8, 10, 34, 35, 47, 48, 53, 64, 65
Bovine serum albumin 120, 248, 251, 255, 294
2-Bromoethanol 161

Cell transformations 151ff
Cellulose, hydrophobically modified hydroxyethyl (HMHEC) 6, 16, 64-69
Charge-transfer complexes 185
α-Chymotrypsin 161
Circular dichroism 175ff
Cloud curve 140
CMC *see* Critical micelle concentration
Coacervation 148, 285
Colloid titration 305
Complex, coacervation 285
-, protein-polyelectrolyte 229, 247, 301
-, water-soluble 247
Conductimetry 135
Conductivity, specific 142
Critical micelle concentration 24, 35, 75, 96, 216
- polymer concentration 54
- salt concentration 54
- shear deformation concentration 222

Debye plot 123, 124
Design enzyme 161
Dextran sulfate 249
Diphasic systems 135
2,6-Diphenyl-4-(2,4,6-triphenyl-1-pyridino) phenoxide 28-30
DNA-containing IPECs 170
Drag reduction 216, 217
Dynamic light scattering, ionomer solutions 349

Egg white proteins 274
Electrochromic displays 341
Electrodes, modified 341
Electrophoresis 125, 132

Enzyme, design 161
EPDM, sulfonated 330
ESR, ionomer solutions 337
1-Ethyl-4-vinylpyridinium/vinylpyridine 154
Ethylbromide 161

Flory-Huggins model 145
Fluorescence probe studies, poly(acrylic acid) 215
- studies, ionomer solutions 336
Fluorophobically modified polymers 7, 100-103, 109

Galacturonan 178
Gel filtration 223
- permeation chromatography 135, 139, 140
Gelatin (type B) 295
Glutamate-dehydrogenase 166
GPC 135, 139, 140
Guinier plot 129
Guluronate 178

Halatotelechelic ionomers 330
Hemoglobin, bovine 301
-, human 248, 304
Holtzer plot 123
Hückel theory 145
Huggins constant 41, 42, 45, 46, 59, 61
Hydrogen-bonding complex, solid properties 200
Hydrophobe-assisted rigidity 222
Hydrophobically modified ethoxylated polyurethanes (HEUR) 7-9, 19
- - hydroxyethyl cellulose (HMHEC) 6, 16, 64-69

Immunochemical reagents 166
Interpolyelectrolyte complexes 151
Interpolymer complexes, ion-conduction 183
Ionomer aggregates 348, 355
- solutions 331ff
- -, aggregation behavior 348
- -, fluorescence studies 336
Ionomers 329
-, halatotelechelic 330
-, perfluorinated 341

Kratky plot 123

Lattice formalism 145
Light scattering 253-257
- -, ionomers 333
Lipopolyamines 167
Local chain rigidity 220
Lysozyme 274

Methylmethacrylate 137
Microencapsulation 285
Multienzyme IPEC systems 166

Nafion 340
Nucleic acids 167

Oligo(ethyleneimine)s 152
Oligo(phosphate)s 152
2-Oxoglutaric acid 166

P4VP quartenary salts 168
PAA 197
PAM-Surf 5
P(AM-AMPS) 142
P(AM-AMS) 142
Pectate-polylysine 176f
Penicillin amidase 161
PEO 121, 220
-, segmental motion 204
Phase diagrams 135, 139
PMA, pyrenyl-tagged 154
PMAA/POE 206
PMMA 121, 137, 185
Poly(acrylamide) 122, 124
Poly(acrylic acid) 120, 183, 249, 274
-, fluorescence probe studies 215
Poly(*tert*-butylaziridine) 157
Poly(diallyldimethylammonium chloride) 248, 251, 255, 303
Poly(dimethylaminoethyl methacrylate) 152
Poly(dimethyldiallylammonium chloride) 120
Poly(ethylene oxide) 121, 220
Poly(ethyleneimine) 120, 157
Poly(methacrylic acid) 183
-/Poly(ethyene oxide) 121
Poly(methylate) polyanions 152
Poly(methylmethacrylate) 121, 137, 185
Poly(oxyethylene) 183, 201
Poly(potassium vinyl sulfate) 157
Poly(sodium acrylates), hydrophobically modified (HMPAA) 9, 11, 52, 54
Poly(vinyl alcohol) 121, 197

Poly(vinyl alcohol sulfate), potassium 248, 253, 303
Poly(4-vinyl-*N*-ethyl pyridinium bromide) 120
Poly(4-vinylpyridine) 248, 250
Poly(*N*-vinyl-2-pyrrolidone) 197
Polyacrylamide 139
- derivatives 135
Polyacrylamides, hydrophobically modified 22
Polyacrylate polyanions 152
Polyampholytes 132
Polydisperse spheres, Debye plot 124
Polydispersity 123
Polyelectrolyte/cationic surfactant 147
- complexes 119
- -, cyanide ion exchanger 321
- -, immobilized enzymes 268, 321
- -, protein precipitation 265, 273
Polyelectrolytes 247, 273
Polyion addition reactions 154
- exchange reactions 154
- substitution reactions 154
Poly-L-lysine 168
Polymer overlap concentration 15, 16, 27, 35, 36, 45, 46
Polymer-polymer affinity 144
Polymethacrylate 152, 153
Polynucleotides 183
Polypeptides 175
Polysaccharide-polylysine complexation 175
Polysoaps, hydrophobically modified 6, 51
Polystyrene 137
-, sulfonated, ionomer solutions 331ff
Polytrimethyl benzyl ammonium chloride 136
Polyurethanes, hydrophobically modified ethoxylated (HEUR) 7-9, 19
Potassium pectate 177
Protease 316
Protein carriers, interpolyelectrolyte complexes 161ff
Protein-electrolyte complexes 247, 308
Protein-polyelectrolyte conjugates 167
Proteins, polyions 159
Pyrene 16, 19, 25, 53, 56, 68
-/auramine O 220

Radius of gyration 122
Rayleigh-Debye-approximation 123
Rheological properties, ionomer solutions 343

SANS, ionomer solutions 335, 341, 343, 348, 349
SAXS, ionomer solutions 341
Shear rotating disk flow 216
- thickening, ionomer solutions 336
Sodium methylsulfate 140
- polystyrene sulfonate 136
Static light scattering 119
Stereocomplexes 185
Styrene 137
Superhelix, polysaccharide 175

Ternary polymer mixture, selective complexation 195
Titration 125
-, turbimetric 276
Trypsin, bovine 305
Turbimetric titration 276

Ultrafiltration 231
Urease 161, 166
UV/VIS-spectroscopy 125

Vinylpyridine 137
Viscosimetry 125, 135

Zimm plots 120

Springer-Verlag and the Environment

We at Springer-Verlag firmly believe that an international science publisher has a special obligation to the environment, and our corporate policies consistently reflect this conviction.

We also expect our business partners – paper mills, printers, packaging manufacturers, etc. – to commit themselves to using environmentally friendly materials and production processes.

The paper in this book is made from low- or no-chlorine pulp and is acid free, in conformance with international standards for paper permanency.

MIX
Papier aus verantwortungsvollen Quellen
Paper from responsible sources
FSC® C105338

If you have any concerns about our products,
you can contact us on
ProductSafety@springernature.com

In case Publisher is established outside the EU,
the EU authorized representative is:
Springer Nature Customer Service Center GmbH
Europaplatz 3, 69115 Heidelberg, Germany

Printed by Libri Plureos GmbH
in Hamburg, Germany